DEFINITIONS

newton—force that will give a 1-kg mass an acceleration of 1 m/sec²
joule—work done by a force of 1 N over a displacement of 1 m
1 newton per sq in. (N/m²) = 1 pascal
1 kilogram force (kgf) = 9.807 N
1 gravity acceleration (g) = 9.807 m/sec²
1 are (a) = 100 m²
1 hectare (ha) = 10,000 m²
1 kip (kip) = 1000 lb

Probability Concepts in Engineering Planning and Design

Probability Concepts in Engineering Planning and Design

VOLUME I BASIC PRINCIPLES

Alfredo H-S. Ang
Professor of Civil Engineering
University of Illinois at Urbana-Champaign

Wilson H. Tang
Associate Professor of Civil Engineering
University of Illinois at Urbana-Champaign

John Wiley & Sons, Inc.
New York London Sydney Toronto

Dedicated to Myrtle and Bernadette

Library of Congress Cataloging in Publication Data:

Ang, Alfredo Hua-Sing, 1930–
    Probability concepts in engineering planning and design.

   Includes bibliographical references and indexes.

   CONTENTS: v. 1. Basic principles.

   1. Engineering—Statistical methods.   2. Probabilities.
   I. Tang, Wilson H., joint author.   II. Title.

TA340.A5   620'.004'2   75–5892

ISBN 0-471-03200-X

Printed in the United States of America

10 9 8 7 6 5 4 3

preface

Uncertainties are unavoidable in the design and planning of engineering systems. Properly, therefore, the tools of engineering analysis should include methods and concepts for evaluating the significance of uncertainty on system performance and design. In this regard, the principles of probability (and its allied fields of statistics and decision theory) offer the mathematical basis for modeling uncertainty and the analysis of its effects on engineering design.

The usefulness of probability and statistics in the analysis of sampled data and for quality control purposes is well recognized; however, the significance of probabilistic concepts transcends any specific application. Indeed, because these concepts are necessary and vital to the proper treatment of uncertainty, probability and statistical decision theory have especially significant roles in all aspects of engineering planning and design, including: (1) the modeling of engineering problems and evaluation of systems performance under conditions of uncertainty; (2) systematic development of design criteria, explicitly taking into account the significance of uncertainty; and (3) the logical framework for risk assessment and risk-benefit trade-off analysis relative to decision making. Our principal aim is to emphasize these wider roles of probability and statistical decision theory in engineering, with special attention on problems related to construction and industrial management; geotechnical, structural, and mechanical design; hydrologic and water resources planning; energy and environmental problems; ocean engineering; transportation planning; and problems of photogrammetric and geodetic engineering.

We are concerned mainly with the practical applications and relevance of probability concepts to engineering. The necessary mathematical concepts are developed in the context of engineering problems and through illustrations of probabilistic modeling of physical situations and phenomena. In this regard, only the essential principles of mathematical probability theory are discussed, and these principles are explained in nonabstract terms in order to stress their relevance to engineering. Mathematical rigor, therefore, is minimized in favor of the applied aspects of probability; this is necessary and essential to enhance the appreciation and recognition of the practical significance of probability concepts. For this purpose, the abstract

v

mathematical concepts are presented and illustrated with a variety of engineering-type problems, and a large number of similar problems are included as exercises. These are intended to illustrate and elucidate specific concepts and, therefore, are necessarily idealized; real-world engineering problems are often more complex than those illustrated.

The book is self-contained and thus is suitable for self-study by practicing engineers who desire a reading and working knowledge of the basic concepts and tools of probability. The presentation of mathematical concepts via illustrative engineering problems should be especially helpful. We hope that this approach will motivate engineers and engineering students to realize the potentials of probability concepts and to learn probability as a part of their professional engineering background; we believe that these concepts are essential tools for proper engineering analysis and design.

The subject is covered in two volumes. Volume I deals with the basic concepts and methods of probability and statistics that are esstential for modeling engineering problems under conditions of uncertainty, whereas Volume II presents advanced concepts and applications, including statistical decision theory, extreme-value statistics, risk analysis, reliability analysis and probability-based design, probabilistic network analysis, queuing theory, and Monte Carlo simulation.

The present volume includes nine chapters. Each chapter deals with certain topics that form the bases for subsequent chapters and concludes with a discussion of the highlights of the chapter and its relation to the subsequent chapters. Chapter 1 stresses the need for and significance of probability in many engineering problems. Chapters 2 to 4 then develop the basic concepts and essential analytical models of probability—in every case, the concepts are developed and illustrated with engineering and physical problems. These are followed by three chapters on the inferential methods of statistics (Chapters 5 to 7), a chapter on probability Bayesian (Chapter 8), and a final chapter (Chapter 9) that introduces the elements of quality assurance.

Volume I is designed for a first course in engineering probability and statistics; only knowledge of elementary calculus is required, and thus the material can be taught to undergraduate engineering students at any level, preferably (in our opinion) at the sophomore level. The material (except Chapter 9 and certain starred topics) was originally developed for a course on basic probability in engineering, required of all civil engineering sophomores at the University of Illinois at Urbana-Champaign. It may be used for a course taught either in the engineering departments or offered for engineers by the departments of mathematics and statistics. Although the level of mathematical sophistication may not satisfy a mathematician, it is appropriate for a first course in engineering probability. We think that the first exposure of engineers to probabilistic concepts and methods should be in physically meaningful terms; this is necessary to properly emphasize and motivate the recognition of the applied aspects of the relevant mathematical concepts.

Volume II is intended for an advanced undergraduate course or a first-year graduate course in engineering that deals with risk and decision analysis in systems planning and design.

Both the customary and SI (système internationale) metric systems of units are used throughout the book. When units are required, a problem is developed and discussed entirely in one of the systems.

Numerous present and former colleagues at the University of Illinois who have taught the basic civil engineering course contributed suggestions for improving the presentation of the material and suggestions for problems. These colleagues include Professors M. Amin, A. Chilton, H. M. Karara, N. Khachaturian, C. P. Siess, W. H. Walker, Y. K. Wen, and B. C. Yen. Their suggestions and advice are greatly appreciated. We are also indebted to the many students who learned the subject from earlier drafts (in the form of class notes with their various degrees of imperfection), and whose learning experiences, questions, and comments contributed immeasurably to the development of the material. The constructive criticisms and suggestions of several prepublication reviewers are gratefully acknowledged, these include, in particular, the detailed suggestions of R. Sexsmith of Cornell University, and the thoughtful comments of J. H. Mize of Oklahoma State University, and J. T-P. Yao of Purdue University. Last, but certainly not least, our sincere gratitude to Professor N. M. Newmark, who for many years as Department Head of Civil Engineering at the University of Illinois, provided the encouragement for and the academic atmosphere conducive to research and innovative developments.

Finally, we are thankful to Connie Crispen for typing and revising the final manuscript and Eldon Boatz for preparing the illustrations.

A. H-S. Ang and W. H. Tang

contents

8. *The Bayesian Approach*

9. *Elements of Quality Assurance and Acceptance Sampling*

1. Role of Probability in Engineering

1.1. INTRODUCTION

Quantitative methods of modeling, analysis, and evaluation are the tools of modern engineering. Some of these methods have become quite elaborate and include sophisticated mathematical modeling and analysis, computer simulation, and optimization techniques. However, irrespective of the level of sophistication in the models, including experimental laboratory models, they are predicated on idealized assumptions or conditions; hence, information derived from these quantitative models may or may not reflect reality closely.

In the development of engineering designs, decisions are often required irrespective of the state of completeness and quality of information, and thus must be formulated under conditions of uncertainty, in the sense that the consequence of a given decision cannot be determined with complete confidence. Aside from the fact that information must often be inferred from similar (or even different) circumstances or derived through modeling, and thus may be in various degrees of imperfection, many problems in engineering involve natural processes and phenomena that are inherently random; the states of such phenomena are naturally indeterminate and thus cannot be described with definiteness. For these reasons, decisions required in the process of engineering planning and design invariably must be made, and are made, under conditions of uncertainty.

The effects of such uncertainty on design and planning are important, to be sure; however, the quantification of such uncertainty, and evaluation of its effects on the performance and design of an engineering system, properly, should include concepts and methods of probability. Furthermore, under conditions of uncertainty the design and planning of engineering systems involve risks, and the formulation of related decisions requires risk-benefit trade-offs, all of which are properly within the province of applied probability.

In this light, we see that the role of probability is quite pervasive in engineering; it ranges from the description of information to the development of bases for design and decision making. Specific examples of such

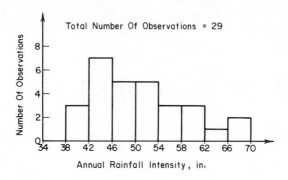

(a) In Number Of Observations

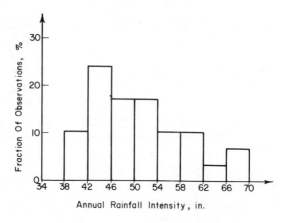

(b) In Fraction Of Total Observations

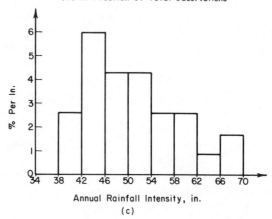

Figure 1.1 Histogram of rainfall intensity (Esopus Creek Watershed, N.Y., 1918–1946) (*a*) In number of observations. (*b*) In fraction of total observations. (*c*) Frequency diagram of rainfall intensity (Esopus Creek Watershed, New York)

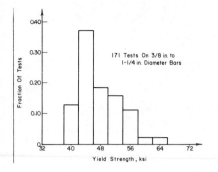

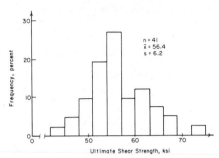

Figure 1.2 Histogram of yield strength of intermediate grade reinforcing bars; data from Julian (1957)

Figure 1.3 Histogram of ultimate shear strength of fillet welds in structural connections; after Kulak (1972)

information and engineering design and decision-making problems are described in the following sections.

1.2. UNCERTAINTY IN REAL-WORLD INFORMATION

1.2.1. Uncertainty associated with randomness

Many phenomena or processes of concern to engineers contain randomness; that is, the actual outcomes are (to some degree) unpredictable. Such phenomena are characterized by experimental observations that are invariably different from one experiment to another (even if performed under apparently identical conditions). In other words, there is usually a range of measured or observed values; moreover, within this range certain values may occur more frequently than others. The characteristics of such experimental data can be portrayed graphically in the form of a *histogram* or *frequency diagram*, such as those shown in Figs. 1.1 through 1.17, which represent information on physical phenomena of significance in engineering. (In some of these figures, specifically Figs. 1.5, 1.6, 1.7, 1.10, 1.13 1.14, and 1.17, theoretical *probability density functions* are also shown; the significance of these theoretical functions and their relations to the experimental frequency diagrams are discussed in Chapters 3 and 6).

A large number of physical phenomena are represented in Figs. 1.1 through 1.17; these are collected here purposely to demonstrate the fact that the natural state of most engineering information contains significant variability.

The histogram, therefore, is a graphic empirical description of the variability of experimental information. For a specific set of experimental data, the corresponding histogram may be constructed as follows.

From the observed range of experimental measurements, choose a

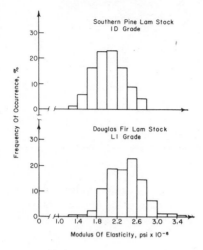

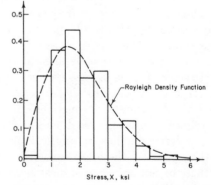

Figure 1.4 Histogram of modulus of elasticity of 2.6 lumber; after Galligan and Snodgrass (1970)

Figure 1.5 Frequency diagram of fatigue lives of 75 S-T aluminum; after Pugsley (1955)

range on the abscissa (for a two-dimensional graph) sufficient to include the largest and smallest observed values, and divide this range into convenient intervals. Then count the number of observations within each interval, and draw vertical bars with heights representing the number of observations in the respective intervals. Alternatively, the heights of the bars may be expressed in terms of the fractions of the total number of observations in

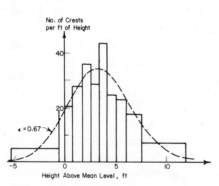

Figure 1.6 Frequency diagram of wave heights above mean sea level; after Cartwright and Longuet-Higgins (1956)

Figure 1.7 Frequency diagram of midship bending stress from one typical record "*S.S. Wolverine State*"; after Hoffman and Lewis (1969)

Table 1.1.

Year	Rainfall intensity (in.)
1918	43.30
1919	53.02
1920	63.52
1921	45.93
1922	48.26
1923	50.51
1924	49.57
1925	43.93
1926	46.77
1927	59.12
1928	54.49
1929	47.38
1930	40.78
1931	45.05
1932	50.37
1933	54.91
1934	51.28
1935	39.91
1936	53.29
1937	67.59
1938	58.71
1939	42.96
1940	55.77
1941	41.31
1942	58.83
1943	48.21
1944	44.67
1945	67.72
1946	43.11

each interval. For example, consider the annual (cumulative) rainfall intensity in the watershed area of the Esopus Creek in New York, recorded between the years 1918 to 1946, as presented in Table 1.1. An examination of these data will reveal that the observed rainfall intensity ranges from 39.91 to 67.72 in. Choosing a uniform interval of 4 in. between 38 and 70 in., the number of observations (and corresponding fraction of total observations) within each interval are shown in Table 1.2.

Plotting the number of observations in a given rainfall interval, we obtain the *histogram* of the rainfall intensity in the Esopus Creek watershed,

Table 1.2.

Interval	Number of observations	Fraction of total observations
38–42	3	0.1034
42–46	7	0.2415
46–50	5	0.1724
50–54	5	0.1724
54–58	3	0.1034
58–62	3	0.1034
62–66	1	0.0345
66–70	2	0.0690
Total = 29		1.0000

as shown in Fig. 1.1*a*, whereas, in terms of the fraction of total observations the same histogram would be as shown in Fig. 1.1*b*.

For the purpose of comparing an empirical frequency distribution (as, for example, portrayed in a histogram) with a theoretical probability density function, the corresponding *frequency diagram* is required. This may be obtained from the histogram by simply dividing the ordinates of the histogram by its total area. In the case of the histogram of Fig. 1.1, we obtain the corresponding frequency diagram by dividing the ordinates in Fig. 1.1*a* by 29 × 4 = 116; or alternatively, by dividing the ordinates in Fig. 1.1*b* by 4 × 1 = 4. The result would be as shown in Fig. 1.1*c*, which is the frequency diagram for the rainfall intensity of the Esopus Creek watershed.

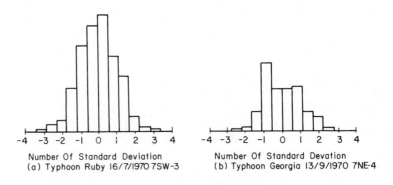

Number Of Standard Deviation
(a) Typhoon Ruby 16/7/1970 7SW-3

Number Of Standard Devation
(b) Typhoon Georgia 13/9/1970 7NE-4

Figure 1.8 Relative dispersions of measured pressure fluctuations on tall buildings during typhoons; after Lam Put (1971)

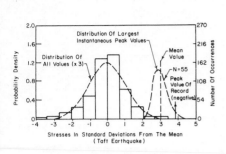

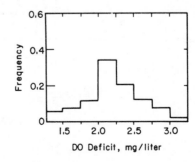

Figure 1.9 Relative dispersion of earthquake-induced shear stresses in soils; after Donovan (1972)

Figure 1.10 Histogram of density of compacted volcanic tuff subgrade; after Pettitt (1967)

The histogram, or frequency diagram, gives a graphic picture of the *relative frequencies* of the various observations or measurements. For most engineering purposes, certain aggregate quantities from the set of observations are more useful than the complete histogram; these include, in particular, the *mean-value* (or *average*) and a *measure of dispersion*. Such quantities may be evaluated from a given histogram; statistically, however, these are usually obtained in terms of the *sample mean* and *sample standard deviation*, as described in Chapter 5.

Clearly, if recorded data of a variable exhibit scatter or dispersion, such as those illustrated in Figs. 1.1 through 1.17, the value of the variable cannot be predicted with certainty. Such a variable is known as a *random variable*, and its value (or range of values) can be predicted only with an associated probability.

When two (or more) random variables are involved, the characteristics of one variable may depend on the value of the other variable (or variables).

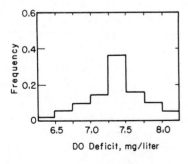

Figure 1.11 Histograms of dissolved oxygen (DO) deficit in Ohio River; after Kothandaraman and Ewing (1969)

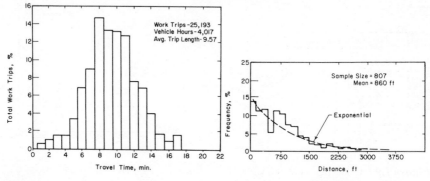

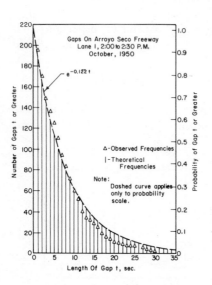

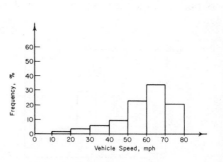

Figure 1.12 O-D trip length frequency distribution, Sioux Falls, S. D., 1956; after U.S. Department of Commerce (1965)

Figure 1.13 Frequency distribution of walking distances for off-street parkers (all trip purposes); after Kanafani (1972)

Figure 1.14 Histogram of gap length between cars on freeway; after Gerlough (1955)

Figure 1.15 Estimated impact speed in 5237 passenger-car single-vehicle accidents; after Viner (1972)

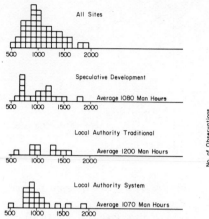

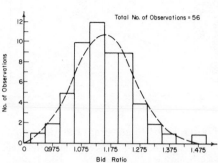

Figure 1.16 Histograms of completion time for house building in England; after Forbes (1969)

Figure 1.17 Highway bid distribution—southwest Texas; after Cox (1969)

Pairs of observed data for the two variables, when plotted on a two-dimensional space, would appear as in Figs. 1.18 through 1.22, which are characterized by scatter or dispersion in the data points, called *scattergrams*. In view of such scatter, the value of one variable, given that of the other, cannot be predicted with certainty. The degree of predictability will depend on the degree of mutual dependency or correlation between the variables, as measured (in the linear case) by the statistical correlation.

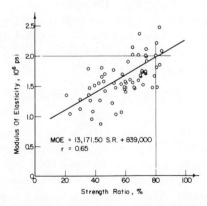

Figure 1.18 Linear regression of modulus of elasticity on strength ratio for 2 × 10 air-dry (16% m.c.) douglas-fir; after Littleford (1967)

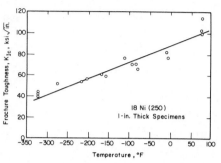

Figure 1.19 Plane-strain fracture-toughness behavior of 18 Ni (250) maraging steel as a function of temperature; after Barsom (1971)

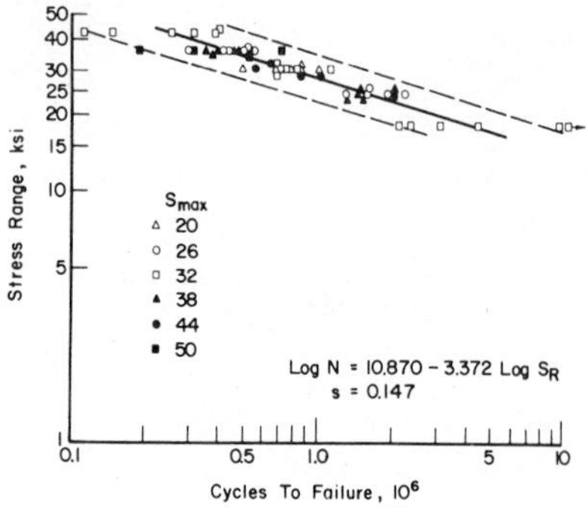

Figure 1.20 Relationship between stress range and fatigue life of welded beams; after Fisher et al. (1970)

Methods for evaluating such correlations are also embodied in statistical analysis.

It should be strongly emphasized that the application of probability is not limited to the description of experimental data, or to the evaluation of the associated statistics (such as the mean, standard deviation, and correlation). Indeed, the much more significant role of probability concepts is in the utilization of this information in the formulation of proper bases for decision making and design. In other words, when we are dealing with information such as that illustrated in Figs. 1.1 through 1.22, which requires probabilistic description, the proper utilization of this information in engineering design and planning will necessarily require concepts and methods of probability. For example, if a design equation involves random variables, such as those described in Figs. 1.1 through 1.22, the quantitative analysis of their effects on, and the formulation of, the design will necessarily involve probabilistic concepts.

1.2.2. Uncertainty associated with imperfect modeling and estimation

Engineering uncertainty, however, is not limited purely to the variability observed in the basic variables. First, the estimated values of a given variable (such as the mean) based on observational data will not be error-free (especially when data are limited). In fact, in some cases, such estimates may not be much better than "educated guesses," based largely on the engineer's judgement. Second, the mathematical or simulation models (for example, formulas, equations, algorithms, computer simulation programs), and even laboratory models, that are often used in engineering analysis and to develop designs are idealized representations of reality; in various degrees, such models are imperfect representations of the real world. Consequently, predictions and/or calculations made on the basis of these models may be inaccurate (to some unknown degree) and thus also contain uncertainty. In certain cases, the uncertainties associated with such prediction or model errors may be much more significant than those associated with the inherent variabilities.

All uncertainties, whether they are associated with inherent variability or with prediction error, may be assessed in statistical terms, and the evaluation of their significance on engineering design accomplished using concepts and methods that are embodied in the theory of probability.

1.3. DESIGN AND DECISION MAKING UNDER UNCERTAINTY

If information is of the type illustrated in Figs. 1.1 through 1.17, in which no single observation is representative, and evaluations and predictions must be based on imperfect models, how should designs be for-

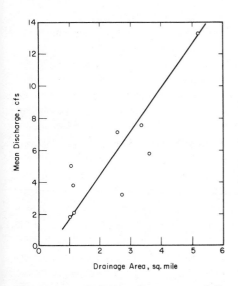

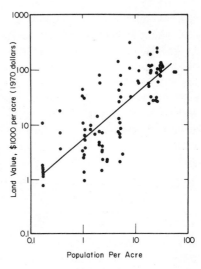

Figure 1.21 Relation of mean annual discharge to drainage area for streams near Honolulu; after Todd and Meyer (1971)

Figure 1.22 Regression line for interstate highway acquisition cost; after Ward (1970)

mulated or decisions affecting a design be resolved? Presumably we may assume consistently worst conditions (for example, specify the highest possible flood, smallest observed fatigue life of materials, and so on) and develop conservative designs on this basis. From the standpoint of system performance and safety, this approach may be suitable; however, the resulting design could be too costly as a consequence of "compounded conservatism." On the other hand, an inexpensive design may not ensure the desired level of performance or safety. Therefore, decisions based on trade-off between cost and benefit (including tangible and intangible factors) are necessary. The most desirable solution is one that is optimal, in the sense of minimum cost and/or maximum benefits. If the available information and evaluative models contain uncertainties, the required trade-off analysis should include the effects of such uncertainties on a given decision.

Such situations are common to many problems of engineering design and planning; in this section we describe several examples illustrating some of these problems. The examples are idealized to simplify the discussions; nevertheless, they serve to illustrate the essence of the decision making aspects of engineering under conditions of uncertainty.

1.3.1. Planning and design of airport pavement

Consider, as the first example, the design of an airport pavement. Among the many factors that have bearing on the design, the thickness of the pavement system (consisting of several layers of subgrade base material and the finished pavement) is one of the principal decision variables. In general, the usable life of the pavement will depend on the thickness of the system; the thicker the pavement system is, the longer its useful life will be. Of course, for the same material and workmanship quality, the cost will also increase with the thickness. On the other hand, a thin system will cost less initially, but the subsequent maintenance and replacement costs will be higher. Therefore the thickness of the pavement system may be determined on the basis of a trade-off between high initial cost with low maintenance, versus low initial cost but high replacement and maintenance costs. For the purpose of such trade-off analysis, the relation between the life of a pavement system and its thickness is required. However, the pavement life is also a function of other variables, including drainage and moisture content, temperature ranges, density and degree of compaction of the subgrade. Since these factors are random (see Fig. 1.10 for the subgrade density), the life of the pavement cannot be predicted with certainty. Hence, the total cost (including initial and maintenance costs) associated with a given pavement thickness cannot be estimated with complete confidence; any meaningful trade-off analysis, therefore, would properly require concepts of probability.

1.3.2. Hydrologic design

Suppose that the protection of a large farm from flooding requires the construction of a main culvert at the junction of a roadway and a stream. A decision on the size (flow capacity) of the culvert is required; clearly, the size will depend on the high stream flow, which is a function of the rainfall intensity in the watershed and the associated runoff. If the culvert is large enough to handle the largest possible flow, there would be no danger to flooding; however, the cost of constructing the culvert may be prohibitively high, and even during the heaviest rainfall the culvert may be used only to a fraction of its capacity—that is, overdesign would be wasteful and costly. On the other hand, if the culvert is too small, its cost may be low, but the farm is likely to be subject to serious flooding every time there is a heavy rainfall, causing damage to crops and erosion of the upstream soil.

The decision would properly require probability considerations, for the following reasons. First, the annual rainfall intensity is highly variable (as illustrated in Fig. 1.1), and the prediction of the stream runoff may be imperfect; consequently, the maximum stream flow (for example, in a year) cannot be predicted with certainty. Assuming that the flow capacity of a given culvert size can be determined accurately, the size of the culvert would depend on the probability of flooding within a given period (for example, ten years). The culvert size then may be determined so that the total expected cost, consisting of the initial cost of constructing the culvert plus the expected loss from flooding and erosion, is minimized. The expected loss is a function of the probability of flooding, and hence the definition of the total expected cost requires probability measures.

1.3.3. Design of structures and machines

In Fig. 1.8 we have an example showing that magnitudes of load (wind pressures in the case of Fig. 1.8) may be described with random variables, whereas in Figs. 1.2 and 1.3 the strengths of structural material and components are shown to be also random, and thus the resulting structural resistance will likewise be random. Even for this idealized situation, the design of a structure (determination of how strong it should be) must consider the question "How safe is safe enough?"—a question that theoretically requires the consideration of risk or probability of failure.

To be specific, consider for example the design of an offshore drilling tower, which is subject to occasional hurricane forces. In such a case, we recognize that aside from the fact that the maximum wind effect during a hurricane is random, the occurrence of hurricanes in a given coastal region is also unpredictable. Hence, in determining the safety level for the design of the tower, the probability of occurrence of strong hurricanes within the specified useful life of the structure must be considered, in addition to the survival probability of the structure during a hurricane. The higher the hurricane force gets, the less frequent will be its occurrence; therefore, if a very strong hurricane is specified for the design, there may be almost no chance of its occurring during the useful life of the drilling tower. Consequently, what level of hurricane force should be used in the design, and what level of protection would be adequate during a hurricane, are decisions that clearly require trade-offs between cost and level of protection in terms of risk or failure probability within the lifetime of the structure.

In structural or machine components that are subject to repeated or cyclic loads, the fatigue life (that is, the number of load cycles until fatigue failure or fracture) of the component is also random, even at constant-amplitude stress cycle as illustrated in Fig. 1.5. For this reason, the useful life of the component is, to some degree, unpredictable. A design will depend on the required life and desired level of reliability; for a given design, the shorter is the required service life, the higher will be its reliability against possible breakdowns within the specified service life. Fatigue life is also a function of the applied stress level; generally, the higher the stress, the shorter the fatigue life. Consequently, if a desired life is specified, the components could be designed to be massive so that the maximum stresses will be low and thus assure long life. This appraoch will, of course, be expensive in terms of material. In contrast, if the parts are under-designed, high stresses may be induced, resulting in short life and thus requiring frequent replacements.

The optimal life may be determined on the basis of minimizing the total expected cost, which would include the initial cost, the expected cost of replacement (a function of the *reliability* or probability of no failure), and the expected cost associated with the loss of revenue incurred during a repair (also a function of reliability). Having decided on the desired design life, the components may then be proportioned accordingly.

1.3.4. Geotechnical design

Properties of soil material are inherently heterogeneous, and natural earth deposits are characterized by irregular layers of various material (for example, clay, silt, sand, gravel, or combinations thereof) with wide ranges of density, moisture content, and other properties that affect the strength and compressibility of the deposit. On the other hand, rock formations are often characterized by irregular systems of geologic faults and fissures that significantly affect the load-bearing capacity of the rock.

In designing foundations to support structures and facilities, the capacity of the *in situ* subsoil and/or rock deposit must be determined. Invariably, however, this determination has to be made on the basis of available geologic information and data from site exploration with limited soil-sampling results.

Because of the natural heterogeneity and irregularity of soil and rock deposits, the capacity of the subsoil could vary widely over a foundation site; moreover, because the required load-bearing capacity must invariably be estimated on the basis of very limited information, such estimates are subject to considerable uncertainty. As a consequence, an estimate may run some risk of overestimating the actual capacity of the soil deposit at a site; in view of this fact, the safety of a foundation designed on the basis of such estimates may not be assured with complete confidence, unless a sufficient margin of safety is provided. On the other hand, an excessively large safety margin may yield an unnecessarily costly support system. Therefore the required safety margin for design may be viewed as a problem involving the trade-off between cost and tolerable risk or probability of failure.

1.3.5. Construction planning and management

Many factors in the construction industry are subject to variability and uncertainty, some of which cannot be controlled. For example, the required durations of various activities in a construction project will depend on the availability of resources, including labor and equipment and their respective productivity, on weather conditions, and on availability of material. None of these factors are completely predictable, and thus the durations of the individual activities as well as the project duration cannot be estimated with much precision or certainty (see for example Fig. 1.16); such durations must be described as random variables. Therefore, in preparing a bid for a project, if conservative (or pessimistic) time estimates are used, the bid price may be too high, thus reducing the chances of winning the bid. On the other hand, if the bid is based on an optimistic estimate of the project time, the contractor may lose money in a successful bid. What degree of conservativeness should the contractor exercise to maximize his profit potential? Realistically, this decision may be based on a consideration of probability—the bid price may be based on a target time corresponding to a specified probability of completion.

1.3.6. Photogrammetric, geodetic, and surveying measurements

All practical measurements are subject to errors, which can be classified as random and systematic errors. Systematic errors can be eliminated or minimized by evaluating them and applying corrections. However, the magnitude and propagation of random errors, inherent in making measurements, can be established and analyzed only on the basis of probability theory. Such a statistical approach is indeed the only reliable means for determining accuracy once measurements are refined beyond instrument capabilities.

The accuracy and precision of measurements can be improved by using instruments capable of keener measurements and by adopting more refined observational procedures. Depending on the importance and cost of a project, the additional cost of increased accuracy and precision may or may not be justified.

The method of least-squares is used widely in photogrammetry, geodesy, and surveying; it is used, for example, in the adjustment of photogrammetric blocks, geodetic networks, and leveling circuits. *A priori* estimation of the accuracy of geodetic coordinates, for example, is essential before finalizing the selection of the configuration of triangulation and trilateration projects.

In conjunction with photogrammetrically produced digital terrain models, least-square fitting using polynomials, potential functions, or trigonometric functions is often used to mathematically describe the surface of the object under study. The object may be a terrain under consideration, such as a possible airport site, an animal for which the surface area is to be determined, or a trileaflet heart valve under study to develop prosthetic valves.

In remote sensing, statistics is used extensively in pattern recognition techniques where the objective is to classify the image spectrally. Sample spectral data from the scene is statistically clustered into distinct groups. This grouping is then often extended to the entire image through the application of discriminate function analysis.

1.4. CONTROL AND STANDARDS

In order to assure some minimum level of quality, or performance, of engineering products or systems, inspections and standards of acceptance are necessary. Clearly, if the standard is too stringent, it may unnecessarily increase product cost or its adherence and enforcement may be difficult; on the other hand, if the standard is too lax, the quality of the product may be overly compromised. Also, if the control variables or design variables are random, what constitutes a stringent or nonstringent standard is not immediately clear; in these cases, the acceptance standards ought to be developed also on the basis of probability considerations.

For example, in constructing an earth embankment, practical standards for acceptability of the compaction should recognize the variability in the density of compacted material, as illustrated in Fig. 1.10. Acordingly, an acceptance sampling plan may be developed based on probability considerations and taking into account such variability.

To control the quality of streams, the parameter most commonly used as a measure of pollution is the concentration of dissolved oxygen (DO) in the stream water, which is random as illustrated in Fig. 1.11. Among environmentalists there is a growing realization of the need for probabilistic standards of stream quality (for example, Loucks and Lynn, 1966; Thayer, 1966). Loucks and Lynn (1966) proposed the following as an example probabilistic stream standard:

The dissolved oxygen concentration in the stream during any 7 consecutive day period must be such that: (i) the probability of its being less than 4 mg/l for any one day is less than 0.20; and (ii) the probability of its being less than 2 mg/l for any one day is less than 0.1 and for two or more consecutive days is less than 0.05.

To ensure the quality of concrete material in reinforced concrete construction, the Building Code of the American Concrete Institute (ACI 318-71) requires the following.

The strength level of the concrete will be considered satisfactory if the averages of all sets of three consecutive strength test results equal or exceed the required f_c' and no individual strength test result falls below the required f_c' by more than 500 psi. Each strength test result shall be the average of two cylinders from the same sample tested at 28 days or the specified earlier age.

These statements imply the need for probability and statistics in the assurance of quality concrete.

1.5 CONCLUDING REMARKS

This chapter has emphasized the importance and role of probability concepts and methods in engineering. The examples enumerated and described in Sections 1.2 to 1.4 should serve to emphasize the pervasiveness of such concepts in engineering planning and design. In particular, it should be stressed that the description of statistical information and estimation of statistics, such as means and variances, are not the only applications of probability theory; the much more significant role of probability concepts, in fact, lies in its unified framework for the quantitative analysis of uncertainty and assessment of associated risk, and in the formulation of trade-off studies relative to decision making, planning, and design.

The many examples presented also serve to illustrate, with real data and realistic engineering problems, that randomness of real-world phenomena and imperfections of engineering predictions and estimations are facts of life. Consequently, uncertainties associated with such randomness and imperfections are unavoidable in engineering planning and design.

Finally, it is important to allay any misconception that extensive data are required to apply probability concepts; the usefulness and relevance of such concepts are equally significant, irrespective of the amount of data or quality of information. Probability is the conceptual and theoretical basis for modeling and analyzing uncertainty. The availability of data and quality of information will affect the degree of uncertainty; however, the lack of sufficient data should not lessen the usefulness of probability as the proper tool for the analysis of such uncertainty and for the evaluation of its effects on engineering performance and design.

In the ensuing chapters of this volume, as well as those in Volume II, the probabilistic concepts and methods necessary for these purposes are developed.

2. Basic Probability Concepts

2.1. EVENTS AND PROBABILITY

2.1.1. Characteristics of probability problems

It may be recognized from the discussions in Chapter 1 that when we speak of probability, we are referring to the occurrence of an event relative to other events; in other words, there is (implicitly at least) more than one possibility, since otherwise the problem would be deterministic. For quantitative purposes, therefore, probability can be considered as a numerical measure of the likelihood of occurrence of an event relative to a set of alternative events.

Accordingly, the first requirement in the formulation of a probabilistic problem is the identification of the set of all possibilities (that is, the *possibility space*) and the event of interest. Probabilities are then associated with specific events within a particular possibility space.

To illustrate the various aspects of a probabilistic problem, as characterized above, consider the following examples.

EXAMPLE 2.1

A contractor is planning the purchase of equipment, including bulldozers, needed for a new project in a remote area. Suppose that from his previous experience, he figures there is a 50% chance that each bulldozer can last at least 6 months without any breakdown. If he purchased 3 bulldozers, what is the probability that there will be only 1 bulldozer left operative in 6 months?

First we observe that at the end of 6 months, the number of operating bulldozers may be 0, 1, 2, or 3; therefore this set of numbers constitutes the possibility space of the number of operational bulldozers after 6 months. However, the probability of the various possible outcomes cannot be readily determined from the information that each bulldozer has a 50% chance of remaining operative after 6 months. For this purpose, the possibility space must be derived in terms of the possible status of each bulldozer after 6 months, as follows.

If we denote the condition of each bulldozer after 6 months as G for *good* and B for *bad* conditions, the possible statuses of the three bulldozers would be

GGG—all three bulldozers in good condition

GGB—first and second bulldozers good, and third one bad

GBB

BBB—all three bulldozers in bad condition

BGG

BBG

GBG

BGB

In this case, therefore, there are a total of 8 possibilities. Since the condition of a bulldozer is equally likely to be good or bad, the 8 possible statuses of the 3 bull-dozers are also equally likely to occur. It is worth noting that among the 8 possible outcomes, only one of them can be realized at the end of 6 months; this means that the different possibilities are *mutually exclusive* (we shall say more on this point in Section 2.2.2).

Among the 8 possible statuses of the 3 bulldozers, the realization of GBB, BGB, or BBG is tantamount to the event "only one bulldozer is operational." And since each possibility is equally likely to occur, the probability of the event within the above possibility space is 3/8.

EXAMPLE 2.2

In designing a left-turn lane for eastbound traffic at a highway intersection, as shown in Fig. E2.2, the probability of 5 or more cars waiting for left turns may be needed to determine the length of the left-turn lane. For this purpose, suppose that over a period of 2 months 60 observations were made (during periods of heavy traffic) of the number of eastbound cars waiting for left turns at this intersection, with the following results.

No. of cars	No. of observations	Relative frequency
0	4	4/60
1	16	16/60
2	20	20/60
3	14	14/60
4	3	3/60
5	2	2/60
6	1	1/60
7	0	0
8	0	0
.	.	.
.	.	.
.	.	.

Conceivably, or theoretically, the number of cars waiting for left turns, during heavy traffic hours, could be any positive integer number; however, in the light of the above traffic count, the possibility of 7 or more cars waiting for left turns is not likely to occur at this intersection.

On the basis of the foregoing results, the observed *relative frequency* (tabulated in the third column above) may be used as the probability of a particular number of

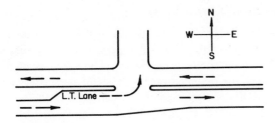

Figure E2.2

cars waiting for left turns. Then the probability of the event "5 or more cars waiting"
is $2/60 + 1/60 = 3/60$.

EXAMPLE 2.3

In the simply supported beam AB shown in Fig. E2.3, the load of 100 kg can be
placed anywhere along the beam. In this case, the reaction at the support A, R_A,
clearly can be any value between 0 and 100 kg; therefore any number between 0
and 100 is a possible reaction value for R_A, and thus is its possibility space.

An event of interest then may be that the reaction is in some specified interval;
for example, ($10 \leq R_A \leq 20$ kg) or ($R_A > 50$ kg). Therefore, if a particular value of
R_A is realized, the event (defined by an interval) containing this value of R_A has
occurred, and we can speak of the probability that R_A will, or will not, be in a given
interval. For instance, if we assume that the 100-kg load is equally likely to be placed
anywhere on the beam, then the probability that the value of R_A will be in a given
interval is proportional to the interval; for example, $P(10 \leq R_A \leq 20) = \frac{10}{100} =$
0.10, and $P(R_A > 50) = \frac{50}{100} = 0.50$.

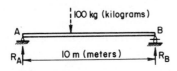

Figure E2.3

From the foregoing examples, the following special characteristics of
probabilistic problems may be observed.

1.   Every problem is defined with reference to a specific possibility space
      (containing more than one possibility), and events are composed of
      one or more possible outcomes within this possibility space.
2.   The probability of an event depends on the probabilities of the in-
      dividual outcomes within a given possibility space.

In Sections 2.2 and 2.3, we shall present the mathematical tools pertinent
to and useful for each of these purposes.

2.1.2.   Calculation of probability

From the examples discussed above, it can be observed that in calculating the probability of an event, a basis for assigning probability measures to the various possible outcomes is necessary. The assignment may be based on prior conditions (or deduced on the basis of prescribed assumptions), or based on the results of empirical observations, or both.

In Examples 2.1 and 2.3, the probabilities of the possible outcomes were based on prior assumptions. In the case of Example 2.1, each of the possible statuses of the 3 bulldozers was assumed to be equiprobable, each equal to $\frac{1}{8}$ (consistent with the prior information that each bulldozer is equally likely to be operative or nonoperative after 6 months) ; whereas, in Example 2.3 the probability that the reaction R_A will be in a given interval was assumed to be proportional to the interval length (consistent with the assumption that the position of the 100-kg load is equally likely to be anywhere along the beam). However, in Example 2.2 the probability of the number of cars waiting for left turns is based on the corresponding observed relative frequency, which is determined from empirical observations.

It should be emphasized that we shall treat probability as a measure necessary and useful in problems where more than one event or outcome is possible. In particular, we shall avoid the philosophical question of the meaning of a probability measure, and concern ourselves merely with the utilitarian aspects of probability and its mathematical theory (see Section 2.3) for modeling problems under conditions of uncertainty, in the same sense that we use the factor of safety to effect engineering design without worrying about its real meaning, or employing Newton's second law of motion without being concerned about the meaning of mass and force.

The usefulness of a calculated probability, however, will depend on the appropriateness of the basis for its determination. In this regard, we observe that the validity of the *a priori* basis for calculating probability depends on the reasonableness of the underlying assumptions, whereas the empirical *relative frequency* basis must rely on a large amount of observational data. When data are limited, the relative frequency by itself may have limited usefulness.

A third basis for calculating probability involves the combination of intuitive or subjective assumptions with experimental observations; the proper vehicle for this combination is Bayes' theorem (see Section 2.3.4), and the result is known as the *Bayesian probability* (see Chapter 8).

2.2.   ELEMENTS OF SET THEORY

Many of the characteristics of a probabilistic problem can be defined formally and modeled succinctly using elementary notions of sets and the mathematical theory of probability. In this and the following section (Section 2.3), we present the basic elements of the theories of sets and probability, as they relate to and are useful for the formulation of probabilistic problems.

2.2.1. Definitions

In the terminology of set theory, the set of all possibilities in a probabilistic problem is collectively a *sample space*, and each of the individual possibilities is a *sample point*. An *event* then is defined as a *subset of the sample space*.

Sample spaces may be *discrete* or *continuous*. In the discrete case, the sample points are individually discrete entities and countable; in the continuous case, the sample space is composed of a continuum of sample points.

A discrete sample space may be *finite* (that is, composed of a finite number of sample points) or *infinite* (that is, with a countably infinite number of sample points). The possible status of the three bulldozers in Example 2.1 is an example of a finite discrete sample space; each of the possible statuses is a sample point, and the eight possibilities collectively constitute the corresponding sample space. Other examples of finite sample spaces are as follows. (1) The winner in a competitive bidding for a construction project will be among the firms submitting bids for the project. The sample space then consists of all the possible bid winners, which are the finite number of firms involved in the bidding; in this case, each of the firms is a sample point. (2) The number of days in a year with freezing temperature in Juneau, Alaska, is limited to 365 days; each day of the year then is a sample point, and collectively all the days of the year constitute the sample space. Example 2.2 is an illustration of a discrete sample space with countably infinite number of sample points; the number of cars waiting for left turns could, theoretically, be any integer number from zero to infinity. Other examples are the following: (1) the number of flaws in a given length of weld, and (2) the number of cars crossing a toll bridge until the next accident on the bridge. In each case we have an infinite number of discrete possibilities. There may be none or only a few flaws in the weld, or the number of flaws could be very large; similarly, an accident may occur with the first car crossing the toll bridge, or there may never be an accident on the bridge.

In a continuous sample space, the number of sample points is effectively always infinite. For example, (1) in considering the location on a toll bridge where a traffic accident may occur, each of the possible locations is a sample point, and the sample space would be the continuum of points on the bridge; and (2) if the bearing capacity of a clay deposit is between 1.5 tsf (tons per sq ft) to 4.0 tsf, then any value within the range 1.5 to

4.0 is a sample point, and the entire continuum of values in this range constitutes the sample space.

Whether a sample space is discrete or continuous, however, an event is always a subset of the appropriate sample space; therefore an event always contain one or more sample points (unless it is an impossible event), and *the realization of any of these sample points constitutes the occurrence of the corresponding event.* Finally, when we speak of probability, we are always referring to an event within a particular sample space.

The following example clarifies the preceding notions in more definitive terms.

EXAMPLE 2.4

Consider again a simply supported beam AB (Fig. E2.4a).

(a) If a concentrated load of 100 lb can be placed only at any of the 2-ft interval points on the beam, the sample space of the reaction R_A will be as shown in Fig. E2.4b. In this case, the sample space of R_A consists of distinct sample points.

Let us also consider the sample space of R_A and R_B (that is, all possible pairs of values of R_A and R_B); in this case, any pair of values of R_A and R_B such that $R_A + R_B = 100$ belongs to the sample space, which is shown in Fig. E2.4c.

(b) If the load can be placed anywhere along the beam, the sample space of R_A can be represented by the line between 0 and 100 (Fig. E2.4d), whereas the corresponding sample space of R_A and R_B is the straight line shown in Fig. E2.4e.

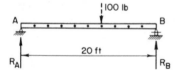

Figure E2.4a

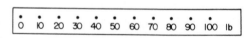

Figure E2.4b   Sample space of R_A

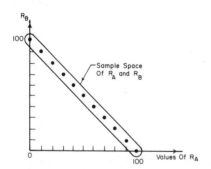

Figure E2.4c

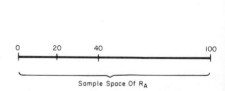

Figure E2.4d   Sample space of R_A

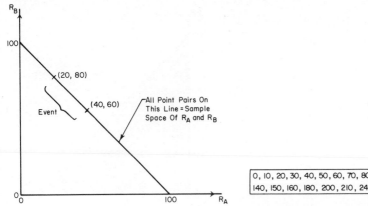

Figure E2.4e

Figure E2.4f Sample Space of R_A

We can then speak of the event that R_A will be between, say, 20 and 40; or that (R_A, R_B) will be between (20, 80) and (40, 60).

(c) Next consider that the load can be 100 lb, 200 lb, or 300 lb, and its position can be at any 2-ft interval on the beam. The sample space of R_A then contains the values listed in Fig. E2.4f, whereas the sample space of R_A and R_B is represented by the two-dimensional coordinates of the points shown in Fig. E2.4g.

However, if the load can be placed anywhere along the beam, then the sample space of R_A and R_B would be described by the three lines shown in Fig. E2.4h.

(d) If the load can be any value between 100 and 300 lb, then the sample space of R_A contains all values between 0 and 300 lb, as represented by the line in Fig. E2.4i, whereas the sample space of R_A and R_B would be the shaded area shown in Fig. E2.4j.

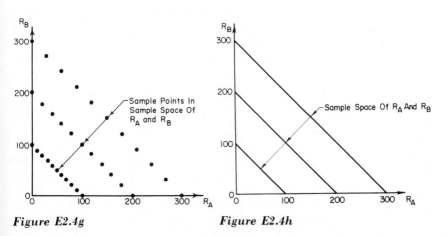

Figure E2.4g

Figure E2.4h

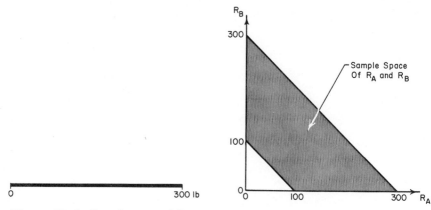

Figure E2.4i Sample space of R_A **Figure E2.4j**

Special events. We define the following special events and adopt the notations indicated below.

1. *Impossible event*, denoted ϕ, is the event with no sample point. It is therefore an *empty* set in a sample space.
2. *Certain event*, denoted S, is the event containing all the sample points in a sample space; that is, it is the sample space itself.
3. *Complementary event* $\bar{E}$. For an event E in a sample space S, the complementary event, denoted $\bar{E}$, contains all the sample points in S that are not in E.

The Venn diagram. A sample space and the events within it can be represented pictorially with the *Venn diagram*—a sample space is represented by a rectangle; an event E is then represented symbolically by a closed region within this rectangle, and the part of the rectangle outside this closed region is the corresponding *complementary event* $\bar{E}$. See Fig. 2.1. In other words, the event E contains all the sample points within the closed

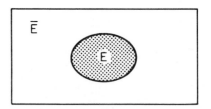

Figure 2.1 A Venn diagram

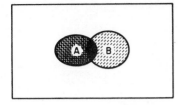 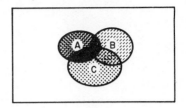

Figure 2.2 Venn Diagram with several events. (*a*) Two events, *A* and *B*. (*b*) Three events, *A*, *B*, and *C*

region, whereas $\bar{E}$ contains all the sample points outside of E. A Venn diagram with two (or more) events would appear as in Fig. 2.2.

2.2.2. Combination of events

In many practical problems, the event of interest may be some combination of other events. For instance, in Example 2.1 the event of *at least 2 bulldozers operative after 6 months* may be of interest. This can be considered as the combination of *2 bulldozers operative* or *3 bulldozers operative*. Such an event is the "union" of the two individual events.

There are two basic ways that events may be combined or derived from other events: by the *union* or *intersection*. Consider two events E_1 and E_2:

The *union of E_1 and E_2*, denoted $E_1 \cup E_2$, is another event that means the occurrence of E_1 or E_2, or both. In other words, $E_1 \cup E_2$ is the subset of sample points that belong to E_1 or E_2 (In set theory, *or* is used in an *inclusive* sense, which means *and/or*).

Examples: (1) In describing the state of supply of construction material, if E_1 represents the shortage of concrete and E_2 represents the shortage of steel, then the union $E_1 \cup E_2$ is the shortage of concrete or steel, or both. (2) In a 20-mile length of an oil pipeline, if E_1 stands for leakage in mile 0 to 15 and E_2 stands for leakage in mile 10 to 20, then $E_1 \cup E_2$ means leakage anywhere along the entire 20-mile pipeline.

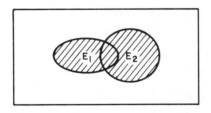

Figure 2.3 Venn diagram for union of events E_1 and E_2

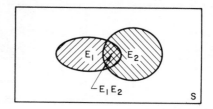

Figure 2.4   Venn diagram of intersection of events E_1 and E_2

The Venn diagram for the union of two events E_1 and E_2 would be the shaded region shown in Fig. 2.3. It follows, therefore, that the portion of the rectangle outside of the shaded region in Fig. 2.3 is the complementary event $\overline{E_1 \cup E_2}$; that is, the complement of $E_1 \cup E_2$.

The union of 3 or more events means the occurrence of at least one of them. For example, transportation between Chicago and New York may be by air, highway, or railway. If we denote the availability of these modes of transport, respectively, as A, H, and R, the available means of transporting material between these two cities can be denoted as $(A \cup H \cup R)$.

The *intersection of E_1 and E_2*, denoted $E_1 \cap E_2$ (or simply E_1E_2), is also an event that means the joint occurrence of E_1 and E_2; in other words, E_1E_2 is the subset of sample points belonging to both E_1 and E_2.

Examples: Referring to the examples described above, (1) E_1E_2 means the shortage of concrete and steel; (2) E_1E_2 means the leakage in mile 10 to 15 along the pipeline; whereas, AHR means all three methods of transport between Chicago and New York are available.

In terms of the Venn diagram, the intersection of two events E_1 and E_2 would be as shown in Fig. 2.4.

EXAMPLE 2.5

In Example 2.2 the sample space is the set $\{0, 1, 2, 3, \ldots\}$; that is, theoretically the sample space contains all non-negative integers.

> If E_1 = the event of more than two cars waiting for left turns;
>   that is, the subset $\{3, 4, 5, \ldots\}$

and

> E_2 = the event between two to four cars waiting for left turns;
>   that is, the subset $\{2, 3, 4\}$

then the union $E_1 \cup E_2$ is the subset $\{2, 3, 4, \ldots\}$ whereas the intersection E_1E_2 is the subset $\{3, 4\}$.

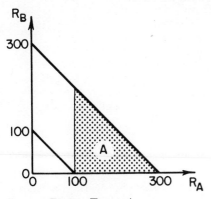

Figure E2.6a Event A

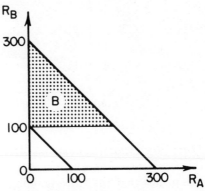

Figure E2.6b Event B

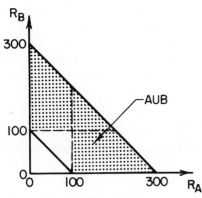

Figure E2.6c Union $A \cup B$

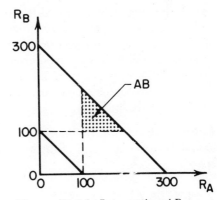

Figure E2.6d Intersection AB

EXAMPLE 2.6

In the last case of Example 2.4, where the load can range between 100 and 300 lb, the sample space of the reactions R_A and R_B is shown in Fig. E2.4*j*.

$$\text{If } A = \text{event } \{R_A > 100 \text{ lb}\}$$

and

$$B = \text{event } \{R_B > 100 \text{ lb}\}$$

the events A and B would be the subsets containing all point pairs of R_A and R_B shown in Figs. E2.6*a* and E2.6*b*, respectively. Observe that the events A and B are defined within the sample space of R_A and R_B. Then the union $A \cup B$ contains all point pairs in the shaded region of Fig. E2.6*c*; whereas the intersection AB is the shaded region shown in Fig. E2.6*d*. In the present example, Figs. E2.6*a* through E2.6*d* serve also as the corresponding Venn diagrams.

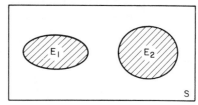

Figure 2.5   Mutually exclusive events E_1 and E_2

Mutually exclusive events.   If the occurrence of one event precludes the occurrence of another event, then the two events are *mutually exclusive*; this means that the corresponding subsets will have no overlap, as shown in the Venn diagram of Fig. 2.5. That is, the subsets are "disjoint." The intersection of two mutually exclusive events E_1 and E_2, therefore, is an impossible event; that is, $E_1 E_2 = \phi$. Examples of mutually exclusive events are (1) making right turn and left turn at a street intersection; (2) flood and drought of a river at a given instant of time; (3) failure and survival of a structure to a strong motion earthquake.

Three or more events are mutually exclusive if the occurrence of one precludes the occurrence of all others. For example, if there are three possible locations for a new airport, then the choices among the three sites are mutually exclusive.

Collectively exhaustive events.   Two or more events are *collectively exhaustive* if the union of all these events constitute the underlying sample space.

EXAMPLE 2.7

Two construction companies *a* and *b* are bidding for jobs. Let *A* denote the event that Company *a* gets a job and *B* denote the event that Company *b* gets a job. Draw the Venn diagrams for the sample spaces of the following:
(a) Company *a* is submitting a bid for one job and Company *b* is submitting a bid for another job.
(b) Companies *a* and *b* are submitting bids to the same job, and there are more than 2 bidders for the job.
(c) Companies *a* and *b* are the only two bidders competing for the same job.

(a) Since companies *a* and *b* may each win a job, the Venn diagram is as shown in Fig. E2.7*a*. The overlapping region indicates that both companies *a* and *b* win jobs. In this case, events *A* and *B* are not mutually exclusive.

(b) Company *a* may win the job, or company *b* may win the job, or some other bidder will win the job. But, if company *a* wins the job, then event *B* will never occur. Therefore event *A* precludes the occurrence of event *B*, and vice versa; hence events *A* and *B* are mutually exclusive. There is no overlapping region in the Venn diagram for events *A* and *B*, as shown in Fig. E2.7*b*. In this case, the complementary event of $(A \cup B)$ means that neither company *a* nor company *b* wins the job.

(c) In this case, the sample space only contains the two events *A* and *B*. If event

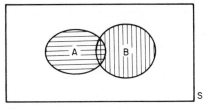

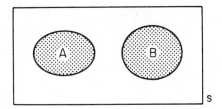

'igure E2.7a   Figure E2.7b

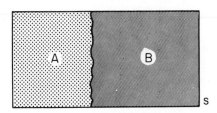

Figure E2.7c

A does not occur, that is, if company *a* loses, then we know definitely that *B* has occurred. Events *A* and *B* are again mutually exclusive; also, *A* and *B* are collectively exhaustive, that is $A \cup B = S$. Hence the corresponding Venn diagram would appear as in Fig. E2.7c.

2.2.3. Operational rules

Sets and the relationships among sets are governed by certain operational rules. In this connection, we adopt the following symbols to designate sets or their associated operations:

$\cup$ union
$\cap$ intersection
$\subset$ belongs to, or is contained in
$\supset$ contains
$\bar{E}$ complement of E

We have seen in Section 2.2.2 that two or more sets can be combined in two ways—through *union* and *intersection*. These and the process of *taking the complement* constitute the basic operations on sets. The rules that govern these operations are the following:

Equality of sets.   Two sets are *equal* if and only if both sets contain exactly the same sample points. On this basis, we immediately observe that

$$A \cup \phi = A$$

$$A \cap \phi = \phi \qquad (2.1a)$$

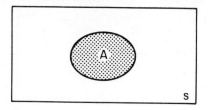

Figure 2.6 Event A

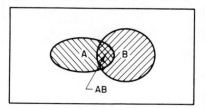

Figure 2.7 Venn diagram of two sets A and B

Also, referring to Fig. 2.6, we have

$$A \cup A = A$$
$$A \cap A = A \tag{2.1b}$$

Furthermore,

$$A \cup S = S$$
$$A \cap S = A \tag{2.1c}$$

Complementary sets. From Fig. 2.1, we observe the following relative to an event E and its complement $\bar{E}$:

$$E \cup \bar{E} = S$$
$$E \cap \bar{E} = \phi \tag{2.2}$$
$$(\bar{\bar{E}}) = E$$

(that is, the complement of the complementary event is the original event).

Commutative rule. Union and intersection of sets are *commutative*; that is,

$$A \cup B = B \cup A$$
$$AB = BA$$

From the Venn diagram of Fig. 2.7, we see that $A \cup B$ and $B \cup A$ clearly contain the same set of sample points and, therefore, are equal subsets within S. Similarly, the same is true of AB and BA.

Associative rule. Union and intersection of sets are *associative*; that is,

$$(A \cup B) \cup C = A \cup (B \cup C)$$
$$(AB)C = A(BC)$$

The equality of the sets $(A \cup B) \cup C$ and $A \cup (B \cup C)$ is clear from the

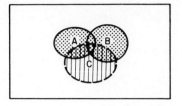 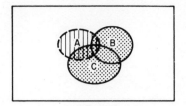

Figure 2.8   Venn diagrams for $(A \cup B) \cup C$ and $A \cup (B \cup C)$

Venn diagram of Fig. 2.8, whereas from Fig. 2.9, we see that $(AB)C = A(BC)$.

Distributive rule.   Union and intersection of sets are *distributive*; that is,

$$(A \cup B)C = AC \cup BC$$

$$(AB) \cup C = (A \cup C)(B \cup C)$$

In this case, the two equalities of the sets are verified by the Venn diagrams of Figs. 2.10 and 2.11, respectively.

These operational rules imply that the rules governing the addition and multiplication of numbers apply to the union and intersection of sets. By assuming the following equivalences—*union for addition* and *intersection for multiplication* (that is, $\cup \rightarrow +$ and $\cap \rightarrow \times$)—the rules of conventional algebra then apply to operations of sets and events. Therefore, in accordance with the hierarchy of algebraic operations, intersection takes precedence over union of events, unless parenthetically indicated otherwise. It should be emphasized, however, that conventional algebraic operations, such as addition and multiplication, have no meaning relative to sets and events. Moreover, there are operations and operational rules that apply to sets that have no counterparts in conventional algebra of num-

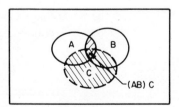 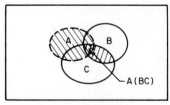

Double Hatched Region = (AB) C   Double Hatched Region = A (BC)

Figure 2.9   Venn diagrams for $(AB)C$ and $A(BC)$

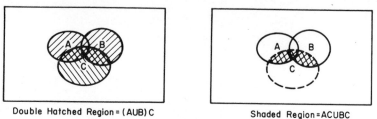

Figure 2.10   Venn diagrams for $A(B \cup C)$ and $AB \cup AC$

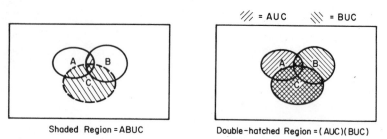

Figure 2.11   Venn diagrams for $AB \cup C$ and $(A \cup C)(B \cup C)$

bers. For example, $A \cup A = A$ and $A \cap A = A$. Another case in point is the second of the distributive rules described above, which says that

$$(A \cup C)(B \cup C) = AB \cup AC \cup CB \cup CC$$

$$= AB \cup C$$

whereas, in conventional algebra, we have

$$(a + c)(b + c) = ab + ac + cb + c^2 \neq ab + c$$

Finally, another rule that also has no counterpart in conventional algebra is the de Morgan's rule, described below.

de Morgan's rule.   Another rule in set theory is the *de Morgan's rule*, which relates sets and their complements. For two events E_1 and E_2, this rule says that

$$\overline{E_1 \cup E_2} = \bar{E}_1 \cap \bar{E}_2$$

To prove this relation, consider the two events E_1 and E_2 as shown in Fig. 2.12. The unshaded region in Fig. 2.12a is clearly $\overline{E_1 \cup E_2}$. The Venn diagrams with $\bar{E}_1$ and $\bar{E}_2$ are individually shown in Fig. 2.12b, the intersection of which is the double-hatched region in Fig. 2.12c. Comparing Figs. 2.12a and 2.12c, we have the above relation, $\overline{E_1 \cup E_2} = \bar{E}_1\bar{E}_2$.

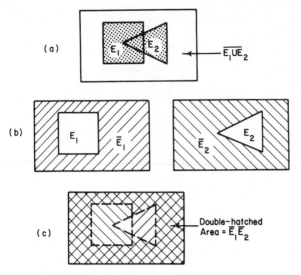

Figure 2.12 Venn diagram for de Morgan's rule

Stated in general, de Morgan's rule is

$$\overline{E_1 \cup E_2 \cup \ldots \cup E_n} = \bar{E}_1 \bar{E}_2 \ldots \bar{E}_n \qquad (2.3a)$$

Applying Eq. 2.3a to $\bar{E}_1, \bar{E}_2, \ldots, \bar{E}_n$, we have

$$\overline{\bar{E}_1 \cup \bar{E}_2 \cup \ldots \cup \bar{E}_n} = E_1 E_2 \ldots E_n$$

Hence taking the complements of both sides of this equation, de Morgan's rule can be stated also as

$$\overline{E_1 E_2 \ldots E_n} = \bar{E}_1 \cup \bar{E}_2 \ldots \cup \bar{E}_n \qquad (2.3b)$$

In view of Eqs. 2.3a and 2.3b we have the following *duality relation.*

The complement of unions and intersections is equal to the intersections and unions of the respective complements. For example,

$$\overline{A \cup BC} = \bar{A} \cap \overline{BC} = \bar{A}(\bar{B} \cup \bar{C}) = \bar{A}\bar{B} \cup \bar{A}\bar{C}$$

$$\overline{(A \cup B)C} = \overline{(A \cup B)} \cup \bar{C} = (\bar{A}\bar{B}) \cup \bar{C}$$

$$\overline{(\overline{E_1 E_2} \cup E_3)}(\bar{E}_1 \cup \bar{E}_3) = \overline{E_1 E_2 \bar{E}_3 \cup E_1 E_3} = \overline{E_1 E_2 \bar{E}_3} \cap \overline{E_1 E_3}$$

EXAMPLE 2.8

A chain consists of two links, as shown in Fig. E2.8. Clearly, the chain will fail if either link breaks; thus, if E_1 = breakage of link 1, and E_2 = breakage of link 2, then

$$\text{Failure of chain} = E_1 \cup E_2$$

Figure E2.8 A two-link chain

and no failure of the chain is, therefore, $\overline{E_1 \cup E_2}$. However, no failure of the chain also means that both links survive; that is,

$$\text{No failure of chain} = \bar{E}_1 \cap \bar{E}_2$$

Therefore

$$\overline{E_1 \cup E_2} = \bar{E}_1\bar{E}_2$$

which is an illustration of de Morgan's rule.

EXAMPLE 2.9

The water supply for a city C comes from two sources A and B. The water is transported by a pipeline consisting of branches 1, 2, and 3, as shown in Fig. E2.9. Assume that either source alone is sufficient to supply the water for the city.

Denote E_1 = failure of branch 1

E_2 = failure of branch 2

E_3 = failure of branch 3

Then shortage of water in the city would be caused by $E_1E_2 \cup E_3$. Therefore, by de Morgan's rule, no shortage means that

$$\overline{E_1E_2 \cup E_3} = (\bar{E}_1 \cup \bar{E}_2)\bar{E}_3$$

in which $(\bar{E}_1 \cup \bar{E}_2)$ means the availability of water at the junction, and $\bar{E}_3$ means no failure of branch 3.

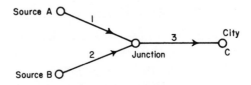

Figure E2.9 Water-supply system

2.3. MATHEMATICS OF PROBABILITY

2.3.1. Basic axioms of probability; addition rule

It may be pointed out that in all of our discussions thus far, we have tacitly assumed that a nonnegative measure, called *probability*, is associated with every event. Implicitly, we have also assumed that such measures possess certain properties and follow certain operational rules. Formally, these properties and rules are embodied in the *mathematical theory of probability*. As in other branches of mathematics, the theory of probability is based on certain fundamental assumptions, or axioms, as follows.

For every event E in a sample space S, there is a probability

$$P(E) \geq 0 \tag{2.4}$$

Secondly, the probability of the *certain event* S is

$$P(S) = 1.0 \tag{2.5}$$

Finally, for two events E_1 and E_2 that are *mutually exclusive*,

$$P(E_1 \cup E_2) = P(E_1) + P(E_2) \tag{2.6}$$

Equations 2.4 through 2.6 then constitute the basic axioms of probability theory. These are essential assumptions and therefore are not subject to proof. However, these axioms and the resulting theory must be consistent with and useful for real-world problems. In this latter regard, we observe that in essence, the probability of an event is a relative measure (that is, relative to other events in the same sample space); for this purpose, therefore, it is convenient or natural to assume such a measure to be nonnegative as prescribed in Eq. 2.4. Moreover, because an event E is always defined within a prescribed sample space S, it is convenient to normalize the probability of an event with respect to S (the certain event), as specified in Eq. 2.5. On the basis of Eqs. 2.4 and 2.5, it follows that the probability of an event E is bounded between 0 and 1.0; that is,

$$0 \leq P(E) \leq 1.0$$

With regard to the third axiom, Eq. 2.6, we observe that from a relative frequency standpoint, if an event E_1 occurs n_1 times among n repetitions of an experiment, and another event E_2 occurs n_2 times (in which E_1 and E_2 are mutually exclusive), then E_1 or E_2 will have occurred $(n_1 + n_2)$ times. Thence, on the basis of relative frequency, we have

$$P(E_1 \cup E_2) = \frac{n_1 + n_2}{n} = \frac{n_1}{n} + \frac{n_2}{n}$$
$$= P(E_1) + P(E_2)$$

It should be emphasized that the mathematical theory of probability is concerned with the logical bases for the relationships among probability measures. All such relationships and the deductive character of the theory can be developed entirely on the basis of the three assumptions described in Eqs. 2.4 through 2.6.

Applying Eq. 2.6 to E and its complement $\bar{E}$, we have

$$P(E \cup \bar{E}) = P(E) + P(\bar{E})$$

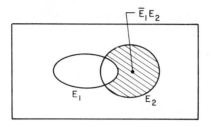

Figure 2.13   Union of E_1 and E_1E_2

but since $E \cup \bar{E} = S$, we have, on the basis of Eq. 2.5,

$$P(E \cup \bar{E}) = P(S) = 1.0$$

Hence the useful relation

$$P(\bar{E}) = 1 - P(E) \tag{2.7}$$

More generally, if E_1 and E_2 are not mutually exclusive, then

$$P(E_1 \cup E_2) = P(E_1) + P(E_2) - P(E_1E_2) \tag{2.8}$$

Equation 2.8 follows from Eq. 2.6 by observing from Fig. 2.13 that $E_1 \cup E_2 = E_1 \cup \bar{E}_1E_2$, where the events E_1 and $\bar{E}_1E_2$ are mutually exclusive; thus, according to Eq. 2.6,

$$P(E_1 \cup \bar{E}_1E_2) = P(E_1) + P(\bar{E}_1E_2)$$

But

$$\bar{E}_1E_2 \cup E_1E_2 = SE_2 = E_2$$

and E_1E_2 and $\bar{E}_1E_2$ are mutually exclusive; hence

$$P(\bar{E}_1E_2) = P(E_2) - P(E_1E_2)$$

thus obtaining Eq. 2.8.

EXAMPLE 2.10

A contractor is starting on two new projects—jobs 1 and 2. There is some uncertainty on the completion time for each job; in one year, each job may be *definitely completed, completion questionable,* or *definitely incomplete.* Let us denote these situations as A, B, and C, respectively, for each job. Describe the sample space for the state of completion of the two jobs; that is, describe all the possible situations of jobs 1 and 2 after one year.

If each of the possibilities for the two jobs is equally likely to occur at the end of one year, what is the probability that exactly one job will be definitely completed in one year?

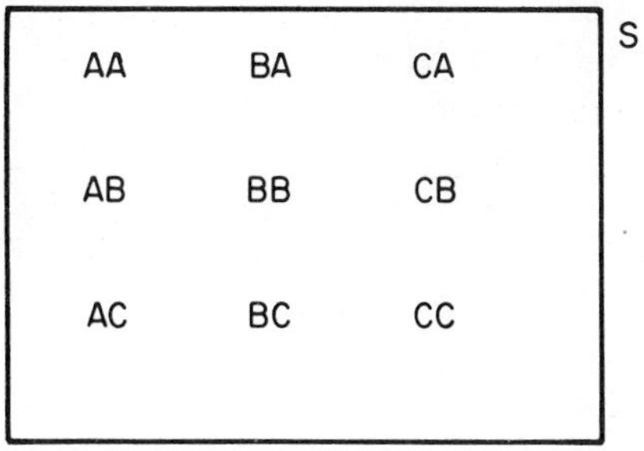

Figure E2.10a   Sample space 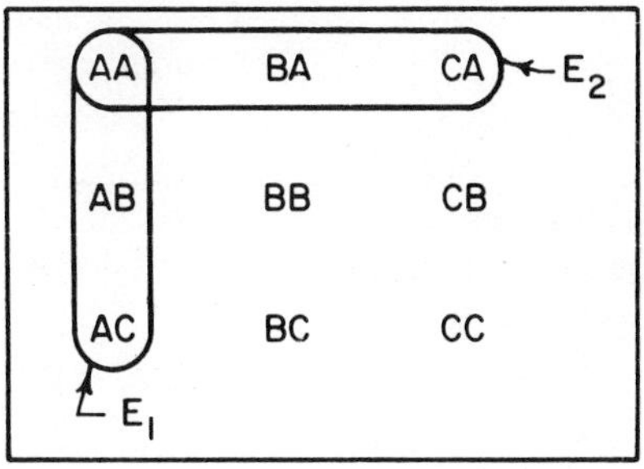   **Figure E2.10b**

Sample space is shown in Fig. E2.10a. Since the event of *exactly one job completed* contains the four sample points AB, AC, BA, and CA, this probability is equal to $4 \times 1/9 = 4/9$.

In this problem, if we let E_1 be the event that job 1 is definitely completed, and E_2 that job 2 is definitely completed, then

$$E_1 \supset \{AA, AB, AC\}$$
$$E_2 \supset \{AA, BA, CA\}$$

The Venn diagram with events E_1 and E_2 will appear as in Fig. 2.10b. If the sample points are equally likely to occur, then $P(E_1) = 3/9$, $P(E_2) = 3/9$, and according to Eq. 2.8

$$P(E_1 \cup E_2) = 3/9 + 3/9 - 1/9 = 5/9$$

which can be verified since $(E_1 \cup E_2) \supset \{AA, AB, AC, BA, CA\}$.

EXAMPLE 2.11

For the purpose of designing the left-turn lane (for eastbound traffic) in Example 2.2, the 60 observations (made at random) of the number of cars waiting for left turns at the intersection, yielded the following results:

No. of cars	No. of observations	Relative frequency
0	4	4/60
1	16	16/60
2	20	20/60
3	14	14/60
4	3	3/60
5	2	2/60
6	1	1/60
7	0	0
8	0	0
.	.	.
.	.	.
.	.	.

Let
$$E_1 = \text{more than 2 cars waiting for left turns}$$
$$E_2 = \text{2 to 4 cars waiting for left turns}$$

Since the number of cars waiting for left turns are mutually exclusive events, a simple extension of Eq. 2.6 (see Eq. 2.6a, pg. 41) and using the above relative frequencies to represent the corresponding probabilities, we obtain

$$P(E_1) = \frac{14}{60} + \frac{3}{60} + \frac{2}{60} + \frac{1}{60} = \frac{20}{60}$$

whereas

$$P(E_2) = \frac{20}{60} + \frac{14}{60} + \frac{3}{60} = \frac{37}{60}$$

Also, in terms of the number of cars waiting for left turns,

$$E_1 E_2 \supset \{3, 4\}$$

and thus

$$P(E_1 E_2) = \frac{14}{60} + \frac{3}{60} = \frac{17}{60}$$

Then, according to Eq. 2.8,

$$P(E_1 \cup E_2) = \frac{20}{60} + \frac{37}{60} - \frac{17}{60} = \frac{40}{60}$$

In this case, we also observe that

$$\bar{E}_1 \cup E_2 \supset \{2, 3, 4, \ldots\}$$

Hence

$$P(E_1 \cup E_2) = \frac{20}{60} + \frac{14}{60} + \frac{3}{60} + \frac{2}{60} + \frac{1}{60} = \frac{40}{60}$$

Thus, verifying the result obtained above using Eq. 2.8.

EXAMPLE 2.12

In Example 2.6, events are represented by areas in the sample space of R_A and R_B, as shown in the Venn diagram of Fig. E2.12. If the probability of an event is proportional to its "area" (this corresponds to the assumption that the sample points are equally likely), we obtain the following.

$$\text{Total "area" of sample space} = \frac{1}{2}[(300)^2 - (100)^2]$$

$$= 40,000$$

Then, referring to Fig. E2.12, we have

$$P(A) = \frac{\frac{1}{2}(200)^2}{40,000} = \frac{1}{2}$$

Similarly,

$$P(B) = \frac{1}{2}$$

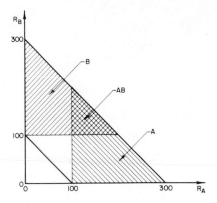

Figure E2.12

whereas

$$P(AB) = \frac{\frac{1}{2}(100)^2}{40{,}000} = \frac{1}{8}$$

and

$$P(A \cup B) = \frac{40{,}000 - \frac{1}{2}(100)^2}{40{,}000} = \frac{7}{8}$$

By Eq. 2.8, we also obtain

$$P(A \cup B) = \frac{1}{2} + \frac{1}{2} - \frac{1}{8} = \frac{7}{8}$$

For three events E_1, E_2, E_3,

$$
\begin{aligned}
P(E_1 \cup E_2 \cup E_3) &= P[(E_1 \cup E_2) \cup E_3] \\
&= P(E_1 \cup E_2) + P(E_3) - P[(E_1 \cup E_2)E_3] \\
&= P(E_1) + P(E_2) + P(E_3) - P(E_1E_2) - P(E_1E_3) \\
&\quad - P(E_2E_3) + P(E_1E_2E_3)
\end{aligned}
\tag{2.9}
$$

The preceding procedure may be extended to the union of any number of events; however, for n events, the probability of the union may be obtained more conveniently using de Morgan's rule, as follows:

$$
\begin{aligned}
P(E_1 \cup E_2 \cup \ldots \cup E_n) &= 1 - P\,(\overline{E_1 \cup E_2 \cup \ldots \cup E_n}) \\
&= 1 - P(\bar{E}_1 \bar{E}_2 \ldots \bar{E}_n)
\end{aligned}
\tag{2.10}
$$

However, if the n events are mutually exclusive, extension of the third proposition (Eq. 2.6) yields

$$P(E_1 \cup E_2 \cup \ldots \cup E_n) = \sum_{i=1}^{n} P(E_i) \tag{2.6a}$$

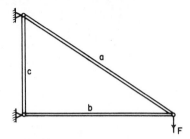

Figure E2.13

EXAMPLE 2.13

Under the load F, the probabilities of failure of the individual members a, b, and c of the truss shown in Fig. E2.13 are 0.05, 0.04, and 0.03, respectively. The failure of any member(s) will constitute failure of the truss.

Assuming that failures of the individual members are *statistically independent*, so that the failure probability of two or more members is equal to the product of the respective member probabilities (see Eq. 2.15, pg. 47), determine the failure probability of the truss.

Denoting the failure events of the three members as A, B, and C, we have $P(A) = 0.05$, $P(B) = 0.04$, and $P(C) = 0.03$. And with the assumption of statistical independence,

$$P(AB) = (0.05)(0.04) = 0.0020$$
$$P(AC) = (0.05)(0.03) = 0.0015$$
$$P(BC) = (0.04)(0.03) = 0.0012$$

and

$$P(ABC) = (0.05)(0.04)(0.03) = 0.00006$$

Then, according to Eq. 2.9,

$$P(\text{failure of truss}) = P(A \cup B \cup C)$$
$$= 0.05 + 0.04 + 0.03$$
$$-0.0020 - 0.0015 - 0.0012$$
$$+0.00006$$
$$= 0.11536$$

This probability may also be obtained (more conveniently) with Eq. 2.10 as follows:

$$P(A \cup B \cup C) = 1 - P(\bar{A}\bar{B}\bar{C})$$

In the present case (see Eq. 2.16), we have

$$P(\bar{A}\bar{B}\bar{C}) = P(\bar{A})P(\bar{B})P(\bar{C})$$
$$= (1 - 0.05)(1 - 0.04)(1 - 0.03) = 0.88464$$

Hence

$$P(\text{failure}) = 1 - 0.88464 = 0.11536$$

2.3.2. Conditional probability; multiplication rule

The probability of an event may depend on the occurrence (or non-occurrence) of another event. If this dependence exists, the associated probability is a *conditional probability.*

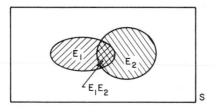

Figure 2.14 Reconstituted sample space E_2

In the sample space of Fig. 2.14, the conditional probability of E_1 assuming E_2 has occurred, denoted $P(E_1 \mid E_2)$, means the likelihood of realizing a sample point in E_1 assuming that it belongs to E_2. Effectively, in other words, we are interested in the event E_1 within the reconstituted sample space E_2. Hence, with the appropriate normalization, we obtain the conditional probability of E_1 *given* E_2 as

$$P(E_1 \mid E_2) = \frac{P(E_1 E_2)}{P(E_2)} \tag{2.11}$$

To clarify this concept, consider the following examples.

EXAMPLE 2.14

Consider a 100-km (kilometer) highway, and assume that the road condition and traffic volume are uniform throughout the 100-km distance, so that accidents are equally likely to occur anywhere on the highway. Define the events

A = an accident in kilometers 0 to 30

B = an accident in kilometers 20 to 60

Since accidents are equally likely anywhere along the highway, it may be assumed that the probability of an accident in a given interval of the highway is proportional to the distance of the interval. Therefore, if an accident occurs on this 100-km highway,

$$P(A) = \frac{30}{100} \quad \text{and} \quad P(B) = \frac{40}{100}$$

Now let us pose the question: "if an accident occurs in the interval (20, 60), what is the probability of the event A?" In this case, we are interested in the probability of A on the condition that B has occurred; this is simply the proportion of the distance that belongs to B within which A is also realized. Clearly, from Fig.

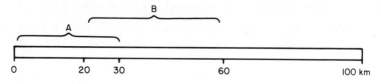

Figure E2.14

E2.14, this conditional probability is

$$P(A \mid B) = \frac{10}{40} = \frac{10/100}{40/100}$$

But, in this case, $10/100 = P(AB)$, and $40/100 = P(B)$, thus illustrating Eq. 2.11.

EXAMPLE 2.15

Consider again the problem of three bulldozers, described earlier in Example 2.1 Let

F = event that the first bulldozer is operational after 6 months

E = 2 bulldozers are operational after 6 months

If the sample points are all equally likely, then referring to the Venn diagram shown in Fig. E2.15, the conditional probability of E given F is

$$P(E \mid F) = \frac{2}{4}$$

This is simply the ratio of the number of sample points in EF relative to those in F, illustrating therefore the notion that F is taken as the new "sample space." Similarly, the conditional probability of F given E would be

$$P(F \mid E) = \frac{2}{3}$$

However, if the sample points are not equally likely, then the associated probability measures must be used in the calculation of the conditional probability. For example, if the probability of a bulldozer's operating at least 6 months is 80%, then (assuming

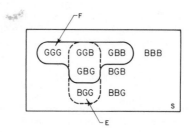

Figure E2.15

statistical independence; see Eq. 2.15, pg. 47) the probabilities of the various sample points will be as follows:

$$P(GGG) = 0.512$$
$$P(GGB) = 0.128$$
$$P(GBB) = 0.032$$
$$P(BBB) = 0.008$$
$$P(BGG) = 0.128$$
$$P(BBG) = 0.032$$
$$P(GBG) = 0.128$$
$$P(BGB) = 0.032$$

In this case, $P(E \mid F)$ must reflect the probabilities of the sample points in EF relative to those of the sample points in F. Accordingly, we have

$$P(E \mid F) = \frac{P(GGB \cup GBG)}{P(GGG \cup GGB \cup GBG \cup GBB)} = \frac{P(EF)}{P(F)}$$

$$= \frac{0.128 + 0.128}{0.512 + 0.128 + 0.128 + 0.032} = \frac{0.256}{0.800} = 0.32$$

It may be emphasized that the conditional probability is merely a generalization of the probability of an event. When we speak of the probability of an event E, it is implicitly conditioned on the sample space S. [This is illustrated in Example 2.14; the probabilities $P(A)$ and $P(B)$ are based on the condition that an accident occurs in the 100-km highway.] To be more explicit, $P(E)$ should be written

$$P(E \mid S) = \frac{P(ES)}{P(S)}$$

But since $ES = E$, and $P(S) = 1.0$,

$$P(E \mid S) = P(E)$$

In other words, conditioning on the sample space S is presumed to be understood; however, when the probability is conditioned on an event other than the original sample space, the reconstituted "sample space" must be made explicit.

We observe also that

$$P(E_1 \mid E_2) + P(\bar{E}_1 \mid E_2) = \frac{P(E_1 E_2)}{P(E_2)} + \frac{P(\bar{E}_1 E_2)}{P(E_2)}$$

$$= \frac{1}{P(E_2)} \left[P\{ (E_1 \cup \bar{E}_1) E_2 \} \right]$$

$$= \frac{P(E_2)}{P(E_2)} = 1.0$$

Therefore

$$P(\bar{E}_1 \mid E_2) = 1 - P(E_1 \mid E_2) \qquad (2.12)$$

which is a generalization of Eq. 2.7. It is important to recognize that in Eq. 2.12 the conditioning event E_2 is the reconstituted sample space; for this reason we must make sure, when applying Eq. 2.12, that the event (for example, E_1) and its complement refer to the same reconstituted sample space E_2. For example, observe the following:

$$P(E_1 \mid \bar{E}_2) \neq 1 - P(E_1 \mid E_2)$$
$$P(\bar{E}_1 \mid \bar{E}_2) \neq 1 - P(E_1 \mid E_2)$$

EXAMPLE 2.16

It has been observed that vehicles approaching a certain intersection in a given direction are twice as likely to go *straight ahead* than to make a *right turn*; also, *left turns* are only half as likely as right turns.

Assume that these conditions are valid for any vehicle. Then if a vehicle approaches the intersection in the indicated direction, we can ask the following.

(a) What are all the possibilities (that is, the different directions for the vehicle to take)?

$$\text{Straight ahead} = E_1$$
$$\text{Turn right} = E_2$$
$$\text{Turn left} = E_3$$

(b) What are the respective probabilities?

$$P(E_1) = \frac{4}{7} \qquad P(E_2) = \frac{2}{7} \qquad P(E_3) = \frac{1}{7}$$

(c) What is the probability of a right turn if a car is definitely going to make a turn?

$$P(E_2 \mid E_2 \cup E_3) = \frac{P[E_2(E_2 \cup E_3)]}{P(E_2 \cup E_3)} = \frac{P(E_2)}{P(E_2 \cup E_3)} = \frac{\frac{2}{7}}{\frac{3}{7}} = \frac{2}{3}$$

On the other hand, if a vehicle is definitely turning at the intersection, the probability that it will *not* turn right is

$$P(\bar{E}_2 \mid E_2 \cup E_3) = 1 - P(E_2 \mid E_2 \cup E_3)$$
$$= 1 - \frac{2}{3} = \frac{1}{3}$$

Statistical independence.   If the occurrence (or nonoccurrence) of one event does not affect the probability of another event, the two events are *statistically independent*. Therefore, if E_1 and E_2 are statistically independent,*

$$P(E_2 \mid E_1) = P(E_2)$$

and   (2.13)

$$P(E_1 \mid E_2) = P(E_1)$$

Multiplication rule.   From Eq. 2.11, the probability of the joint event E_1E_2 is

$$P(E_1E_2) = P(E_1 \mid E_2) \, P(E_2)$$

or   (2.14)

$$P(E_1E_2) = P(E_2 \mid E_1) \, P(E_1)$$

If E_1 and E_2 are statistically independent events, then this multiplication rule becomes

$$P(E_1E_2) = P(E_1) \, P(E_2) \tag{2.15}$$

For three events, the multiplication rule is

$$P(E_1E_2E_3) = P(E_1 \mid E_2E_3) \, P(E_2 \mid E_3) \, P(E_3) \tag{2.14a}$$

and if the events are statistically independent,

$$P(E_1E_2E_3) = P(E_1) \, P(E_2) \, P(E_3) \tag{2.15a}$$

We would expect that if E_1 and E_2 are statistically independent, their complements $\bar{E}_1$ and $\bar{E}_2$ would also be statistically independent. This can be verified in the case of two events as follows:

$$
\begin{aligned}
P(\bar{E}_1\bar{E}_2) &= P(\overline{E_1 \cup E_2}) = 1 - P(E_1 \cup E_2) \\
&= 1 - [P(E_1) + P(E_2) - P(E_1) \, P(E_2)] \\
&= [1 - P(E_1)][1 - P(E_2)] \\
&= P(\bar{E}_1) \, P(\bar{E}_2) \tag{2.16}
\end{aligned}
$$

Finally, we should emphasize that all the mathematical rules pertaining to probability apply equally to conditional probabilities defined within the same reconstituted sample space, including specifically the following:

$$P(E_1 \cup E_2 \mid A) = P(E_1 \mid A) + P(E_2 \mid A) - P(E_1E_2 \mid A) \tag{2.17}$$

$$P(E_1E_2 \mid A) = P[(E_1 \mid E_2) \mid A] P(E_2 \mid A) \tag{2.18}$$

* This way of defining statistical independence is intuitively more direct. Although this is somewhat unconventional, because statistical independence is usually defined mathematically in the form of Eq. 2.15, the Mathematical Association of America (1972) suggested the use of the conditional definition of statistical independence—that is, Eq. 2.13.

whereas if E_1 and E_2 are statistically independent,

$$P(E_1E_2 \mid A) = P(E_1 \mid A) \, P(E_2 \mid A) \qquad (2.18a)$$

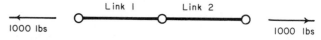

Figure E2.17

EXAMPLE 2.17

Consider again a chain system consisting of two links (Fig. E2.17). If the applied force is 1000 lb, it is obvious that any link in the chain will fail if its strength is less than 1000 lb. Suppose that the probability of this happening to either link is 0.05. What is the probability of failure of the chain?

Let E_1 and E_2 denote the failure of links 1 and 2, respectively. Therefore the failure of the chain is

$$P(E_1 \cup E_2) = P(E_1) + P(E_2) - P(E_1E_2)$$
$$= 0.05 + 0.05 - P(E_2 \mid E_1)P(E_1)$$
$$= 0.10 - 0.05 \, P(E_2 \mid E_1)$$

We observe that the conditional probability $P(E_2 \mid E_1)$ is required; this will depend on the degree of mutual dependence between E_1 and E_2. For example, if the links are randomly selected from two suppliers, then E_1 and E_2 may be assumed to be statistically independent; thus $P(E_2 \mid E_1) = P(E_2) = 0.05$. In such a case,

$$P(E_1 \cup E_2) = 0.1 - 0.05 \times 0.05 = 0.0975$$

Conversely, if the two links were fabricated from the same steel bar by the same manufacturer, the characteristics of the two links can be expected to be quite similar. In the extreme case, the strength of the links may be assumed to be identical; in this case, $P(E_2 \mid E_1) = 1.0$. Thence

$$P(E_1 \cup E_2) = 0.10 - 0.05 \times 1.0 = 0.05$$

which is the same as the failure probability of one link.

The failure probability of the chain system, therefore, ranges between 0.05 (which is the failure probability of a single link) and 0.0975, depending on the conditional probability $P(E_2 \mid E_1)$, which is a function of the degree of correlation between the strengths of the two links.

EXAMPLE 2.18

Two power generating units a and b operate in parallel to supply the power requirements of a small city. The demand for power is subject to considerable fluctuation, and it is known that each unit has a capacity so that it can supply the city's full power requirement 75% of the time in case the other unit fails. The probability of failure of each unit is 0.10, whereas the probability that both units will fail is 0.02.

If there is failure in the power generation, what is the probability that the city will have its supply of full power?

Let

$$A = \text{event unit } a \text{ fails}$$
$$B = \text{event unit } b \text{ fails}$$

Then

$$P(A) = P(B) = 0.10$$
$$P(AB) = 0.02$$

hence

$$P(A \mid B) = P(B \mid A) = \frac{0.02}{0.10} = 0.20$$

The conditional probability that when there is failure, only one of the two units failed is

$$P(A\bar{B} \cup \bar{A}B \mid A \cup B) = P(A\bar{B} \mid A \cup B) + P(\bar{A}B \mid A \cup B)$$

$$= \frac{P[A\bar{B}(A \cup B)]}{P(A \cup B)} + \frac{P[\bar{A}B(A \cup B)]}{P(A \cup B)}$$

$$= \frac{P(A\bar{B})}{P(A \cup B)} + \frac{P(\bar{A}B)}{P(A \cup B)}$$

$$= \frac{P(\bar{B} \mid A)P(A) + P(\bar{A} \mid B)P(B)}{P(A) + P(B) - P(B \mid A)P(A)}$$

$$= \frac{0.8 \times 0.1 + 0.8 \times 0.1}{0.1 + 0.1 - 0.2 \times 0.1}$$

$$= 2\left(\frac{0.08}{0.18}\right) = 0.89$$

Thus the probability that the city will have supply of full power, when there is failure in the power generation, is $0.89 \times 0.75 = 0.67$.

EXAMPLE 2.19

Before a section (say $\frac{1}{10}$ mile long) of a pavement is accepted by the State Highway Department, the thickness of an 8-in. pavement is inspected for specification compliance by ultrasonics reading (see Fig. E2.19). This is done at every $\frac{1}{10}$-mile point of the pavement; each $\frac{1}{10}$-mile section will be accepted if the measured thickness is at least 7.5 in.; otherwise the entire section will be rejected.

Suppose, from past experience, that 90% of all sections constructed by the contractor were found to be in compliance with specifications. However, the ultrasonics thickness determination is only 80% reliable; that is, there is a 20% chance that a conclusion based on ultrasonics test may be erroneous.

(a) What is the probability that a particular section of the pavement is well constructed (that is, at least 7.5-in. thick) *and* will be accepted by the Highway Department?

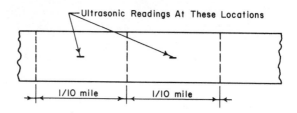

Figure E2.19

Let
$$G = \text{actual thickness of pavement is at least 7.5 in.}$$
$$A = \text{measured thickness} \geq 7.5 \text{ in.}$$

The statement "reliability of the ultrasonics test is 80%" may be interpreted to mean

$$P(G \mid A) = 0.80$$

and

$$P(\bar{G} \mid \bar{A}) = 0.80$$

Hence

$$P(G \mid \bar{A}) = 1 - 0.80 = 0.20$$

Based on the contractor's past record, we may assume that 90% of his work will have satisfactory ultrasonics readings; hence

$$P(A) = 0.90$$

The event of interest is GA; its probability, therefore, is

$$P(GA) = P(G \mid A)P(A)$$
$$= (0.80)(0.90) = 0.72$$

(b) What is the probability that a section is poorly constructed (that is, has thickness less than 7.5 in.) but will be accepted on the basis of the ultrasonics test?

In this case, we have

$$P(\bar{G}A) = P(\bar{G} \mid A)P(A)$$
$$= (0.20)(0.90) = 0.18$$

EXAMPLE 2.20

The settlement problem of a steel frame may be idealized as follows. A and B represent two footings resting on soil (Fig. E2.20). Each footing may either remain at the original level or settle 5 cm. The probability of settlement in each footing is 0.1. However, the probability that a footing will settle, given that the other has settled, is 0.8.

(a) The possible conditions of the two footings are as follows:

$$AB \qquad A \text{ settles, } B \text{ settles}$$
$$\bar{A}B \qquad A \text{ does not settle, } B \text{ settles}$$
$$A\bar{B} \qquad A \text{ settles, } B \text{ does not settle}$$
$$\bar{A}\bar{B} \qquad A \text{ does not settle, } B \text{ does not settle}$$

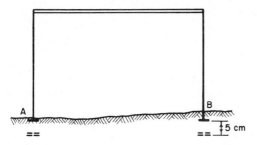

Figure E2.20

(b) The probability of settlement (that is, either A or B will settle) is

$$P(A \cup B) = P(A) + P(B) - P(AB)$$
$$= P(A) + P(B) - P(A)P(B \mid A)$$
$$= 0.1 + 0.1 - 0.1 \times 0.8 = 0.12$$

(c) If we are interested in the event E that differential settlement (that is, a difference in the level of the two footings) occurs, the event will consist of $\bar{A}B$ and $A\bar{B}$. Since these two events are mutually exclusive,

$$P(E) = P(\bar{A}B) + P(A\bar{B})$$
$$= P(B)P(\bar{A} \mid B) + P(A)P(\bar{B} \mid A)$$
$$= (0.1)[1 - P(A \mid B)] + (0.1)[1 - P(B \mid A)]$$
$$= (0.1)(0.2) + (0.1)(0.2) = 0.04$$

EXAMPLE 2.21

The foundation of a tall building may fail either from bearing capacity, or by excessive settlement. Let B and S represent the respective failure modes. If $P(B) = 0.001$ and $P(S) = 0.008$, and $P(B \mid S) =$ probability of failure in bearing capacity given that it has excessive settlement $= 0.1$, determine (a) the probability of failure of the foundation; (b) the probability that the building has excessive settlement but no failure in bearing capacity.

(a)
$$P(F) = P(B \cup S) = P(B) + P(S) - P(B \cap S)$$
$$= P(B) + P(S) - P(B \mid S)P(S)$$
$$= 0.001 + 0.008 - (0.1)(0.008)$$
$$= 0.009 - 0.0008 = 0.0082$$

(b)
$$P(S \cap \bar{B}) = P(\bar{B} \mid S)P(S)$$
$$= [1 - P(B \mid S)]P(S)$$
$$= (1 - 0.1)(0.008) = 0.9 \times 0.008 = 0.0072$$

In this problem, the conditional probability $P(B \mid S)$ cannot be larger than $1/8$; can you explain why?

EXAMPLE 2.22

There are two streams flowing past an industrial plant. The dissolved oxygen, DO, level in the water downstream is an indication of the degree of pollution caused by the waste dumped from the industrial plant. Let A denote the event that stream a is polluted, and B the event that stream b is polluted. From measurements taken on the DO level of each stream over the last year, it was determined that in a given day

$$P(A) = \frac{2}{5} \quad \text{and} \quad P(B) = \frac{3}{4}$$

and the probability that at least one stream will be polluted in any given day is $P(A \cup B) = 4/5$.

(a) Determine the probability that stream a is also polluted given that stream b is polluted.

(b) Determine the probability that stream b is also polluted given that stream a is polluted.

First, we compute the probability that both streams are polluted. Since

$$P(A \cup B) = P(A) + P(B) - P(A \cap B)$$

we have

$$P(A \cap B) = P(A) + P(B) - P(A \cup B)$$

$$= \frac{2}{5} + \frac{3}{4} - \frac{4}{5} = \frac{7}{20}$$

Therefore

$$P(A \mid B) = \frac{P(A \cap B)}{P(B)} = \frac{7/20}{3/4} = \frac{7}{15}$$

and

$$P(B \mid A) = \frac{P(A \cap B)}{P(A)} = \frac{7/20}{2/5} = \frac{7}{8}$$

In other words, stream b is very likely to be polluted when stream a is polluted, whereas chances are less than 50% that stream a will be polluted when stream b is polluted.

2.3.3.   Theorem of total probability

Sometimes the probability of an event A cannot be determined directly. However, its occurrence is always accompanied by the occurrence of other events E_i, $i = 1, 2, \ldots, n$, such that the probability of A will depend on which of the events E_i has occurred. In such a case the probability of A will be an expected probability (that is, the average probability weighted by those of E_i). Such problems require the *theorem of total probability*. By way of introduction, consider the following example.

EXAMPLE 2.23

Suppose that there is considerable uncertainty concerning the fate of the U.S. supersonic transport (SST) project. Whether or not the United States will have a commercial SST by 1980 will depend on the result of the presidential election in 1976. Suppose also that if the Democrats win the election, the probability of an SST by 1980 is only 20%, whereas if the Republicans win in 1976, this probability will be 70%.

Clearly, without knowing the party that will win the 1976 election, we cannot say whether the required probability will be 20% or 70%. However, if the two major parties have equal chances of winning in 1976, this probability would be the average of 0.20 and 0.70; or

$$P(\text{SST by 1980}) = 0.2(0.5) + 0.7(0.5) = 0.45$$

If the Republicans are favored by 3 to 2 to win in 1976, it would be reasonable to weigh the preceding probabilities by the respective odds of winning the election; thus

$$P(\text{SST by 1980}) = 0.2(0.4) + 0.7(0.6) = 0.50$$

whereas, if the Democrats are favored 3 to 2 to win the election, the corresponding probability would be

$$P(\text{SST in 1980}) = 0.2(0.6) + 0.7(0.4) = 0.40$$

Formally, consider n mutually exclusive and collectively exhaustive events $E_1, E_2, \ldots , E_n$; that is, $E_1 \cup E_2 \cup \ldots \cup E_n = S$. Then if A is an event also in the same sample space (see Fig. 2.15), we have

$$A = AS$$

$$= A (E_1 \cup E_2 \cup \ldots \cup E_n)$$

$$= AE_1 \cup AE_2 \cup \ldots \cup AE_n$$

where $AE_1, AE_2, \ldots , AE_n$ are also mutually exclusive, as can be seen in the Venn diagram of Fig. 2.15. Then

$$P(A) = P(AE_1) + P(AE_2) + \cdots + P(AE_n)$$

and by virtue of the multiplication rule, Eq. 2.14, we obtain the *total probability* theorem

$$P(A) = P(A \mid E_1) P(E_1) + P(A \mid E_2) P(E_2) + \cdots + P(A \mid E_n) P(E_n)$$

$$(2.19)$$

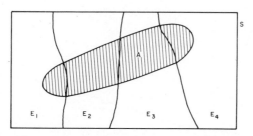

Figure 2.15   Venn diagram with events A and $E_1, E_2 \ldots , E_n$

In applying the total probability theorem, it is important to observe that the events E_i, $i = 1, 2, \ldots, n$, must be mutually exclusive and collectively exhaustive.

EXAMPLE 2.24

Figure E2.24 shows one direction of two interstate highways I_1 and I_2 merging into I_3. Assume that I_1 and I_2 have equal capacities; the rush-hour traffic, however, is somewhat different, so that during rush hours

$$P(I_1) = P(\text{excessive traffic in } I_1) = 10\%$$
$$P(I_2) = P(\text{excessive traffic in } I_2) = 20\%$$

Also, denoting $P(I_1 \mid I_2)$ as the probability of excessive traffic in I_1, given excessive traffic in I_2, we have

$$P(I_1 \mid I_2) = 50\%$$

and

$$P(I_2 \mid I_1) = 100\%$$

(a) If the capacity of I_3 is the same as that of I_1 or I_2, what is the probability of excessive traffic in I_3? Assume that when I_1 and I_2 are carrying less than their traffic capacities, I_3 may be exceeded with probability 20%.

First, we observe that this probability will depend on the traffic conditions in I_1 and I_2, which may be $I_1 I_2$, $\bar{I}_1 I_2$, $I_1 \bar{I}_2$, or $\bar{I}_1 \bar{I}_2$, with respective probabilities as follows:

$$
\begin{aligned}
P(I_1 I_2) &= P(I_1 \mid I_2)P(I_2) = 0.5 \times 0.2 = 0.10 \\
P(\bar{I}_1 I_2) &= P(\bar{I}_1 \mid I_2)P(I_2) \\
&= [1 - P(I_1 \mid I_2)]P(I_2) = 0.5(0.2) = 0.10 \\
P(I_1 \bar{I}_2) &= P(\bar{I}_2 \mid I_1)P(I_1) \\
&= [1 - P(I_2 \mid I_1)]P(I_1) = 0 \\
P(\bar{I}_1 \bar{I}_2) &= 1 - [P(I_1 I_2) + P(\bar{I}_1 I_2) + P(I_1 \bar{I}_2)] \\
&= 1 - (0.1 + 0.1 + 0) = 0.80
\end{aligned}
$$

Clearly, the traffic in I_3 will be excessive when the traffic in I_1 or I_2, or both, is excessive. Also, we have $P(I_3 \mid \bar{I}_1 \bar{I}_2) = 0.20$.

Then

$$
\begin{aligned}
P(I_3) &= P(I_3 \mid I_1 I_2)P(I_1 I_2) + P(I_3 \mid \bar{I}_1 I_2)P(\bar{I}_1 I_2) + P(I_3 \mid I_1 \bar{I}_2)P(I_1 \bar{I}_2) \\
&\quad + P(I_3 \mid \bar{I}_1 \bar{I}_2)P(\bar{I}_1 \bar{I}_2) \\
&= 1.00(0.10) + 1.00(0.10) + 1.00(0) + 0.20(0.80) \\
&= 0.36
\end{aligned}
$$

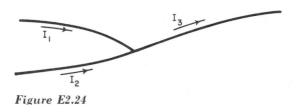

Figure E2.24

(b) If the capacity of I_3 is *twice* that of I_1 or I_2, what is the probability of excessive traffic in I_3? Assume that if only I_1 or I_2 has excessive traffic, the capacity of I_3 may be exceeded with probability of 15%.

Therefore $P(I_3 | I_1 \bar{I}_2) = P(I_3 | \bar{I}_1 I_2) = 0.15$. Furthermore, it is obvious that I_3 will have excessive traffic when I_1 and I_2 both have excessive traffic. Then, in this case,

$$P(I_3) = P(I_3 | I_1 I_2)P(I_1 I_2) + P(I_3 | \bar{I}_1 I_2)P(\bar{I}_1 I_2) + P(I_3 | I_1 \bar{I}_2)P(I_1 \bar{I}_2)$$
$$P(I_3 | \bar{I}_1 \bar{I}_2)P(\bar{I}_1 \bar{I}_2)$$
$$= 1.00(0.10) + 0.15(0.10) + 0.15(0) + 0(0.80)$$
$$= 0.115$$

EXAMPLE 2.25

Suppose that in any given year, the probability of damaging storms (that is, storms with wind speed exceeding say 60 mph) in the county of Champaign is 0.20. During such a storm, if not accompanied by tornadoes, the probability of structural failures in the city of Urbana (which is in Champaign County) is 0.10.

When a storm occurs in the county, the probability that it will be accompanied by a tornado is 0.25, and the probability that this tornado will hit the city of Urbana is 0.05. Assume that tornadoes occur only during a storm, and when the city is hit by a tornado it is certain to cause structural failures, whereas the probability of structural failures in the city when a tornado occurs in the county but does not hit the city is 0.10.

Calculate the probability of structural failures in the city of Urbana in a period of one year.

Define the following events:

$$F = \text{failure of structures in city of Urbana}$$
$$S = \text{storm in Champaign County}$$
$$T = \text{tornado in Champaign County}$$
$$H = \text{tornado hitting city of Urbana}$$

Clearly, the events ST, $S\bar{T}$, $\bar{S}T$, and $\bar{S}\bar{T}$ are mutually exclusive and collectively exhaustive; hence the probability of structural failures in the city is

$$P(F) = P(F | ST)P(ST) + P(F | S\bar{T})P(S\bar{T}) + P(F | \bar{S}T)P(\bar{S}T)$$
$$+ P(F | \bar{S}\bar{T})P(\bar{S}\bar{T})$$

where

$$P(F | ST) = P[(F | ST) | H]P(H) + P[(F | ST) | \bar{H}]P(\bar{H})$$
$$= 1.00(0.05) + 0.10(0.95) = 0.145$$
$$P(F | S\bar{T}) = 0.10$$
$$P(F | \bar{S}T) = \text{unknown; not needed in this problem}$$
$$P(F | \bar{S}\bar{T}) = 0$$

Also

$$P(ST) = P(T | S)P(S) = 0.25(0.20) = 0.05$$
$$P(S\bar{T}) = P(\bar{T} | S)P(S) = 0.75(0.20) = 0.15$$
$$P(\bar{S}T) = 0$$
$$P(\bar{S}\bar{T}) = 0.80$$

Therefore

$$P(F) = 0.145(0.05) + 0.10(0.15) + (?)(0) + 0(0.80)$$
$$= 0.00725 + 0.015 = 0.0222$$

2.3.4. Bayes' theorem

In the situation underlying the total probability theorem (see Section 2.3.3), if the event A occurred, what is the probability that a particular E_i also occurred? This may be considered as a "reverse" probability.

Applying Eq. 2.14 to the joint event AE_i, we have

$$P(A \mid E_i)P(E_i) = P(E_i \mid A)P(A)$$

Therefore we obtain the desired probability

$$P(E_i \mid A) = \frac{P(A \mid E_i)P(E_i)}{P(A)} \tag{2.20}$$

which is known as *Bayes' theorem*. If $P(A)$ is expanded using the total probability theorem, Eq. 2.20 becomes

$$P(E_i \mid A) = \frac{P(A \mid E_i)P(E_i)}{\sum\limits_{j=1}^{n} P(A \mid E_j)P(E_j)} \tag{2.20a}$$

EXAMPLE 2.26

Referring again to the pavement problem of Example 2.19, we might ask, "What is the probability that if a section is well constructed, it will be accepted on the basis of the ultrasonics test?"

This means $P(A \mid G)$, which according to Eq. 2.20, is given by

$$P(A \mid G) = \frac{P(G \mid A)P(A)}{P(G)}$$

From Example 2.19, we have

$$P(G \mid A) = 0.80 \quad \text{and} \quad P(A) = 0.90$$

To determine $P(G)$, we observe that A and $\bar{A}$ are mutually exclusive and collectively exhaustive; hence, according to Eq. 2.19,

$$P(G) = P(G \mid A)P(A) + P(G \mid \bar{A})P(\bar{A})$$
$$= 0.80(0.90) + (0.20)(0.10)$$
$$= 0.74$$

Therefore the required probability is

$$P(A \mid G) = \frac{0.80(0.90)}{0.74} = 0.973$$

whereas

$$P(\bar{A} \mid G) = 1 - 0.973 = 0.027$$

which is the probability that a well-constructed section may be rejected on the basis of the ultrasonics test. These latter probabilities should be compared and contrasted with $P(G \mid A)$ and $P(\bar{G} \mid A)$ of Example 2.19; the difference in meaning is not trivial.

EXAMPLE 2.27

The air pollution in a city is caused mainly by industrial and automobile exhausts. In the next 5 years, the chances of successfully controlling these two sources of pollution are, respectively, 75% and 60%. Assume that if only one of the two sources is successfully controlled, the probability of bringing the pollution below acceptable level would be 80%.

(a) What is the probability of successfully controlling air pollution in the next 5 years?

(b) If, in the next 5 years, the pollution level is not sufficiently controlled, what is the probability that it is *entirely* caused by the failure to control automobile exhaust?

Assuming statistical independence between controlling industrial (I) and automobile (A) exhausts, we have

$$P(AI) = 0.75 \times 0.60 = 0.45$$
$$P(A\bar{I}) = 0.25 \times 0.60 = 0.15$$
$$P(\bar{A}I) = 0.75 \times 0.40 = 0.30$$
$$P(\bar{A}\bar{I}) = 0.25 \times 0.40 = 0.10$$

Then, denoting E as the event of controlling air pollution,

(a) $P(E) = 1.00(0.45) + 0.80(0.15) + 0.80(0.30) + 0(0.10)$
$$= 0.81$$

(b) $P(\bar{A}\bar{I} \mid \bar{E}) = \dfrac{P(\bar{E} \mid \bar{A}\bar{I}) P(\bar{A}\bar{I})}{P(\bar{E})} = \dfrac{0.20 \times 0.30}{0.19}$
$$= 0.32$$

(c) A related question: If pollution is not controlled, what is the probability that control of automobile exhaust was not successful? This calls for $P(\bar{A} \mid \bar{E})$; but

$$
\begin{aligned}
P(\bar{A} \mid \bar{E}) &= P(\bar{A}I \cup \bar{A}\bar{I} \mid \bar{E}) \\
&= P(\bar{A}I \mid \bar{E}) + P(\bar{A}\bar{I} \mid \bar{E}) \\
&= \frac{P(\bar{E} \mid \bar{A}I)P(\bar{A}I)}{P(\bar{E})} + \frac{P(\bar{E} \mid \bar{A}\bar{I})P(\bar{A}\bar{I})}{P(\bar{E})} \\
&= \frac{0.20(0.30)}{0.19} + \frac{1.00(0.10)}{0.19} \\
&= 0.84
\end{aligned}
$$

whereas

$$
\begin{aligned}
P(\bar{I} \mid \bar{E}) = P(\bar{I}A \cup \bar{I}\bar{A} \mid \bar{E}) &= \frac{P(\bar{E} \mid \bar{I}A)P(\bar{I}A)}{P(\bar{E})} + \frac{P(\bar{E} \mid \bar{I}\bar{A})P(\bar{I}\bar{A})}{P(\bar{E})} \\
&= \frac{0.20(0.15) + 1.00(0.10)}{0.19} = \frac{0.13}{0.19} = 0.68
\end{aligned}
$$

EXAMPLE 2.28

Aggregates for construction are ordered from two different companies. Company A delivers 600 loads each day, out of which 3% do not satisfy the specified quality.

Company B supplies 400 loads each day, out of which only 1% are substandard.

(a) What is the probability that a load of aggregate picked at random came from company A?

(b) What is the probability that a load of aggregate picked at random will not pass the specified standard?

(c) If a load of aggregate was found to be substandard, what is the probability that it came from company A?

Solutions:

(a) Since there are altogether 1000 loads, out of which 600 came from company A, the probability that a load picked at random comes from company A is

$$P(A) = \frac{600}{1000} = 0.6$$

(b) The substandard aggregate may come from either company A or company B. We may apply the theorem of total probability to compute the probability of the event E, that is, picking a load of substandard aggregate:

$$P(E) = P(E \mid A)P(A) + P(E \mid B)P(B)$$
$$= 0.03 \times \frac{600}{1000} + 0.01 \times \frac{400}{1000}$$
$$= 0.018 + 0.004 = 0.022$$

(c) If the load of aggregate picked at random is substandard, the probability that it comes from company A is no longer 0.6 as in (a), because the sample space is changed. Instead of 1000 loads, the new sample space consists of only substandard aggregate loads which is

$$(0.03 \times 600 + 0.01 \times 400) = 18 + 4 = 22 \text{ loads}$$

out of which only 18 are from company A. Hence

$$P(A \mid \text{the aggregate is substandard}) = \frac{0.03 \times 600}{0.03 \times 600 + 0.01 \times 400}$$
$$= \frac{18}{22} = 0.818$$

Since the aggregate of company A is of poorer quality than that of B, the additional information that a load of aggregate is substandard increases the probability that such a load comes from A.

Bayes' theorem is useful for revising or updating the calculated probability as more data and information become available. The following examples will serve to illustrate this, including how prior information (which may be based on judgmental assumptions) is combined with test results to update the calculated probability.

EXAMPLE 2.29

Consider a pile foundation, in which pile groups are used to support the individual column footings. Each of the pile group is designed to support a load of 200 tons.

Under normal condition, this is quite safe. However, on rare occasions the load may reach as high as 300 tons. The foundation engineer wished to know the probability that a pile group can carry this extreme load of up to 300 tons.

Based on previous experience with similar pile foundations, supplemented with blow counts and soil tests, the engineer estimated a probability of 0.70 that any pile group can support a 300-ton load. Also, among those that have capacity less than 300 tons, 50% failed at loads less than 280 tons.

To improve the estimated probability, the foundation engineer ordered one pile group to be proof-loaded to 280 tons. If the pile group survives the specified proof load, the probability of the pile group supporting a load of 300 tons can be updated as follows.

Let

A = event that the capacity of pile group $\geq$ 300 tons

T = event of a successful proof load.

Then according to the information given above, $P(\bar{T} \mid A) = 0.5$, and $P(A) = 0.70$; and clearly $P(T \mid A) = 1.0$. Bayes' theorem then gives

$$P(A \mid T) = \frac{P(T \mid A)P(A)}{P(T \mid A)P(A) + P(T \mid \bar{A})P(\bar{A})}$$

$$= \frac{(1.00)(0.70)}{1.00(0.70) + 0.5(0.3)} = 0.833$$

Therefore, if the proof test is successful, the required probability is increased from 0.7 to 0.833.

EXAMPLE 2.30

Aggregates for a highway pavement are extracted from a gravel pit. Based on experience with the material from this area, it is known that the probabilities are

$$P(G) = P \text{ (good-quality aggregate)} = 0.70$$
$$P(\bar{G}) = P \text{ (poor-quality aggregate)} = 0.30$$

In order to improve this prior information, the engineer tested a sample of the aggregate. However, the test method is not perfectly reliable—the probability that a perfectly good-quality aggregate will pass the test is 80%, whereas, the probability of a poor-quality aggregate passing the test is 10%.

Let T_1 denote the event that a sample passes the test. Then, if a sample does indeed pass the test, the updated probability is

$$P(G \mid T_1) = \frac{P(T_1 \mid G)P(G)}{P(T_1 \mid G)P(G) + P(T_1 \mid \bar{G})P(\bar{G})}$$

$$= \frac{(0.8)(0.7)}{(0.8)(0.7) + (0.1)(0.3)} = 0.95$$

Therefore, with a positive test result, the probability of good-quality aggregate is increased significantly—from 70% to 95%.

Suppose that the engineer is not satisfied with just one sample test, and another sample is tested. If this additional sample also passes the test, the probability is updated further as follows:

$$P(G \mid T_2) = \frac{P(T_2 \mid G)P(G)}{P(T_2 \mid G)P(G) + P(T_2 \mid \bar{G})P(\bar{G})}$$

$$= \frac{(0.8)(0.95)}{(0.8)(0.95) + (0.1)(0.05)} = 0.993$$

This updating is performed sequentially. The updating may also be performed in a single step using the two test results together. In this latter case, we have

$$P(G \mid T_1 T_2) = \frac{P(T_1 T_2 \mid G)P(G)}{P(T_1 T_2 \mid G)P(G) + P(T_1 T_2 \mid \bar{G})P(\bar{G})}$$

$$= \frac{(0.8)(0.8)(0.7)}{(0.8)(0.8)(0.7) + (0.1)(0.1)(0.3)} = 0.993$$

which is clearly the same as the result obtained sequentially above, as it should be.

2.4. CONCLUDING REMARKS

In this chapter, we learn that a probabilistic problem involves the determination of the probability of an event within an exhaustive set of possibilities (or possibility space). Two things are paramount in the formulation and solution of such problems: (1) the definition of the possibility space and the identification of the event within this space; and (2) the evaluation of the probability of the event. The relevant mathematical bases useful for these purposes are the theory of sets and the theory of probability. In this chapter, the basic elements of both theories are developed in elementary and nonabstract terms, and are illustrated with physical problems.

Defined in the context of sets, events can be combined to obtain other events via the operational rules of sets and subsets; basically, these consist of the *union* and *intersection* of two or more events including their complements. Similarly, the operational rules of the theory of probability provide the bases for the deductive relationships among probabilities of different events within a given possibility space; specifically, these consist of the *addition rule*, the *multiplication rule*, the *theorem of total probability*, and *Bayes' theorem*.

In essence, the concepts developed in this chapter constitute the fundamentals of applied probability. In Chapters 3 and 4, additional analytical tools will be developed based on these fundamental concepts.

PROBLEMS

Sections 2.1 & 2.2

2.1 The possible settlements for the three supports of a bridge shown in Fig. P2.1 are as follows:

support *A*—0 in., 1 in., 2 in.
support *B*—0 in., 2 in.
support *C*—0 in., 1 in., 2 in.

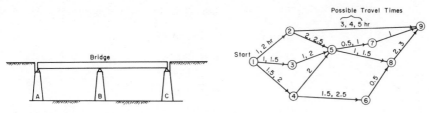

Figure P2.1   **Figure P2.2**

(a) Identify the sample space representing all possible settlements of the three supports; for example $(1, 0, 2)$ means A settles 1 in., B settles 0 in., and C settles 2 in.

(b) If E is the event of 2 in. differential settlement between any adjacent supports of the bridge, determine the sample points of E.

2.2 Figure P2.2 shows a network of highways connecting the cities $1, 2, \ldots, 9$.

(a) Identify the sample space representing all possible routes between cities 1 and 9.

(b) The possible travel times between any two connecting nodes are as indicated in Fig. P2.2 (for example, from 2 to 9, the possible travel times are 3, 4, 5 hr). What are the possible travel times between 1 and 9 through route ① → ② → ⑨? How about through route ① → ④ → ⑥ → ⑧ → ⑨?

2.3 A 6 m × 48 m apartment building may be divided into 1-, 2-, or 3-bedroom units, or combinations thereof (Fig. P2.3). If 1-bedroom units are each 6 m × 6 m, 2-bedroom units are each 6 m × 12 m, and 3-bedroom units are each 6 m × 16 m, how may the apartment building be subdivided into one or more types of units?

2.4 A left-turn pocket of length 60 ft is planned at a street intersection. Assume that only two types of vehicles will be using it; a type-A vehicle will occupy 15 ft of the pocket, whereas a type-B vehicle will occupy 30 ft.

(a) Identify all the possible combinations of types A and B vehicles waiting for left turns from the pocket.

(b) Group these possibilities into events of 1, 2, 3, and 4 vehicles waiting for left turns.

2.5 Strong wind at a particular site may come from any direction between due east $(\theta = 0°)$ and due north $(\theta = 90°)$. All values of wind speed V are possible.

(a) Sketch the sample space for wind speed and direction.

(b) Let $A = \{V > 20 \text{ mph}\}$
$B = \{12 \text{ mph} < V \leq 30 \text{ mph}\}$
$C = \{\theta \leq 30°\}$

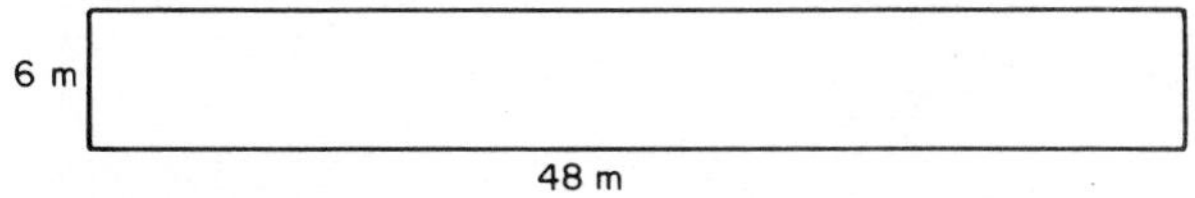

6 m ⌈   ⌉
48 m

Figure P2.3

Identify the events A, B, C, and $\bar{A}$ in the sample space sketched in part (a).

(c) Use new sketches to identify the following events:

$$D = A \cap C$$
$$E = A \cup B$$
$$F = A \cap B \cap C$$

(d) Are the events D and E mutually exclusive? How about events A and C?

2.6 The possible values of the water height H, relative to mean water level, at each of the two rivers A and B are as follows (in meters):

$$H = -3, -2, -1, 0, 1, 2, 3, 6$$

(a) Consider river A and define the following events:

$$A_1 = \{H_A > 0\}, \qquad A_2 = \{H_A = 0\}, \qquad A_3 = \{H_A \le 0\}$$

List all pairs of mutually exclusive events among A_1, A_2, and A_3.

(b) At each river, define
Normal water, $N = \{-1 \le H \le 1\}$
Drought, $D = \{H < -1\}$
Flood, $F = \{H > 1\}$
Use the ordered pair (h_A, h_B) to identify sample points relating to joint water levels in A and B, respectively; thus $(3, -1)$ defines the condition $h_A = 3$ and $h_B = -1$ simultaneously. Determine the sample points for the events

(i) $N_A \cap N_B$ (ii) $(F_A \cup D_A) \cap N_B$

2.7 The sequence of main activities in the construction of two structures is shown in Fig. P2.7. The construction of the superstructures A and B can start as soon as their common foundation has been completed.

The possible times of completion for each phase of construction are indicated in Fig. P2.7; for example, the foundation phase may take 5 or 7 months.

(a) List the possible combinations of times for each phase of the project; for example, (5, 3, 6) denotes the event that it takes 5 months for foundation, 3 months for superstructure A, and 6 months for superstructure B.

(b) What are the possible *total* completion times for structure A alone? For structure B alone?

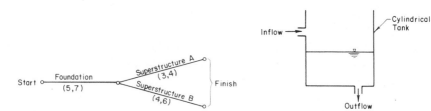

Figure P2.7   **Figure P2.8**

(c) What are the possible *total* completion times for the project?

(d) If the possibilities in part (a) are equally likely, what is the probability that the complete project will be finished within 10 months?

2.8 A cylindrical tank is used to store water for a town (Fig. P2.8). The available supply is not completely predictable. In any one day, the inflow is equally likely to fill 6, 7, or 8 ft of the tank. The demand for water is also variable, and may (with equal likelihood) require an amount equivalent to 5, 6, or 7 ft of water in the tank.

(a) What are the possible combinations of inflow and outflow in a day?

(b) Assuming that the water level in the tank is 7 ft at the start of a day, what are the possible water levels in the tank at the end of the day? What is the probability that there will be at least 9 ft of water remaining in the tank at the end of the day?

2.9 A power plant has two generating units, numbered 1 and 2. Because of maintenance and occasional machine malfunctions, the probabilities that, in a given week, units No. 1 and 2 will be out of service (these two events are denoted by E_1 and E_2) are 0.01 and 0.02, respectively.

During a summer week there is a probability of 0.10 that the weather will be extremely hot (say average temperature $> 85°F$; this event is denoted by H) so that demand for power for air-conditioning will increase considerably. The performance of the power plant in terms of its potential ability to meet the demand in a given week can be classified as

(i) satisfactory S, if *both* units are functioning *and* the average temperature is below $85°F$

(ii) poor P, if one of the units is out of service *and* the average temperature is above $85°F$

(iii) marginal M, otherwise.

Assume H, E_1, and E_2 are statistically independent.

(a) Define the events S, P, and M in terms of H, E_1, and E_2.

(b) What is the probability that *exactly one unit* will be out of service in any given week?

(c) Find $P(S)$, $P(P)$, and $P(M)$.

Section 2.3

2.10 A cantilever beam has 2 hooks where weights ① and ② may be hung (Fig. P2.10). There can be as many as two weights or no weight at each hook. In order to design this beam, the engineer needs to know the fixed-end moment at A, that is, M_A.

(a) What are all the possible values of M_A?

(b) Let

$$E_1 \text{ denote the event that } M_A > 600 \text{ ft-lb}$$
$$E_2 \text{ denote the event that } 200 \leq M_A < 800 \text{ ft-lb}$$

Are events E_1 and E_2 mutually exclusive? Why?

(c) Are events E_1 and E_3 mutually exclusive? Where $E_3 = \{0, 100, 400\}$.

(d) With the following information:

$$\text{Probability that weight ① hangs at } B = 0.2$$
$$\text{Probability that weight ① hangs at } C = 0.7$$
$$\text{Probability that weight ② hangs at } B = 0.3$$
$$\text{Probability that weight ② hangs at } C = 0.5$$

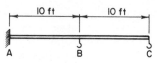

Figure P2.10

What are the probabilities associated with each sample point in part (a)? Assume that the location of weight ① does not affect the probability of the location of weight ②.

(e) Determine the probabilities of the following events:

$$E_1, E_2, E_1 \cap E_2, E_1 \cup E_2, \bar{E}_2$$

2.11 In a building construction project, the completion of the building requires the successive completion of a series of activities. Define

E = excavation completed on time; and $P(E) = 0.8$

F = foundation completed on time; and $P(F) = 0.7$

S = superstructure completed on time; and $P(S) = 0.9$

Assume statistical independence among these events.

(a) Define the event {project completed on time} in terms of E, F, and S. Compute the probability of on-time completion.

(b) Define, in terms of E, F, S and their complements, the following event:
G = excavation will be on time and at least one of the other two operations will not be on time
Calculate $P(G)$.

(c) Define the event
H = only one of the three operations will be on time

2.12 The waste from an industrial plant is subjected to treatment before it is ejected to a nearby stream. The treatment process consists of three stages, namely: primary, secondary, and tertiary treatments (Fig. P2.12). The primary treatment may be rated as good (G_1), incomplete (I_1) or failure (F_1). The secondary treatment may be rated as good (G_2) or failure (F_2), and the tertiary treatment may also be rated as good (G_3) or failure (F_3). Assume that the ratings in each treatment are *equally likely* (for example, the primary treatment will be equally likely to be good or incomplete or failure). Furthermore, the performances of the three stages of treatment are statistically independent of one another.

(a) What are the possible combined ratings of the three treatment stages? (for example, G_1, F_2, G_3 denotes a combination where there is a good primary and tertiary, but a failure in the secondary treatment). What is the probability of each of these combinations (or sample points)?

(b) Suppose the event of satisfactory overall treatment requires at least two stages of good treatment. What is the probability of this event?

(c) Suppose:

E_1 = good primary treatment

E_2 = good secondary treatment

E_3 = good tertiary treatment

Determine

$$P(\bar{E}_1), \qquad P(E_1 \cup E_2), \qquad P(E_2 E_3)$$

(d) Express in terms of E_1, E_2, E_3 the event of satisfactory overall treatment as defined in part (b). (*Hint.* $E_1 E_2$ is part of this event.)

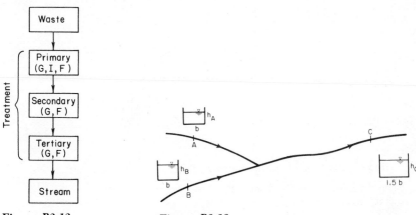

Figure P2.12   **Figure P2.13**

2.13 The cross-sections of the rivers at A, B, and C are shown in Fig. P2.13 and the flood levels at A and B, above mean flow level, are as follows:

Flood level at A (ft)	Probability
0	0.25
2	0.25
4	0.25
6	0.25

Flood level at B (ft)	Probability
0	0.20
2	0.20
4	0.20
6	0.20
8	0.20

Assume that the flow velocities at A, B, and C are the same. What is the probability that the flood at C will be higher than 6 ft above the mean level? Assume statistical independence between flood levels at A and B. *Ans. 0.3.*

2.14 Figure P2.14 is a plot of test results showing the degree of subgrade compac-

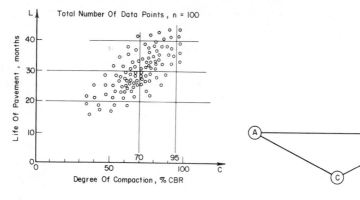

Figure P2.14   **Figure P2.15**

tion C versus the life of pavement L. Determine the following:

(a) $P(20 < L \leq 40 \mid C \geq 70)$

(b) $P(L > 40 \mid C \leq 95)$

(c) $P(L > 40 \mid 70 < C \leq 95)$

(d) $P(L > 30$ and $C < 70)$

2.15 The highway system between cities A and B is shown in Fig. P2.15. Travel between A and B during the winter months is not always possible because some parts of the highway may not be open to traffic, because of extreme weather condition. Let E_1, E_2, E_3 denote the events that highway AB, AC, and CB are open, respectively.

On any given day, assume

$$P(E_1) = 2/5 \qquad P(E_3 \mid E_2) = 4/5$$
$$P(E_2) = 3/4 \qquad P(E_1 \mid E_2 E_3) = 1/2$$
$$P(E_3) = 2/3$$

(a) What is the probability that a traveler will be able to make a trip from A to B if he has to pass through city C? *Ans. 0.6.*

(b) What is the probability that he will be able to get to city B? *Ans. 0.7.*

(c) Which route should he try first in order to maximize his chance of getting to B?

2.16 A contractor is submitting bids to two jobs A and B. The probability that he will win job A is $P(A) = \frac{1}{4}$ and that for job B is $P(B) = \frac{1}{3}$.

(a) Assuming that winning job A and winning job B are independent events, what is the probability that the contractor will get at least a job?

(b) What is the probability that the contractor got job A if he has won at least one job?

(c) If he is also submitting a bid for job C with probability of winning it $P(C) = 1/4$, what is the probability that he will get at least one job? Again assume statistical independence among A, B, and C. What is the probability that the contractor will not get any job at all?

2.17 Cities 1 and 2 are connected by route A, and route B connects cities 2 and 3

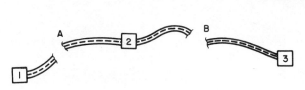

Figure P2.17

(Fig. P2.17). Denote the eastbound lanes as A_1 and B_1, and the westbound lanes as A_2 and B_2, respectively.

Suppose that the probability is 90% that a lane in route A will not require major repair work for at least 2 years; the corresponding probability for a lane in route B is only 80%.

  (a) Determine the probability that route A will require major repair work in the next two years. Do the same for route B.

      Assume that if one lane of a route needs repair, the chances that the other lane will also need repair is 3 times its original probability. *Ans. 0.17; 0.28.*

  (b) Assuming that the need for repair works in routes A and B are independent of each other, what is the probability that the road between cities 1 and 3 will require major repair in two years? *Ans. 0.40.*

2.18 The water supply system for a city consists of a storage tank and a pipe line supplying water from a reservoir some distance away (Fig. P2.18). The amount of water available from the reservoir is variable depending on the precipitation in the watershed (among other things). Consequently, the amount of water stored in the tank would be also variable. The consumption of water also fluctuates considerably.

To simplify the problem, denote

$$A = \text{available water supply from the reservoir is low}$$
$$B = \text{water stored in the tank is low}$$
$$C = \text{level of consumption is low}$$

and assume that

$$P(A) = 20\%$$
$$P(B) = 15\%$$
$$P(C) = 50\%$$

The reservoir supply is regulated to a certain extent to meet the demand, so

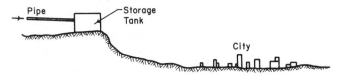

Figure P2.18

that
$$P(\bar{A} \mid \bar{C}) \equiv P(\text{reservoir supply is high} \mid \text{consumption is high})$$
$$= 75\%$$

Also, $P(B \mid A) = 50\%$, whereas the amount of water stored is independent of the demand.

Suppose that a water shortage will occur when there is high demand (or consumption) for water but either the *reservoir supply is low* or the *stored water is low*. What is then the probability of a water shortage? Assume that $P(AB \mid \bar{C}) = 0.5 \, P(AB)$.

2.19 The time T (in minutes) that it takes to load crushed rocks from a quarry onto a truck varies considerably. From a record of 48 loadings, the following were observed.

Loading time T (minutes)	No. of observations
0 to 1^-	0
1 to 2^-	5
2 to 3^-	12
3 to 4^-	15
4 to 5^-	10
5 to 6^-	6
≥ 6	0
	Total $= \overline{48}$

(a) Sketch the histogram for the above data.
(b) Based on these data, what is the probability that the loading time T for a truck will be at least 4 minutes?
(c) What is the probability that the total time for loading 2 consecutive trucks will be less than 6 minutes? Assume the loading times for any two trucks to be statistically independent.
(d) In order to make a conservative estimate of the loading time, it is assumed that loading a truck will require at least 3 minutes; on this assumption, what will be the probability that the loading time for a truck will be less than 4 minutes?

2.20 A gravity retaining wall may fail either by sliding (A) or overturning (B) or both (Fig. P2.20). Assume:
   (i) Probability of failure by sliding is twice as likely as that by overturning; that is, $P(A) = 2P(B)$.
   (ii) Probability that the wall also fails by sliding, given that it has failed by overturning, $P(A \mid B) = 0.8$
   (iii) Probability of failure of wall $= 10^{-3}$
   (a) Determine the probability that sliding will occur.   *Ans. 0.00091.*
   (b) If the wall fails, what is the probability that *only* sliding has occurred? *Ans. 0.546.*

2.21 Two cables are used to lift a load W (Fig. P2.21). However, normally only cable A will be carrying the load; cable B is slightly longer than A, so nor-

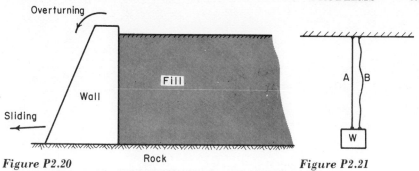

Overturning

Wall

Sliding

Figure P2.20   Rock   *Figure P2.21*

A | B

W

mally it does not participate in carrying the load. But if cable A breaks, then B will have to carry the full load, until A is replaced.

The probability that A will break is 0.02; also, the probability that B will fail if it has to carry the load by itself is 0.30.

(a) What is the probability that both cables will fail?

(b) If the load remains lifted, what is the probability that none of the cables have failed?

2.22   The preliminary design of a bridge spanning a river consists of four girders and three piers as shown in Fig. P2.22. From consideration of the loading and resisting capacities of each structural element the failure probability for each girder is 10^{-5} and each pier is 10^{-6}. Assume that failures of the girders and piers are statistically independent. Determine:

(a) The probability of failure in the girder(s).

(b) The probability of failure in the pier(s).

(c) The probability of failure of the bridge system.

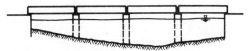

Figure P2.22

2.23   The town shown in Fig. P2.23 is protected from floods by a reservoir dam that is designed for a 50-year flood; that is, the probability that the reservoir will overflow in a year is 1/50 or 0.02. The town and reservoir are located in an active seismic region; annually, the probability of occurrence of a destructive earthquake is 5%. During such an earthquake, it is 20% probable that the dam will be damaged, thus causing the reservoir water to flood the town. Assume that the occurrences of natural floods and earthquakes are statistically independent.

(a) What is the probability of an earthquate-induced flood in a year?

(b) What is the probability that the town is free from flooding in any one year?

(c) If the occurrence of an earthquake is assumed in a given year, what is the probability that the town will be flooded that year?

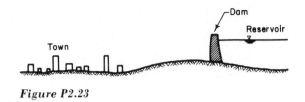

Figure P2.23

2.24 From a survey of 1000 water-pipe systems in the United States, 15 of them are reported to be contaminated by bacteria alone whereas 5 are reported to have an excessive level of lead concentration and among these 5, there are 2 that are found to contain bacteria also.

  (a) What is the probability that a pipe system selected at random will contain bacteria? *Ans. 0.017.*
  (b) What is the probability that a pipe system selected at random is contaminated? *Ans. 0.02.*
  (c) Suppose that a pipe system is found to contain bacteria. What is the probability that its lead concentration is also excessive? *Ans. 2/17.*
  (d) Assume that the present probability of contamination as computed in part (b) is not satisfactory, and it is proposed that it should not exceed 0.01. Suppose that it is difficult to control the lead contamination, but it is possible to reduce the likelihood of bacteria contamination. What should be the permissible probability of bacteria contamination? Assume that the value of the conditional probability computed in part (c) still applies. *Ans. 0.00567.*

2.25 The structural component shown in Fig. P2.25 has welds to be inspected for flaws. From experience, the likelihood of detecting flaws in a foot of weld provided by the manufacturer is 0.1; and the probability of detecting flaws in a weld of length L ft is given by

$$P(F_L) = 0.1L \quad \text{for} \quad 0 \le L \le 2 \text{ ft}$$

In general, the quality between sections of welds in a structural component is

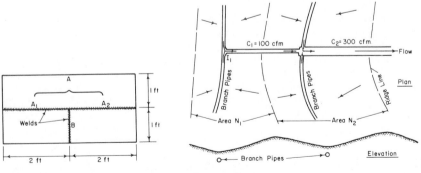

Figure P2.25   **Figure P2.26**

correlated. Assume the following:
   (i) If flaws are detected in section A_1, the probability of flaws being detected in A_2 will be three times its original probability.
   (ii) If flaws are detected in section A, the probability of detecting flaws in section B will be doubled.

Let F_{A_1}, F_{A_2}, F_A, and F_B be the events of flaws detected in weld sections A_1, A_2, A, and B, respectively.
   (a) What is the probability of detecting flaws in A?   *Ans. 0.28.*
   (b) What is the probability of detecting flaws in the structural component? *Ans. 0.324.*
   (c) If flaws are detected in the structural component, what is the probability that they are found *only* in A?   *Ans. 0.692.*

2.26 The storm drainage in a residential subdivision can be divided into watershed areas N_1 and N_2 as shown in Fig. P2.26. The drainage system consists of the main sewers with capacities $C_1 = 100$ cfm (cubic feet per minute) and $C_2 = 300$ cfm, respectively. The amounts of drainage from N_1 and N_2 are variable, depending on the rainfall intensities in the subdivision (assume that whenever it rains the entire subdivision is covered); in any given year, the maximum flow, I_1 and I_2, and their corresponding probabilities are as follows.

I_1 (cfm)	Probability	I_2 (cfm)	Probability
80	0.60	100	0.50
120	0.40	210	0.30
		250	0.20

Neglect the possibility of flooding in N_1 caused by the overflow of pipe C_2.
   (a) What is the probability of flooding in area N_1? Flooding occurs only when the drainage exceeds the capacity of the main sewer.
   (b) What is the probability of flooding in area N_2?
   (c) What is the probability of flooding in the subdivision?

2.27 In order to study the parking problem of a college campus, an average worker in office building D, say Mr. X, is selected and his chance of getting a parking space each day is studied. (Assume that Mr. X will check the parking lots A, B, C in that sequence and will park his car as soon as a space is found.) Assume that there are only three parking lots available, of which A and B are free, whereas C is metered (Fig. P2.27). No other parking facilities (say street parking) are allowed. From statistical data, the probabilities of getting a parking space each week day morning in lots A, B, C are 0.2, 0.1, 0.5, respectively. However, if lot A is full, the probability that Mr. X will find a space in B is only 0.04. Also, if lots A and B are full, Mr. X will only have a probability of 40% of getting a parking space in C. Determine the following:
   (a) The probability that Mr. X will not be able to secure a free space on a weekday morning.   *Ans. 0.768.*
   (b) The probability that Mr. X will be able to park his car on a weekday morning.   *Ans. 0.539.*
   (c) If Mr. X has successfully parked his car one morning, what is the probability that it will be free of charge?   *Ans. 0.43.*

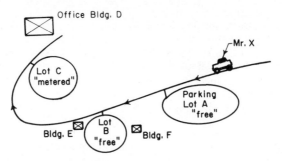

Figure P2.27

2.28 Pollution is becoming a problem in cities I and II. City I is affected by both air and water pollution, whereas city II is subjected to air pollution only. A three-year plan has been put into action to control these sources of pollution in both cities. It is estimated that the air pollution in city I will be successfully controlled is 4 times as likely as that in city II. However, if air pollution in city II is controlled, then air pollution in city I will be controlled with 90% probability.

The control of water pollution in city I may be assumed to be independent of the control of air pollution in both cities. In city I, the probability that pollution will be completely controlled (that is, both sources are controlled) is 0.32, whereas it is also estimated that water pollution is only half as likely to be controlled as the air pollution in that city. Let

A_I be the event "air pollution in city I is controlled"

A_{II} be the event "air pollution in city II is controlled"

W_I be the event "water pollution in city I is controlled"

Determine:
  (a) Probability that air pollution will be controlled in both cities. *Ans. 0.18.*
  (b) Probability that pollution in both cities will be completely controlled. *Ans.* 0.072.
  (c) Probability that at least one city will be free of pollution.   *Ans. 0.448.*

2.29 A form of transportation is to be provided between two cities that are 200 miles apart. The alternatives are highway (*H*), railway (*R*), or air transport

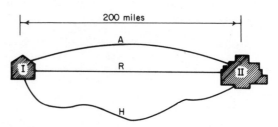

Figure P2.29

(A); the last one meaning the construction of airports in both cities. (See Fig. P2.29.) Because of the relative merits and costs, the odds that a Committee of Planners will decide on R, H, or A are 1 to 2 to 3. Only one of these three means can be constructed.

However, if the committee decides on building a railroad (R), the probability that it will be completed in one year is 50%; if it decides on a highway (H), the corresponding probability is 75%; and if it decides on air travel, there is a probability of 90% that the airports will be completed in one year.

(a) What is the probability that the two cities will have a means of transportation in one year?

(b) If some transportation facility between the two cities is completed in one year, what is the probability that it will be air transport (A)?

(c) If the committee decides in favor of land facilities, what is the probability that the final decision will be for a highway (H)?

2.30 "Liquefaction of sand" denotes a phenomenon in foundation engineering, in which a mass of saturated sand suddenly loses its bearing capacity because of rapid changes in loading conditions—for example, resulting from earthquake vibrations. When this happens, disastrous effects on structures built on the site may follow.

For simplicity, rate earthquake intensities into low (L), medium (M), and high (H). The likelihoods of liquefaction associated with earthquakes of these intensities are, respectively, 0.05, 0.20, and 0.90.

Assume that the relative frequencies of occurrence of earthquakes of these intensities are, respectively, 1, 0.1, and 0.01 per year.

(a) What is the probability that the next earthquake is of low intensity? *Ans. 0.9.*

(b) What is the probability of liquefaction of sand at the site during the next earthquake? *Ans. 0.07.*

(c) What is the probability that the sand will survive the next three earthquakes (that is, no liquefaction)? Assume the conditions between earthquakes are statistically independent. *Ans. 0.80.*

2.31 There are three modes of transporting material from New York to Florida, namely, by land, sea, or air. Also land transportation may be by rail or highway. About half of the materials are transported by land, 30% by sea, and the rest by air.

Also, 40% of all land transportation is by highway and the rest by rail shipments. The percentages of damaged cargo are, respectively, 10% by highway, 5% by rail, 6% by sea, and 2% by air.

(a) What percentage of all cargoes may be expected to be damaged?

(b) If a damaged cargo is received, what is the probability that it was shipped by land? By sea? By air?

2.32 The amount of stored water in a reservoir (Fig. P2.32a) may be idealized into three states: full (F), half-full (H), and empty (E). Because of the probabilistic nature of the inflowing water into the reservoir, as well as the outflow from the reservoir to meet uncertain demand for water, the amount of water stored may shift from one state to another during each season. Suppose that these transitional probabilities from one state to another are as indicated in Fig. P2.32b. For example, in the beginning of a season, if the water storage is empty, the probability that it will become half-full at the end of the season is 0.5 and the probability that it will remain empty is 0.4, and so on. Assume that the water level is full at the start of the season.

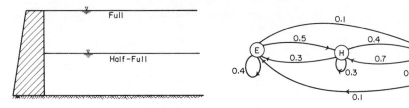

Figure P2.32a   *Figure P2.32b*

(a) What is the probability that the reservoir will be full at the end of one season? What is the probability that the reservoir will contain water at the end of one season?   *Ans. 0.2; 0.9.*

(b) What is the probability that the reservoir will be full at the end of the second season?   *Ans. 0.33.*

(c) What is the probability that the reservoir will contain water at the end of the second season?   *Ans. 0.73.*

2.33 At a quarry, the time required to load crushed rocks onto a truck is equally likely to be either 2 or 3 minutes (Fig. P2.33). Also the number of trucks in a queue waiting to be loaded at any time varies considerably, as reflected in the following set of 30 observations taken at random. The time required to

No. of trucks in queue	No. of observations	Relative frequency
0	6	0.2
1	3	0.1
2	9	0.3
3	9	0.3
4	3	0.1
5	0	0.0
	Total = $\overline{30}$	

load a truck is statistically independent of the queue size.

(a) If there are two trucks in the queue when a truck arrives at the quarry, what is the probability that its "waiting time" will be less than 5 minutes?   *Ans. 0.25.*

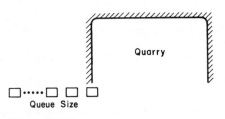

Figure P2.33

(b) Before arriving at the quarry (and thus not knowing the size of the queue), what is the probability that the waiting time of a particular truck will be less than 5 minutes? *Ans. 0.375.*

2.34 A chemical plant produces a variety of products using four different processes; the available labor is sufficient only to run one process at a time. The plant manager knows that the discharge of dangerous pollution into the plant waste water system and thence into a nearby stream is dependent on which process equipment is in operation. The probability that a particular process will be producing dangerous pollution products is as shown below:

$$
\begin{array}{ll}
\text{process } A & 40\% \\
\text{process } B & 5\% \\
\text{process } C & 30\% \\
\text{process } D & 10\%
\end{array}
$$

All other processes in the plant are considered harmless.

In a typical month the relative likelihoods of processes A, B, C, and D operating through the month are $2:4:3:1$, respectively.

(a) What is the probability that there will be *no* dangerous pollution discharged in a given month?

(b) If dangerous pollution is detected in the plant discharge, what is the probability that process A was operating?

(c) The pollution products that are discharged by the various processes have different probabilities of producing a fish kill in the stream that the plant uses for disposal, as follows.

Process	Probability of fish kill
A	0.9
B	0.1
C	0.8
D	0.3

Based on these assumptions what is the probability that fish will be killed by pollution in the stream in a given month?

(d) Of the four processes, which is the most fruitful one (in terms of minimizing the likelihood of fish kill) to select for clean-up if only one can be improved?

2.35 The probability of occurrence of fire in a subdivision has been estimated to be 30% for one occurrence and 10% for two occurrences in a year. Assume that the chance for three or more occurrences is negligible. In a fire, the probability that it will cause structural damage is 0.2. Assume that structural damages between fires are statistically independent.

(a) What is the probability that there will be no structural damage caused by fire in a year? *Ans. 0.904.*

(b) If a small town consists of two such subdivisions, what is the probability that there will be some structural damage caused by fire in the town in a year? Assume that the events of fire-induced structural damage in the two subdivisions are statistically independent. *Ans. 0.183.*

2.36   At a construction project, the amount of material (say lumber for falsework) available for any day is variable, and can be described with the frequency diagram of Fig. P2.36. The amount of material used in a day's construction is either 150 units or 250 units, with corresponding probabilities 0.70 and 0.30.

(a) What is the probability of shortage of material in any day? Shortage occurs whenever the available material is less than the amount needed for that day's construction.

(b) If there is a shortage of material, what is the probability that there were fewer than 200 units available?

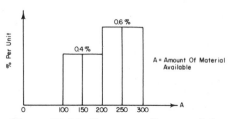

Figure P2.36   Frequency diagram of *A*

2.37   The completion time of a construction project depends on whether the carpenters and plumbers working on the project will go on strike. The probabilities of delay (*D*) are 100%, 80%, 40%, and 5% if *both go on strike, carpenters alone go on strike, plumbers alone go on strike*, and *neither of them strikes*, respectively. Also, there is 60% chance that plumbers will strike if carpenters strike, and if plumbers go on strike there is 30% chance that carpenters would follow. It is known that the chance for the plumbers' strike is 10%. Let

$$C = \text{event that carpenters went on strike}$$
$$P = \text{event that plumbers went on strike}$$
$$D = \text{delay in project completion}$$

(a) Determine probability of delay in completion. *Ans. 0.118.*
(b) If there is a delay in completion, determine the following:
    (i) Probability that both carpenters and plumbers strike. *Ans. 0.254.*
    (ii) Probability that carpenters strike and plumbers do not. *Ans. 0.136.*
    (iii) Probability of carpenters' strike. *Ans. 0.390.*

2.38   The water supply for a city comes from two reservoirs, *a* and *b* (Fig. P2.38). Because of variable rainfall conditions each year, the amount of water in each reservoir may *exceed* or *not exceed* the normal capacity. Let *A* denote the event that the water in reservoir *a* exceeds its normal capacity, and let *B* denote that for reservoir *b*. The following probabilities are given: $P(B) = 0.8$, $P(AB) = 0.6$, $P(\bar{A} \mid \bar{B}) = 0.7$. In addition, the probabilities that the city will have satisfactory supply of water *if only one reservoir exceeds, both reservoirs exceed*, and *none of the reservoir exceeds* the normal capacities are

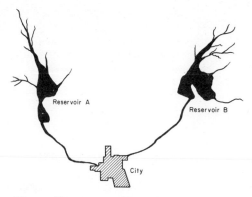

Figure P2.38

0.7, 0.9, and 0.3, respectively. What is the probability that the city will have satisfactory water supply?   *Ans. 0.764.*

2.39 A water tower is located in an active earthquake region. When an earthquake occurs, the probability that the tower will fail depends on the magnitude of the earthquake *and* also on the amount of storage in the tank at the time of shaking of the ground. For simplicity, assume that the tank is either full (F) or half-full (H) with relative likelihoods of 1 to 3. The earthquake magnitude may be assumed to be either strong (S) or weak (W) with relative frequencies 1 to 9.

When a strong earthquake occurs, the tower will definitely collapse regardless of the storage level. However, the tower will certainly survive a weak earthquake if the tank is only half-full. If the tank is full during a weak earthquake, it will have a 50-50 chance of survival.

If the tower collapsed during a recent earthquake, what is the probability that the tank was full at the time of the earthquake?

2.40 For a county in Texas, the probabilities that it will be hit by one or two hurricanes each year are 0.3 and 0.05, respectively. The event that it will be hit by three or more hurricanes in a year may be assumed to have negligible probability.

This county may be subjected to floods each year from the melting of snow in the upstream regions, or from the heavy precipitation brought by hurricanes, or both. Normally, the chance of flood in a year, caused by the melting snow only, is 10%. However, during a hurricane there is a 25% probability of flooding. Assume that floods caused by the melting snow and floods caused by hurricanes are independent events.

What is the probability that there will be flooding in this county in a year?

2.41 Before the design of a tunnel through a rocky region, geological exploration was conducted to investigate the joints and the potential slip surfaces that exist in the rock strata (Fig. P2.41). For economic reasons, only portions of the strata are explored. In addition the measurements recorded by the instruments are not perfectly reliable. Thus the geologist can only conclude that the condition of the rock may be either highly fissured (H), medium fissured (M), or slightly fissured (L) with relative likelihoods of 1:1:8. Based on this

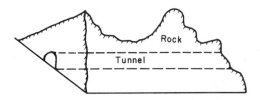

Figure P2.41

information, the engineer designs the tunnel and estimates that if the rock condition is L, the reliability of the proposed design is 99.9%. However, if it turns out that the rock condition is M, the probability of failure will be doubled; similarly, if the rock condition is H, the probability of failure will be 10 times that for condition L.

(a) What is the expected reliability of the proposed tunnel design? *Ans. 0.998.*

(b) A more reliable device is subsequently used to improve the prediction of rock condition. Its results indicate that a highly fissured condition for the rock around the tunnel is practically impossible, but it cannot give better information on the relative likelihood between rock conditions M and L. In light of this new information, what would be the revised reliability of the proposed tunnel design? *Ans. 0.9989.*

(c) If the tunnel collapsed, what should be the updated probabilities of M and L? *Ans. 0.797; 0.203.*

2.42 Three research and development groups, A, B, and C, submitted proposals for a research project to be awarded by a research agency of the government. From past performance records, the respective histograms of completion time relative to the scheduled target time t_0 are shown in Fig. P2.42. It is known that groups A and B have about equal chances of getting the project, whereas C is twice as likely as either A or B to win the contract.

Based on past performance records, determine:

(a) The probability that the project will be completed on schedule. *Ans. 0.60.*

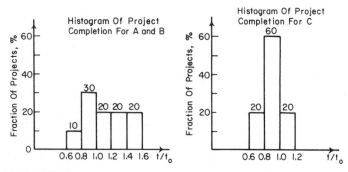

Figure P2.42

(b) If the project completion is delayed, what is the probability that it was originally awarded to C? *Ans. 0.25.*

2.43 Two independent remote sensing devices, A and B, mounted on an airplane are used to determine the locations of diseased trees in a large area of forest land. The detectability of device A is 0.8 (that is, the probability that a group of diseased trees will be detected by device A is 0.8), whereas the detectability of device B is 0.9.

However, when a group of diseased trees has been detected its location may not be pinpointed accurately by either device. Based on a detection from device A alone, the location can be accurately determined with probability 0.7, whereas the corresponding probability with device B alone is only 0.4. If the same group of diseased trees is detected by both devices, its location can be pinpointed with certainty. Determine the following.

(a) The probability that a group of diseased trees will be detected. *Ans. 0.98.*

(b) The probability that a group of diseased trees will be detected by *only one* device. *Ans. 0.26.*

(c) The probability of accurately locating a group of diseased trees. *Ans. 0.848.*

3. Analytical Models of Random Phenomena

3.1. RANDOM VARIABLES

In engineering and the physical sciences many random phenomena of interest are associated with the numerical outcomes of some physical quantity. In the various examples discussed earlier, we were concerned with the number of bulldozers operative after six months, the time required to complete a project, and the flood of a river above mean flow level, all of which are outcomes in numerical terms. However, we also saw examples in which the outcomes are not in numerical terms—for example, the state of completion of a project in one year, the survival or failure of a chain, and the availability of different modes of transportation. Events of this latter type may also be identified numerically by artificially assigning numerical values to each of the possible alternative events; for example, the three states of completion of a project in one year (*definitely completed*, *completion questionable*, and *definitely incomplete*) may be arbitrarily assigned the numbers 1, 2, and 3, respectively.

In other words, the possible outcomes of a random phenomenon can be identified numerically, either naturally or artificially. In any case, an outcome or event may be identified through the value(s) of a function; such a function is a *random variable*, which is usually denoted with a capital letter. The value (or range of values) of a random variable then represents a distinct event; for example, if the values of X represent floods above mean level, then $X > 7$ ft stands for the occurrence of a flood higher than 7 ft, and (referring to the example above) if Y is the state of completion of a project in one year, then $Y = 2$ means that the project's completion is questionable in one year. In short, a random variable is a device (cooked up when necessary) to identify events in numerical terms. Henceforth, we can then say that $(X = a)$, or $(X \leqslant b)$, or $(a < X \leqslant b)$ is an event.

More formally, a random variable may be considered as a rule that maps events in a sample space into the real line. The mapping is one-to-one; also, mutually exclusive events are mapped into nonoverlapping intervals on the real line. In Fig. 3.1 the events E_1, E_2, and so on, from the sample space S

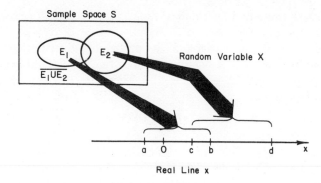

Figure 3.1   Mapping of events into real line through random variable X

are mapped into the real line through the random variable X; these events can then be identified as follows:

$$E_1 = (a < X \leqslant b)$$

$$E_2 = (c < X \leqslant d)$$

$$\overline{E_1 \cup E_2} = (X \leqslant a) \cup (X > d); \quad E_1 E_2 = (c < X \leq b)$$

Consistent with the underlying sample space, a random variable may be *discrete* or *continuous*.

The purpose and advantages of identifying events in numerical terms should be obvious—this will then permit convenient analytical description as well as graphical display of events and their probabilities.

3.1.1. Probability distribution of a random variable

Since the value of a random variable represents an event, it can assume a numerical value only with an associated probability or probability measure. The rule for describing the probability measures associated with all the values of a random variable is a *probability distribution* or "probability law."

If X is a random variable, its probability distribution can always be described by its *cumulative distribution function* (CDF), which is

$$F_X(x) \equiv P(X \leqslant x) \qquad \text{for all } x^* \tag{3.1}$$

Here X is a *discrete* random variable if only certain discrete values of x have positive probabilities. Alternatively, X is a *continuous* random variable if probability measures are defined for any value of x. A random variable may

---

* A standard notation is to denote a random variable with a capital letter, and its value with the corresponding lowercase letter.

also be both discrete and continuous; an example of such a *mixed* random variable is shown in Fig. 3.2c.

For a discrete random variable X, its probability distribution may also be described in terms of a *probability mass function* (PMF), which is simply a function expressing $P(X = x)$ for all x. Therefore, if X is a discrete random variable with PMF $p_X(x_i) \equiv P(X = x_i)$, its distribution function is

$$F_X(x) = P(X \leqslant x) = \sum_{\text{all } x_i \leqslant x} P(X = x_i) = \sum_{\text{all } x_i \leqslant x} p_X(x_i) \quad (3.2)$$

However, if X is continuous, probabilities are associated with intervals on the real line (since events are defined as intervals on the real line); consequently, at a specific value of X, such as $X = x$, only the *density of probability* is defined. Thus, for a continuous random variable, the probability law may also be described in terms of a *probability density function* (PDF), so that if $f_X(x)$ is the PDF of X, the probability of X in the interval $(a, b]$ is

$$P(a < X \leqslant b) = \int_a^b f_X(x) \, dx \quad (3.3)$$

It follows then that the corresponding distribution function is

$$F_X(x) = P(X \leqslant x) = \int_{-\infty}^x f_X(\xi) \, d\xi \quad (3.4)$$

Accordingly, if $F_X(x)$ has a first derivative, then, from Eq. 3.4,

$$f_X(x) = \frac{dF_X(x)}{dx} \quad (3.5)$$

We might reiterate that $f_X(x)$ is not a probability; however, $f_X(x) \, dx = P(x < X \leqslant x + dx)$ is the probability that values of X will be in the interval $(x, x + dx]$.

It should be emphasized that any function used to represent the probability distribution of a random variable must necessarily satisfy the axioms of probability (see Section 2.3.1). For this reason, the function must be nonnegative and the probabilities associated with all possible values of the random variable must add up to 1.0. In other words, if $F_X(x)$ is the distribution function of X, then it must have the following properties:

(a) $F_X(-\infty) = 0$; $F_X(+\infty) = 1.0$

(b) $F_X(x) \geqslant 0$, and is nondecreasing with x.

(c) It is continuous with x.

Conversely, any function possessing these properties is a bona fide cumula-

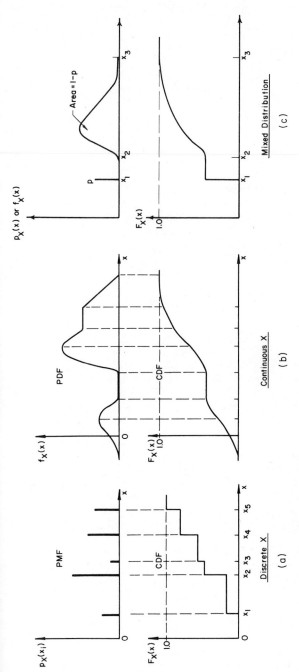

Figure 3.2 Bona fide probability distributions

tive distribution function. By virtue of these properties and Eqs. 3.2 through 3.5, the PMF and PDF are nonnegative functions of x, whereas the probabilities of a PMF add up to 1.0, and the total area under a PDF is also equal to 1.0. Figure 3.2 presents graphic examples of legitimate probability distributions. Figure 3.2 also illustrates the graphical characteristics of the probability distributions of discrete, continuous, and mixed random variables.

We observe that we can write Eq. 3.3 as

$$P(a < X \leqslant b) = \int_{-\infty}^{b} f_X(x) \, dx - \int_{-\infty}^{a} f_X(x) \, dx$$

Similarly, for discrete X, we have

$$P(a < X \leqslant b) = \sum_{\text{all } x_i \leq b} p_X(x_i) - \sum_{\text{all } x_i \leq a} p_X(x_i)$$

Thus, by virtue of Eqs. 3.2 and 3.4,

$$P(a < X \leqslant b) = F_X(b) - F_X(a) \tag{3.6}$$

EXAMPLE 3.1

For an example of a discrete random variable, consider again the problem of bulldozers in Example 2.1.

Using X as the random variable, whose values represent the number of good bulldozers after 6 months, the events in the sample space S are mapped (naturally) into the discrete values of the real line as shown in Fig. E3.1a.

Thus $(X = 0)$, $(X = 1)$, $(X = 2)$, and $(X = 3)$ can be used to identify the respective events of interest.

If the probability that a bulldozer will remain operational after 6 months is $p = 0.8$, then assuming the conditions between bulldozers to be statistically independent, the PMF of X becomes

$$P(X = 0) = (0.2)^3 = 0.008$$
$$P(X = 1) = 3[0.8(0.2)^2] = 0.096$$
$$P(X = 2) = 3[(0.8)^2 0.2] = 0.384$$
$$P(X = 3) = (0.8)^3 = 0.512$$

whereas $P(X = x) = 0$ for all other x. These results can be portrayed graphically as shown in Fig. E3.1b. The corresponding cumulative distribution function (CDF) would appear as in Fig. E3.1c.

Analytically, the PMF described above is given by the binomial distribution (see Section 3.2.3) with $n = 3$ and $p = 0.8$.

EXAMPLE 3.2

To illustrate a continuous random variable, consider the problem described in Example 2.14. If the volume of traffic and road conditions along the 100-km highway

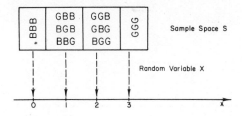

Figure E3.1a

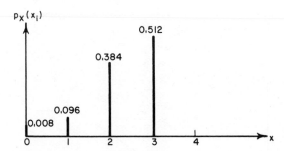

Figure E3.1b PMF of X

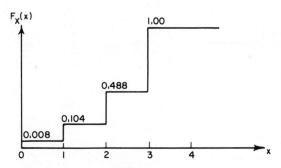

Figure E3.1c CDF of X

are about the same, the likelihood of accidents is roughly uniform over the 100-km distance. If X is a random variable whose values denote the distance (from km 0) at which accidents occur, then the probability density function (PDF) of X is constant between 0 and 100 km; that is

$$f_X(x) = c \qquad 0 \leq x \leq 100$$
$$= 0 \qquad \text{elsewhere}$$

where $c = 1/100$. Graphically, this is shown in Fig. E3.2a. The corresponding distri-

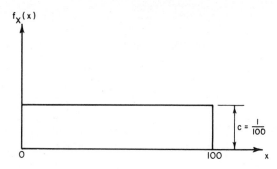

Figure E3.2a   PDF of X

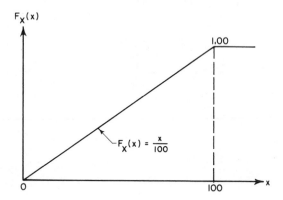

Figure E3.2b   CDF of X

bution function is

$$F_X(x) = \int_0^x c\,dx = cx = \frac{x}{100} \qquad 0 \le x \le 100$$
$$= 1.0 \qquad\qquad\qquad x > 100$$
$$= 0 \qquad\qquad\qquad x < 0$$

and graphically is as shown in Fig. E3.2b. Then, for example, the probability

$$P(20 < X \le 35) = \int_{20}^{35} \frac{1}{100}\,dx = 0.15$$

or, alternatively, using Eq. 3.6,

$$P(20 < X \le 35) = F_X(35) - F_X(20)$$
$$= \frac{35}{100} - \frac{20}{100} = 0.15$$

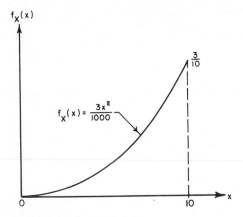

Figure E3.3   PDF of X

EXAMPLE 3.3

Suppose that a random variable X has a PDF of the form

$$f_X(x) = \alpha x^2 \qquad 0 \le x \le 10$$
$$= 0 \qquad \text{elsewhere}$$

(See Fig. E3.3.) Under what condition (i.e. what value of α) is this function a bona fide PDF?

In order to satisfy all the properties of a PDF, we must have

$$\int_0^{10} \alpha x^2 \, dx = 1.0$$

from which

$$\frac{\alpha}{3} (10)^3 = 1.0$$

and

$$\alpha = \frac{3}{1000}$$

The probability

$$P(X > 5) = 1 - P(X \le 5) = 1 - \int_0^5 \frac{3x^2}{1000} \, dx = 1 - \frac{5^3}{1000} = 0.875$$

3.1.2. Main descriptors of a random variable

The probabilistic characteristics of a random variable would be described completely if the form of the distribution function (or equivalently its probability density or mass function) and the associated parameters are specified. In practice, however, the form of the distribution function may not be known; consequently, approximate description of a random variable is often necessary. The probabilistic characteristics of a random variable may be described approximately in terms of certain key quantities or *main descriptors* of the random variable; the most important of these are the *central value* of the random variable, and a *measure of the dispersion* of its values. A *skewness measure* may also be important and useful when the underlying distribution is known to be nonsymmetric.

Moreover, even when the distribution function is known, the principal quantities remain useful, because they convey information on the properties of the random variable that are of first importance in practical applications. Also, the parameters of the distribution may be derived as functions of these quantities, or may be the parameters themselves (see Chapter 5).

Mean or expected value (a central value).   Since there is a range of possible values of a random variable, we would naturally be interested in some central value, such as the *average*. In particular, because the different values of the random variable are associated with different probabilities or probability densities, the "weighted average" would be of special interest; this is known as the *mean value* or the *expected value* of the random variable.

Therefore, if X is a discrete random variable with PMF $p_X(x_i)$, its "weighted" average value, denoted $E(X)$, is

$$E(X) = \sum_{\text{all } x_i} x_i \, p_X(x_i) \tag{3.7a}$$

Similarly, for a continuous random variable X with PDF $f_X(x)$, the mean value is

$$E(X) = \int_{-\infty}^{\infty} x \, f_X(x) \, dx \tag{3.7b}$$

Mathematical expectation.   The notion of a weighted average or expected value can be generalized for a function of X. Given a function $g(X)$, its expected value $E[g(X)]$, obtained as a generalization of Eq. 3.7, is

$$E[g(X)] = \sum_{\text{all } x_i} g(x_i) p_X(x_i) \tag{3.8a}$$

if X is discrete; whereas, if X is continuous,

$$E[g(X)] = \int_{-\infty}^{\infty} g(x) f_X(x) \, dx \tag{3.8b}$$

In either case, $E[g(X)]$ is known as the *mathematical expectation* of $g(X)$.

Other quantities that are used also to designate the central value of a random variable include the *mode* (or modal value) and the *median*.

The *mode* $\tilde{x}$ is the most probable value of a random variable; that is, it is the value of the random variable with the largest probability or the highest probability density.

The *median* is the value of a random variable at which values above and below it are equally probable; that is, if x_m is the median of X, then

$$F_X(x_m) = 0.50 \qquad (3.9)$$

In general, the mean, median, and mode of a random variable are different, especially if the density function is not symmetric. However, if the PDF is symmetric and unimodal (single mode), these three quantities coincide.

Variance and standard deviation (measures of dispersion). Besides the central value, the next most important quantity of a random variable is its measure of dispersion or variability; that is, the quantity that gives a measure of how closely the values of the variate are clustered (or conversely, how widely they are spread) around the central value. Intuitively, such a measure must be a function of the deviations from the central value. However, whether a deviation is above or below the central value should be of no significance; consequently, the function should be an even function of the deviations.

If the deviations are taken with respect to the mean value, then a suitable average measure of dispersion is the *variance*. For a discrete random variable X with PMF $p_X(x_i)$, the variance of X is

$$\mathrm{Var}(X) = \sum_{\text{all } x_i} (x_i - \mu_X)^2 \, p_X(x_i) \qquad (3.10)$$

in which $\mu_X \equiv E(X)$. We observe that this is simply the weighted average of squared deviations, or, in accordance with Eq. 3.8, it is the mathematical expectation of $g(X) = (X - \mu_X)^2$. Therefore, according to Eq. 3.8b, if X is continuous with PDF $f_X(x)$, the variance is

$$\mathrm{Var}(X) = \int_{-\infty}^{\infty} (x - \mu_X)^2 f_X(x) \, dx \qquad (3.11)$$

Expanding the integrand in Eq. 3.11, we have

$$\mathrm{Var}(X) = \int_{-\infty}^{\infty} (x^2 - 2\mu_X x + \mu_X^2) f_X(x) \, dx$$

$$= E(X^2) - 2\mu_X E(X) + \mu_X^2$$

Thus a useful relation for the variance is

$$\mathrm{Var}(X) = E(X^2) - \mu_X^2 \qquad (3.12)$$

In Eq. 3.12, the term $E(X^2)$ is known as the *mean-square* value of X.

Dimensionally, a more convenient measure of dispersion is the square root of the variance, or the *standard deviation* σ; that is,

$$\sigma_X = \sqrt{\mathrm{Var}(X)} \qquad (3.13)$$

It is hard to say, solely on the basis of the variance or standard deviation, whether the dispersion is large or small; for this purpose, the measure of dispersion relative to the central value is more useful. In other words, whether the dispersion is large or small is meaningful only relative to the central value. For this reason, the *coefficient of variation* (COV),

$$\delta_X = \frac{\sigma_X}{\mu_X} \tag{3.14}$$

is often a preferred and convenient nondimensional measure of dispersion or variability.

EXAMPLE 3.1 *(continued)*

Referring back to Example 3.1, the expected number of operating bulldozers at the end of 6 months is

$$E(X) = 0(0.008) + 1(0.096) + 2(0.384) + 3(0.512)$$
$$= 2.40$$

This illustrates the fact that the expected value of a discrete random variable may not be a possible value of the random variable.

The corresponding variance is

$$\text{Var } (X) = 0.008(0 - 2.4)^2 + 0.096(1 - 2.4)^2$$
$$+0.384(2 - 2.4)^2 + 0.512(3 - 2.4)^2$$
$$= 0.48$$

Using Eq. 3.12, we may compute the variance also as

$$\text{Var } (X) = [1^2(0.096) + 2^2(0.384) + 3^2(0.512)] - (2.40)^2$$
$$= 0.48$$

The standard deviation, therefore, is

$$\sigma_X = \sqrt{0.48} = 0.69$$

and coefficient of variation (COV) is

$$\delta_X = \frac{0.69}{2.40} = 0.29$$

EXAMPLE 3.3 *(continued)*

For the random variable X with the density function of Example 3.3, the mean and variance are, respectively,

$$E(X) = \int_0^{10} x\left(\frac{3x^2}{1000}\right) dx$$
$$= \left[\frac{3}{4000} x^4\right]_0^{10} = \frac{3 \cdot 10^4}{4000} = \frac{30}{4} = 7.50$$

$$\text{Var}(X) = \int_0^{10} (x - 7.5)^2 \left(\frac{3x^2}{1000}\right) dx$$

$$= \frac{3}{1000} \int_0^{10} [x^4 - 15x^3 + (7.5)^2 x^2] \, dx = 3.75$$

or, by Eq. 3.12,

$$\text{Var}(X) = \int_0^{10} x^2 \left(\frac{3x^2}{1000}\right) dx - (7.50)^2 = 3.75$$

Therefore the standard deviation is

$$\sigma_X = \sqrt{3.75} = 1.94$$

and the corresponding COV is

$$\delta_X = \frac{1.94}{7.50} = 0.26$$

From Fig. E3.3, the modal value is obviously $\tilde{x} = 10$. To determine the median, Eq. 3.9 yields

$$\int_0^{x_m} \frac{3x^2}{1000} \, dx = 0.50$$

from which we have

$$x_m^3 = 500$$

Thus the median is

$$x_m = 7.94$$

EXAMPLE 3.4

A contractor has an experience record that shows 60% of his projects are completed on schedule. If this performance record prevails, the probability of the number of completions in the next 6 jobs can be described by the binomial distribution (see Section 3.2.3) as follows: if X is the number of jobs completed among 6 future jobs, then

$$P(X = x) = \binom{6}{x}(0.6)^x(0.4)^{6-x} \qquad x = 0, 1, 2, \ldots, 6$$

$$= 0 \qquad\qquad\qquad \text{otherwise}$$

where

$$\binom{6}{x} = \frac{6!}{x!\,(6-x)!}$$

The mean number of jobs completed on schedule, therefore, is

$$E(X) = \sum_{x=0}^{6} x \binom{6}{x}(0.6)^x(0.4)^{6-x}$$

$$= 1\binom{6}{1}(0.6)(0.4)^5 + 2\binom{6}{2}(0.6)^2(0.4)^4$$

$$+ 3\binom{6}{3}(0.6)^3(0.4)^3 + 4\binom{6}{4}(0.6)^4(0.4)^2$$

$$+ 5\binom{6}{5}(0.6)^5(0.4) + 6\binom{6}{6}(0.6)^6$$

$$= 0.03686 + 2(0.13830) + 3(0.27640)$$
$$+ 4(0.31110) + 5(0.18660) + 6(0.04666)$$
$$= 3.60$$

Therefore the average number of jobs among 6 that can be completed on schedule is between 3 and 4.

The corresponding variance is

$$\text{Var}(X) = \sum_{x=0}^{6} (x - 3.60)^2 \left[\binom{6}{x}(0.6)^x(0.4)^{6-x} \right]$$

$$= (-3.60)^2 \binom{6}{0}(0.4)^6 + (-2.60)^2 \binom{6}{1}(0.6)(0.4)^5$$

$$+ (-1.60)^2 \binom{6}{2}(0.6)^2(0.4)^4 + (-0.60)^2 \binom{6}{3}(0.6)^3(0.4)^3$$

$$+ (0.4)^2 \binom{6}{4}(0.6)^4(0.4)^2 + (1.40)^2 \binom{6}{5}(0.6)^5(0.4)$$

$$+ (2.40)^2 \binom{6}{6}(0.6)^6$$

$$= 0.0531 + 0.2482 + 0.3539 + 0.0995$$
$$+ 0.0498 + 0.1626 + 0.2684$$
$$= 1.2355$$

The standard deviation, therefore, is

$$\sigma_X = \sqrt{1.2355} = 1.11$$

and coefficient of variation (COV) is

$$\delta_X = \frac{\sigma_X}{\mu_X} = \frac{1.11}{3.60} = 0.308$$

In this case, $X = 4$ has the highest probability; hence the mode is 4.

EXAMPLE 3.5

Suppose that the useful life T (in hours) of welding machines is not predictable, but can be described with a PDF known as the *exponential* distribution (see Section 3.2.7)

$$f_T(t) = \lambda e^{-\lambda t} \quad t \geq 0$$
$$= 0 \quad t < 0$$

in which λ is a constant parameter. Graphically, this probability density function is shown in Fig. E3.5a. In this case, the corresponding distribution function is

$$F_T(t) = \int_0^t \lambda e^{-\lambda \tau} \, d\tau = 1 - e^{-\lambda t}$$

which is also shown graphically in Fig. E3.5b.

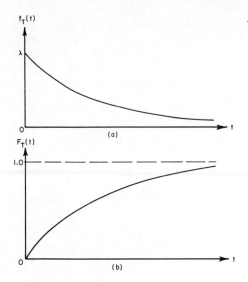

Figure E3.5 PDF and CDF of T

The mean life of a welding machine then is

$$\mu_T = E(T) = \int_0^\infty t \cdot \lambda e^{-\lambda t} \, dt$$

Integration by part will yield $\mu_T = 1/\lambda$. Therefore, for the exponential distribution, the parameter λ is equal to the reciprocal of the mean life; that is, $\lambda = 1/E(T)$.

The mode $\tilde{t}$ in this case is zero, whereas the corresponding median life t_m is obtained as follows. According to the definition of the median, Eq. 3.9,

$$\int_0^{t_m} \lambda e^{-\lambda t} \, dt = 0.50$$

thus obtaining

$$t_m = \frac{-\ln 0.5}{\lambda} = \frac{0.693}{\lambda}$$

Therefore

$$t_m = 0.693 \mu_T$$

The variance of T is

$$\sigma_T^2 = \int_0^\infty \left(t - \frac{1}{\lambda} \right)^2 \lambda e^{-\lambda t} \, dt$$

Integration by parts yields

$$\sigma_T^2 = \frac{1}{\lambda^2}$$

Thus the corresponding standard deviation is

$$\sigma_T = \frac{1}{\lambda} = \mu_T$$

Therefore the COV of the exponential distribution is

$$\delta_T = 1.00$$

Measure of skewness. Another useful property of a random variable is the symmetry or lack of symmetry of its probability distribution, and its associated degree and direction of asymmetry. A measure of this asymmetry or skewness is the *third central moment,* or

$$E(X - \mu_X)^3 = \sum_{\text{all } x_i} (x_i - \mu_X)^3 p_X(x_i) \qquad \text{for discrete } X$$

and

$$E(X - \mu_X)^3 = \int_{-\infty}^{\infty} (x - \mu_X)^3 f_X(x) \, dx \qquad \text{for continuous } X$$

Observe that $E(X - \mu_X)^3$ is zero if the probability distribution is symmetric about μ_X; otherwise it may be positive or negative. It will be positive if the values of X that are greater than μ_X are more widely dispersed than the dispersion of $X < \mu_X$. On the other hand, this third moment about the mean will be negative if the reverse situation is true. Therefore, the skewness of a random variable may be designated as *positive* or *negative* in accordance with the sign of the third moment $E(X - \mu_X)^3$; the magnitude of this third moment gives the corresponding degree of skewness. These properties are illustrated in Fig. 3.3.

A convenient nondimensional measure of skewness is the *skewness coefficient,*

$$\theta = \frac{E(X - \mu)^3}{\sigma^3} \tag{3.15}$$

Analogies with properties of area. The mean value and the variance correspond, respectively, to the *centroidal distance* and the *central moment-of-inertia* of an area. To see this, consider a unit area having a general shape shown in Fig. 3.4.

The centroidal distance x_0 of the area is

$$x_0 = \frac{\int_{-\infty}^{\infty} xf(x) \, dx}{\text{area}} = \int_{-\infty}^{\infty} xf(x) \, dx \tag{3.16}$$

which is also the first moment (about 0) of the irregular-shaped area.

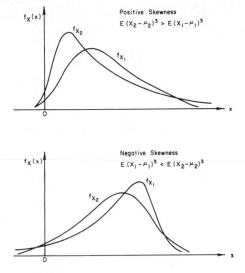

Figure 3.3 Asymmetric PDF

The moment of inertia about the vertical centroidal axis is

$$I_y = \int_{-\infty}^{\infty} (x - x_0)^2 f(x) \; dx \tag{3.17}$$

Comparing Eqs. 3.7*b* and 3.11 with Eqs. 3.16 and 3.17, respectively, we see that the mean value is equivalent to the centroidal distance, whereas the variance is equivalent to the centroidal moment of inertia of an area. In this regard, we can therefore refer to the mean value as the *first*

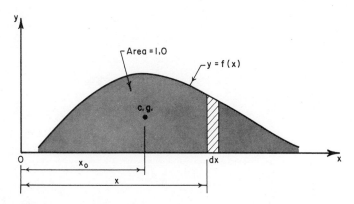

Figure 3.4 An irregular area

moment, and the variance as the *second (central) moment* of a random variable. More generally, extending this terminology, we shall refer to

$$E(X^n) = \int_{-\infty}^{\infty} x^n f_X(x)\, dx \tag{3.18}$$

as the *nth moment* of X.

MOMENT-GENERATING AND CHARACTERISTIC FUNCTIONS*

The approximate description of a random variable (discussed in Section 3.1.2) can be improved with a knowledge of its higher-order moments. Indeed, if all the moments of a random variable are known, its probability distribution would also be completely described. This means that a function through which all the moments can be generated is an alternative way of describing the probability law of a random variable; such a function is a *moment-generating function.* Its complex form is a *characteristic function.*

The moment-generating function of a random variable X, denoted $G_X(s)$, is defined as the expected value of e^{sX}; that is,

$$G_X(s) \equiv E(e^{sX}) \tag{3.19}$$

where s is an auxiliary (deterministic) variable.

Therefore, if X has a PDF $f_X(x)$, the corresponding moment-generating function is

$$G_X(s) = \int_{-\infty}^{\infty} e^{sx} f_X(x)\, dx \tag{3.20a}$$

whereas if X is discrete with PMF

$$G_X(s) = \sum_{\text{all } x_i} e^{sx\hat{i}} p_X(x_i) \tag{3.20b}$$

From Eq. 3.20a we observe that

$$\left.\frac{dG_X(s)}{ds}\right|_{s=0} = \int_{-\infty}^{\infty} x f_X(x)\, dx$$

Therefore

$$\frac{dG_X(0)}{ds} = E(X), \text{ the expected value of } X.$$

Similarly,

$$\frac{d^2 G_X(0)}{ds^2} = \int_{-\infty}^{\infty} x^2 f_X(x)\, dx = E(X^2)$$

and, in general,

$$\frac{d^n G_X(0)}{ds^n} = \int_{-\infty}^{\infty} x^n f_X(x)\, dx = E(X^n) \tag{3.21}$$

Therefore the nth moment of a random variable is given by the nth derivative of its moment-generating function evaluated at $s = 0$.

---

* This section is presented here only for mathematical definition; the material is not necessary for understanding the remainder of the book.

It can also be shown that the variance is given by

$$\text{Var}(X) = \frac{d^2}{ds^2}\ln G_X(0) \tag{3.22}$$

The characteristic function of X is defined as

$$\phi_X(s) \equiv E(e^{isX}) = G_X(is) \tag{3.23}$$

in which $i = \sqrt{-1}$. Therefore

$$\phi_X(s) = \int_{-\infty}^{\infty} e^{isx} f_X(x)\, dx \tag{3.24a}$$

or

$$\phi_X(s) = \sum_{\text{all } x_j} e^{isx_j} \cdot p_X(x_j) \tag{3.24b}$$

In terms of the characteristic function, the nth moment of X is given by

$$E(X^n) = \frac{1}{(i)^n}\frac{d^n\phi_X(0)}{ds^n} \tag{3.25}$$

whereas the special relation for the variance is

$$\text{Var}(X) = \frac{1}{(i)^2}\frac{d^2}{ds^2}\ln\phi_X(0) \tag{3.26}$$

3.2. USEFUL PROBABILITY DISTRIBUTIONS

Any function possessing all the properties cited earlier (in Section 3.1.1) can be used to describe the probability distribution of a random variable. However, there are a number of discrete and continuous functions that are specially useful because of one or more of the following reasons: (1) The function is the result of an underlying physical process and is derived on the basis of certain physically reasonable assumptions; (2) the function is the result of some limiting process; and (3) it is widely known and the necessary statistical information (including probability tables) is available widely. Several of these probability distribution functions are presented and their special properties described in this section.

3.2.1. The normal distribution

Perhaps the best-known and most widely used probability distribution is the *normal distribution*, also known as the *Gaussian distribution*. The normal distribution has a probability density function given by

$$f_X(x) = \frac{1}{\sigma\sqrt{2\pi}}\exp\left[-\frac{1}{2}\left(\frac{x-\mu}{\sigma}\right)^2\right] \qquad -\infty < x < \infty \tag{3.27}$$

where μ and σ are the parameters of the distribution, which are also the *mean and standard deviation*, respectively, of the variate. A short notation for this distribution is $N(\mu, \sigma)$, which we shall adopt.

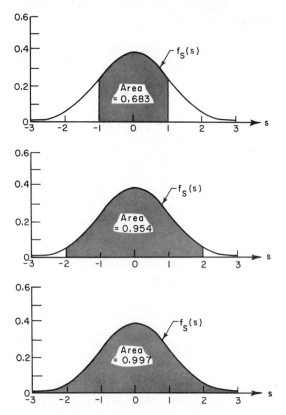

Figure 3.5   Density functions of the standard normal distribution

The standard normal distribution. A Gaussian distribution with parameters $\mu = 0$ and $\sigma = 1.0$ is known as the *standard normal* distribution and is denoted appropriately as $N(0, 1)$. The density function, accordingly, is

$$f_S(s) = \frac{1}{\sqrt{2\pi}} e^{-(1/2)s^2} \qquad -\infty < s < \infty \qquad (3.27a)$$

Several density functions of $N(0, 1)$ are shown graphically in Fig. 3.5; of some interest are the total probabilities within a specified number of standard deviations from the mean (which is zero), as shown in Fig. 3.5. Observe that the density function of $N(0, 1)$ is symmetric about zero.

Because of its wide usage, a special notation $\Phi(s)$ is commonly used to designate the distribution function of the *standard normal variate* S; that is,

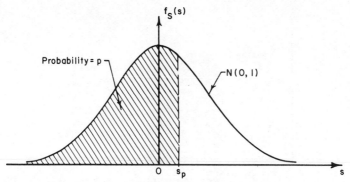

Figure 3.6 The standard normal density function

$\Phi(s) = F_S(s)$, where S has $N(0, 1)$ distribution. Referring to Fig. 3.6, we have

$$\Phi(s_p) = p$$

Conversely, the value of a standard normal variate at a cumulative probability p would be denoted as

$$s_p = \Phi^{-1}(p)$$

This notation will be used throughout this book.

The distribution function of $N(0, 1)$, that is, $\Phi(s)$, is tabulated widely as tables of normal probabilities—for example, Table A.1 of Appendix A. Observe from Table A.1 that the probabilities are given only for positive values of the variate. This is because by virtue of symmetry of the standard normal PDF about zero, the probabilities for negative values of the variate can be obtained as

$$\Phi(-s) = 1 - \Phi(s) \tag{3.27b}$$

By the same token, values of s corresponding to $p < 0.5$ may be obtained as

$$s = \Phi^{-1}(p) = -\Phi^{-1}(1 - p) \tag{3.27c}$$

With the table of $\Phi(s)$, probabilities of any other normal distributions can then be determined readily as follows. Suppose a normal variate X with distribution $N(\mu, \sigma)$; the probability

$$P(a < X \leqslant b) = \frac{1}{\sigma\sqrt{2\pi}} \int_a^b \exp\left[-\frac{1}{2}\left(\frac{x - \mu}{\sigma}\right)^2\right] dx$$

Clearly this is the area under the normal curve between a and b, as shown in Fig. 3.7. Theoretically, the required probability can be obtained by

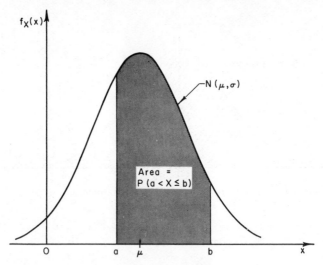

Figure 3.7   PDF for $N(\mu, \sigma)$

evaluating the preceding integral directly; however, this can be done also by making the following change of variable:

$$s = \frac{x - \mu}{\sigma} \quad \text{and} \quad dx = \sigma\, ds$$

Then

$$P(a < X \leqslant b) = \frac{1}{\sigma\sqrt{2\pi}} \int_{(a-\mu)/\sigma}^{(b-\mu)/\sigma} e^{-(1/2)s^2}\, \sigma\, ds$$

$$= \frac{1}{\sqrt{2\pi}} \int_{(a-\mu)/\sigma}^{(b-\mu)/\sigma} e^{-(1/2)s^2}\, ds$$

which may be recognized to be the area of the standard normal density function between $(a - \mu)/\sigma$ and $(b - \mu)/\sigma$, and thus according to Eq. 3.6 can be determined also as

$$P(a < X \leq b) = \Phi\left(\frac{b - \mu}{\sigma}\right) - \Phi\left(\frac{a - \mu}{\sigma}\right) \tag{3.28}$$

EXAMPLE 3.6

Suppose, from historical record, that the total annual rainfall in a catch basin is estimated to be normal $N(60 \text{ in.}, 15 \text{ in.})$.

(a) What is the probability that in future years the annual rainfall will be between 40 and 70 in.?

According to Eqs. 3.28 and 3.27b, this probability is

$$P(40 < X \leq 70) = \Phi\left(\frac{70 - 60}{15}\right) - \Phi\left(\frac{40 - 60}{15}\right)$$

$$= \Phi(0.67) - \Phi(-1.33)$$

$$= \Phi(0.67) - [1 - \Phi(1.33)]$$

From Table A.1 we therefore obtain the probability

$$P(40 < X \leq 70) = 0.748571 - (1 - 0.908241)$$

$$= 0.6568$$

(b) What is the probability that the annual rainfall will be at least 30 in.?

$$P(X \geq 30) = \Phi(\infty) - \Phi\left(\frac{30 - 60}{15}\right)$$

$$= 1 - \Phi(-2.00) = 1 - [1 - \Phi(2.00)] = \Phi(2.00)$$

$$= 0.9772$$

(c) What is the 10-percentile annual rainfall in the basin (that is, the value of the variate at which the cumulative probability is 10%)? In other words, the probability that the annual rainfall will be less than the 10-percentile value is 10%.

In this case, we wish to determine $x_{.10}$ so that

$$P(X \leq x_{.10}) = 0.10$$

Therefore

$$\Phi\left(\frac{x_{.10} - 60}{15}\right) = 0.10$$

Observing from Table A.1 that probabilities less than 0.50 are associated with negative values of the variate, and using Eq. 3.27c, we have

$$\frac{x_{.10} - 60}{15} = \Phi^{-1}(0.10) = -\Phi^{-1}(0.90) = -1.28$$

Hence the 10-percentile annual rainfall is

$$x_{.10} = 60 - 1.28(15) = 40.8 \text{ in.}$$

EXAMPLE 3.7

A shell structure is resting on three supports, A, B, and C, as shown in Fig. E3.7. Even though the loads from the roof transmitted to the three supports can be estimated accurately, the soil conditions under A, B, and C are not completely predictable. Assume that the settlements ρ_A, ρ_B, ρ_C are independent normal variates with means 2, 2.5, 3 cm (centimeter) and coefficients of variation 20%, 20%, 25%, respectively.

(a) What is the probability that the maximum settlement will exceed 4 cm?

(b) If it is known that A and B have settled 2.5 and 3.5 cm, respectively, what is the probability that the maximum differential settlement will not exceed 0.8 cm? That it will not exceed 1.5 cm?

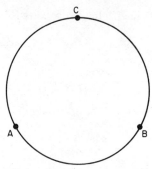

Figure E3.7

Solutions

(a) $P(\max \rho > 4 \text{ cm}) = 1 - P(\max \rho \leq 4 \text{ cm})$

$= 1 - P(\rho_A \leq 4 \cap \rho_B \leq 4 \cap \rho_C \leq 4)$

$= 1 - P(\rho_A \leq 4) P(\rho_B \leq 4) P(\rho_C \leq 4)$

$= 1 - \Phi\left(\dfrac{4-2}{0.4}\right) \Phi\left(\dfrac{4-2.5}{0.5}\right) \Phi\left(\dfrac{4-3}{0.75}\right)$

$= 1 - \Phi(5) \cdot \Phi(3) \cdot \Phi(1.333)$

$= 1 - 1 \times 0.9986 \times 0.9088$

$= 0.0925$

(b) Since the differential settlement between A and B, that is, $\Delta_{AB} = 3.5 \text{ cm} - 2.5 \text{ cm} = 1 \text{ cm}$,

$$P(\Delta_{\max} \leq 0.8 \text{ cm}) = 0$$

regardless of what ρ_C is.

For the event $\Delta_{\max}$ is less than 1.5 cm, we have to know what ρ_C is. If $\rho_C < 1$ cm or $\rho_C > 4$ cm, then $\Delta_{AC} > 1.5$ cm; also if $\rho_C < 2$ cm or $\rho_C > 5$ cm, then $\Delta_{BC} > 1.5$ cm. From these two conditions we see that the acceptable region of ρ_C is (2 cm $\leq \rho_C \leq 4$ cm). Any other values of ρ_C will definitely give rise to a maximum differential settlement exceeding 1.5 cm. Therefore

$$P(\Delta_{\max} \leq 1.5 \text{ cm}) = P(2 \text{ cm} \leq \rho_C \leq 4 \text{ cm})$$

$$= \Phi\left(\frac{4-3}{0.75}\right) - \Phi\left(\frac{2-3}{0.75}\right)$$

$$= \Phi(1.333) - \Phi(-1.333)$$

$$= 0.9088 - 0.0912$$

$$= 0.8176$$

3.2.2. The logarithmic normal distribution

A random variable X has a *logarithmic normal* (or simply log-normal) probability distribution if $\ln X$ (the natural logarithm of X) is normal.

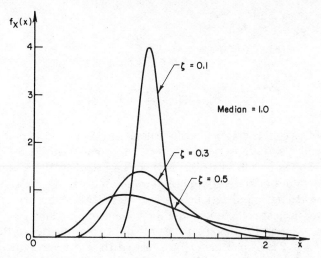

Figure 3.8 Log-normal density functions

In this case, the density function of X is

$$f_X(x) = \frac{1}{\sqrt{2\pi}\,\zeta x} \exp\left[-\frac{1}{2}\left(\frac{\ln x - \lambda}{\zeta}\right)^2\right] \qquad 0 \leq x < \infty \qquad (3.29)$$

where $\lambda = E(\ln X)$ and $\zeta = \sqrt{\mathrm{Var}(\ln X)}$ are, respectively, the mean and standard deviation of $\ln X$, and are the parameters of the distribution. Equation 3.29 is illustrated in Fig. 3.8 for various values of ζ.

Because of its relationship with the normal distribution (that is involving a logarithmic transformation), probabilities associated with a log-normal variate can also be determined using the table of standard normal probabilities. We show this as follows.

On the basis of Eq. 3.3, the probability that X will assume values in an interval $(a, b]$ is

$$P(a < X \leq b) = \int_a^b \frac{1}{\sqrt{2\pi}\,\zeta x} \exp\left[-\frac{1}{2}\left(\frac{\ln x - \lambda}{\zeta}\right)^2\right] dx$$

Let

$$s = \frac{\ln x - \lambda}{\zeta};$$

then $dx = x\zeta \, ds$, and

$$P(a < X \leq b) = \frac{1}{\sqrt{2\pi}} \int_{(\ln a - \lambda)/\zeta}^{(\ln b - \lambda)/\zeta} e^{-(1/2)s^2} \, ds$$

$$= \Phi\left(\frac{\ln b - \lambda}{\zeta}\right) - \Phi\left(\frac{\ln a - \lambda}{\zeta}\right) \qquad (3.30)$$

In view of this convenient facility for calculating probabilities of log-normal variates and also because the values of the random variable are always positive, the log-normal distribution may be useful in those applications where the values of the variate are known to be strictly positive; for example, the strength and fatigue life of material, the intensity of rainfall, the time for project completion, and the volume of air traffic.

We observe from Eq. 3.30 that the probability is a function of the parameters λ and ζ. These parameters are related to the mean μ and standard deviation σ of the variate as follows.

Let $Y = \ln X$, which is $N(\lambda, \zeta)$. It follows that $X = e^Y$ and

$$\mu = E(X) = E(e^Y)$$

$$= \frac{1}{\sqrt{2\pi}\,\zeta} \int_{-\infty}^{\infty} e^y \exp\left[-\frac{1}{2}\left(\frac{y - \lambda}{\zeta}\right)^2\right] dy$$

$$= \frac{1}{\sqrt{2\pi}\,\zeta} \int_{-\infty}^{\infty} \exp\left[y - \frac{1}{2}\left(\frac{y - \lambda}{\zeta}\right)^2\right] dy$$

By completing the square on the exponent, we obtain

$$\mu = \left[\frac{1}{\sqrt{2\pi}\,\zeta} \int_{-\infty}^{\infty} \exp\left\{-\frac{1}{2}\left(\frac{y - (\lambda + \zeta^2)}{\zeta}\right)^2 dy\right\}\right] \exp(\lambda + \tfrac{1}{2}\zeta^2)$$

We recognize that the quantity within the brackets is the total unit area of the Gaussian density function $N(\lambda + \zeta^2, \zeta)$; hence

$$\mu = \exp(\lambda + \tfrac{1}{2}\zeta^2)$$

Thus we have

$$\lambda = \ln \mu - \tfrac{1}{2}\zeta^2 \qquad (3.31)$$

Similarly, we derive the variance of X as follows:

$$E(X^2) = \frac{1}{\sqrt{2\pi}\,\zeta} \int_{-\infty}^{\infty} e^{2y} \exp\left[-\frac{1}{2}\left(\frac{y - \lambda}{\zeta}\right)^2\right] dy$$

$$= \frac{1}{\sqrt{2\pi}\,\zeta} \int_{-\infty}^{\infty} \exp\left[-\frac{1}{2\zeta^2}\left\{y^2 - 2(\lambda + 2\zeta^2)y + \lambda^2\right\}\right] dy$$

By completing the square on the exponent, the above integral yields

$$E(X^2) = \left[\frac{1}{\sqrt{2\pi}\,\zeta} \int_{-\infty}^{\infty} \exp\left\{ -\frac{1}{2}\left(\frac{y - (\lambda + 2\zeta^2)}{\zeta} \right)^2 dy \right\} \right] \exp[2(\lambda + \zeta^2)]$$

$$= \exp[2(\lambda + \zeta^2)].$$

Thence, according to Eq. 3.12, and using Eq. 3.31,

$$\mathrm{Var}(X) = \exp[2(\lambda + \zeta^2)] - \exp[2(\lambda + \tfrac{1}{2}\zeta^2)]$$

$$= \mu^2(e^{\zeta^2} - 1)$$

from which we obtain

$$\zeta^2 = \ln\left(1 + \frac{\sigma^2}{\mu^2} \right) \tag{3.32}$$

If σ/μ is not large, say ≤ 0.30, $\ln[1 + (\sigma^2/\mu^2)] \simeq \sigma^2/\mu^2$. In such cases, therefore,

$$\zeta \simeq \frac{\sigma}{\mu} = \delta, \text{ the COV} \tag{3.32a}$$

The median is often used to designate the central value of a log-normal variate. If x_m is the median, then by definition, Eq. 3.9,

$$P(X \leq x_m) = 0.5$$

or

$$\Phi\left(\frac{\ln x_m - \lambda}{\zeta} \right) = 0.5$$

Thus

$$\frac{\ln x_m - \lambda}{\zeta} = \Phi^{-1}(0.5) = 0$$

Therefore, in terms of the median, the parameter λ is

$$\lambda = \ln x_m \tag{3.33}$$

conversely,

$$x_m = e^{\lambda} \tag{3.33a}$$

Comparing Eqs. 3.31 and 3.33 and using Eq. 3.32, we obtain the relation between the mean and median of a log-normal variate as

$$x_m = \frac{\mu}{\sqrt{1 + \delta^2}} \tag{3.34}$$

This means then that the median of a log-normal variate is always less than its mean value; that is, $x_m < \mu$.

EXAMPLE 3.8

In Example 3.6, suppose that the annual rainfall has a log-normal distribution (instead of normal) with the same mean and standard deviation of 60 and 15 in., respectively. What would then be the answers to the questions raised in Example 3.6? We first obtain the parameters λ and ζ as follows. Using Eq. 3.32a,

$$\zeta \simeq \frac{15}{60} = 0.25$$

and from Eq. 3.31

$$\lambda = \ln 60 - \tfrac{1}{2}(0.25)^2 = 4.09 - 0.03 = 4.06$$

(a) In this case, the probability that the annual rainfall will be between 40 and 70 in. is

$$P(40 < X \le 70) = \Phi\left(\frac{\ln 70 - 4.06}{0.25}\right) - \Phi\left(\frac{\ln 40 - 4.06}{0.25}\right)$$

$$= \Phi(0.75) - \Phi(-1.48)$$

$$= 0.773373 - 0.069437 = 0.7039$$

(b) The probability that the annual rainfall will be at least 30 in. is

$$P(X \ge 30) = 1 - \Phi\left(\frac{\ln 30 - 4.06}{0.25}\right)$$

$$= 1 - \Phi(-2.64) = 0.9958$$

(c) The 10-percentile annual rainfall is

$$\Phi\left(\frac{\ln x_{.10} - 4.06}{0.25}\right) = 0.10$$

$$\frac{\ln x_{.10} - 4.06}{0.25} = -1.28$$

$$\ln x_{.10} = 4.06 - 1.28(0.25) = 3.74$$

Therefore

$$x_{.10} = e^{3.74} = 42.10 \text{ in.}$$

3.2.3. Bernoulli sequence and the binomial distribution

Problems of concern to engineers and engineering planners sometimes require the consideration of the potential occurrence or recurrence of an event in a sequence of repeated "trials." For example, in allocating a fleet of construction equipments for a project, the anticipated conditions of every piece of equipment in the fleet over the project duration would have some bearing on the determination of the required fleet size; whereas, in planning the flood control system for a river basin, the annual maximum flow of the river over a sequence of years would be important in the determination of the design flood. In these cases, the operational conditions of each piece of equipment, and the maximum flow of the river each year relative to a specified flood level, constitute the respective *trials*. These problems are also such that there are only two possible outcomes in each trial, namely, the *occurrence* or *nonoccurrence* of an event—each piece of equipment *may* or *may not* malfunction over the duration of the project; each year, the maximum flow of the river *may* or *may not* exceed some specified flood level.

Problems of the type described above may be modeled by a *Bernoulli sequence*, which is based on the following assumptions:

1. Each trial has only two possible outcomes: the *occurrence* or *nonoccurrence* of an event.
2. The probability of occurrence of the event in each trial is constant.
3. The trials are statistically independent.

Therefore, in the examples cited above, if the operational conditions between equipments are statistically independent and the probability of malfunction for every piece of equipment is the same, then the conditions of the entire fleet of equipments constitute a Bernoulli sequence. Similarly, if the annual maximum floods are statistically independent, and each year the probability of the annual flood's exceeding some specified level is constant, then the annual maximum floods over a series of years also constitute a Bernoulli sequence.

The binomial distribution. If the probability of occurrence of an event in each trial is p (and probability of nonoccurrence is $1 - p$), then the probability of exactly x occurrences among n trials in a Bernoulli sequence is given by the *binomial* PMF as follows:

$$P(X = x) = \binom{n}{x} p^x (1 - p)^{n-x} \qquad x = 0, 1, 2, \ldots, n \qquad (3.35)$$

SAMPLING WITH REPLACEMENT

where n and p are parameters, and $\binom{n}{x} = n!/[x!(n - x)!]$ is the *binomial coefficient* (see Appendix B). The PMF for such a distribution with $p = 0.80$ and $n = 3$ was illustrated earlier (Example 3.1).

We observe that the probability of realizing a particular sequence of exactly x occurrences of the event among n trials is $p^x (1 - p)^{n-x}$. However, the specific sequence of trials in which the event may occur x times can be permuted among the n trials, so that the number of distinct sequences with exactly x occurrences is $\binom{n}{x}$; for example, if there are x breakdowns among a fleet of n pieces of equipments, the x breakdowns may be realized in $\binom{n}{x}$ different sequences. Thus we obtain Eq. 3.35.

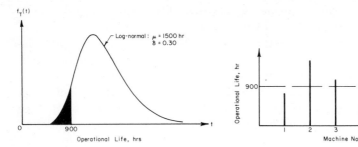

Figure E3.9a   Figure E3.9b

Figure E3.9a   *Figure E3.9b*

EXAMPLE 3.9

Suppose that five road graders are used in a highway project. The operational life T of such equipments is known to have a log-normal distribution with a mean life of 1500 hr and a COV of 30% (see Fig. E3.9a). Among the five machines in use, what is the probability that two of them will malfunction in less than 900 hr of operation? Assume statistical independence between the conditions of the machines.

Each grader may or may not malfunction after 900 hr of operation. The probability of malfunction within this period is determined as follows:

$$\zeta \simeq 0.30$$
$$\lambda = \ln 1500 - \tfrac{1}{2}(0.3)^2 = 7.27$$

Therefore the probability of a machine malfunctioning in 900 hr (see Fig. E3.9a) is

$$p = P(T < 900) = \Phi\left(\frac{\ln 900 - 7.27}{0.30}\right)$$
$$= \Phi(-1.56) = 0.0594$$

For the five machines taken collectively, the actual operational lives may conceivably be as shown in Fig. E3.9b. That is, as illustrated in Fig. E3.9b, machines 1 and 4 have operational lives less than 900 hr. The probability that machines 1 and 4 will malfunction within 900 hr whereas 2, 3, and 5 remain operational is $p^2(1 - p)^3$. But the two malfunctions may happen to any two among the five machines; consequently, if X is the number of machines malfunctioning in 900 hr,

$$P(X = 2) = \binom{5}{2}(0.0594)^2(0.9406)^3$$
$$= \frac{5!}{2!\,3!}(0.0594)^2(0.9406)^3$$
$$= 0.0294$$

The probability of malfunction among the five graders (that is, malfunction in one or more machines) is

$$P(X \geq 1) = 1 - P(X = 0)$$
$$= 1 - (0.9406)^5 = 0.2638$$

In spite of its simplicity, the Bernoulli model is quite useful in many engineering applications. Engineering problems involving situations with only two alternative possibilities are numerous. Aside from those cited and illustrated above, other problems of this nature include the following. In a series of soil borings, each boring may or may not encounter boulders; in monitoring the daily water quality of a river on the downstream side of an industrial plant, the water tested daily may or may not meet the pollution control standards; the individual items produced on an assembly line in an industrial plant may or may not pass the inspection to assure product quality; and a nuclear power plant may or may not be hit by a tornado in a year. In each case, if the situation is repeated, we have a Bernoulli sequence.

It should be emphasized also that in modeling problems with the Bernoulli sequence, the individual trials must be discrete. In spite of this requirement, however, certain continuous problems may be modeled (approximately at least) with the Bernoulli sequence. For example, time and space problems, which are generally continuous, may be modeled with the Bernoulli sequence by discretizing time (or space) into finite intervals and admitting only two possibilities within each interval; what happens in each time (or space) interval then constitutes a trial, and in the series of intervals a Bernoulli sequence. Consider for example the following.

EXAMPLE 3.10

In planning the flood control system for a river, the yearly maximum flood of the river is of concern. Suppose that the probability of the annual maximum flood exceeding some specified design level h_0 is 0.10 in any year; what is the probability that the level h_0 will be exceeded once in the next five years?

In this case, we observe that the natural time interval is one year, and within each year there is only one maximum flood that may or may not exceed the level h_0. Therefore the series of annual floods can be modeled as a Bernoulli sequence. Furthermore, assuming that h_0 is high enough that there is no likelihood of its being exceeded more than once a year, the number of exceedances of level h_0, therefore, has a binomial distribution. On this basis, if X is the number of exceedances of flood level h_0 in the next five years, then we have

$$P(X = 1) = \binom{5}{1}(0.1)^1(0.9)^4 = 0.328$$

The probability that there will be at most one exceedance of level h_0 (that is, one or none) in the next five years is

$$P(X \leq 1) = P(X = 0) + P(X = 1)$$
$$= \binom{5}{0}(0.1)^0(0.9)^5 + \binom{5}{1}(0.1)^1(0.9)^4$$
$$= 0.590 + 0.328 = 0.918$$

3.2.4. The geometric distribution

In a Bernoulli sequence, the number of trials until a specified event occurs for the first time is governed by the *geometric distribution*. We observe that if the first occurrence of the event is realized on the tth trial, then there must be no occurrence of this event in any of the prior $(t - 1)$ trials. Therefore, if T is the appropriate random variable,

$$P(T = t) = pq^{t-1}   t = 1, 2, \ldots \tag{3.36}$$

which is known as the *geometric distribution*.

The return period. In a time (or space) problem that can be modeled as a Bernoulli sequence, the number of time (or space) intervals until the first occurrence of an event is called the *first occurrence time*.

We observe that if the individual trials (or intervals) in the sequence are statistically independent, the first occurrence time must also be the time between any two consecutive occurrences of the same event; that is, the *recurrence time* is equal to the first occurrence time.

The recurrence time, therefore, in a Bernoulli sequence also has a geometric distribution; the *mean recurrence time*, which is popularly known in engineering as the (average) *return period*, therefore, is

$$\bar{T} = E(T) = \sum_{t=1}^{\infty} t \cdot pq^{t-1} = p(1 + 2q + 3q^2 + \cdots)$$

For $q < 1.0$, the infinite series in the parentheses yields

$$\frac{1}{(1 - q)^2} = \frac{1}{p^2}$$

Hence

$$\bar{T} = \frac{1}{p} \tag{3.37}$$

Equation 3.37, therefore, means that *on the average* the time between two consecutive occurrences of an event is equal to the reciprocal of the probability of the event within one time unit. It must be emphasized that the return period is only an average duration between events, and should not be construed as the actual time between occurrences; the actual time is T, which is a random variable.

EXAMPLE 3.11

A radio transmission tower is designed for a "50-year wind," that is, a wind velocity having a return period of 50 years.

(a) What is the probability that the design wind velocity will be exceeded for the first time on the fifth year after completion of the structure?

In this case, the probability of encountering the 50-year wind in any one year is $p = 1/50 = 0.02$. The required probability then is

$$P(T = 5) = (0.02)(0.98)^4 = 0.018$$

(b) What is the probability that the first such wind velocity will occur within 5 years after completion of the structure?

$$P(T \leq 5) = \sum_{t=1}^{5} (0.02)(0.98)^{t-1}$$
$$= 0.02 + 0.0196 + 0.0192 + 0.0188 + 0.0184$$
$$= 0.096$$

It may be recognized that this is the same as the event of *at least* one 50-year wind in 5 years; thus the desired probability may also be obtained as $1 - (0.98)^5 = 0.096$.

However, the above is different from the event of experiencing exactly one 50-year wind in 5 years; the probability in this case would be $\binom{5}{1}(0.02)(0.98)^4 = 0.092$.

EXAMPLE 3.12

An offshore structure is designed for a height of 8 m above the mean sea level (Fig. E3.12). This height corresponds to a 10% probability of being exceeded by sea waves in a year. What is the probability that the structure will be subjected to waves exceeding 8 m within the return period of the design wave height?

The return period of the design wave is

$$\bar{T} = \frac{1}{0.10} = 10 \text{ years}$$

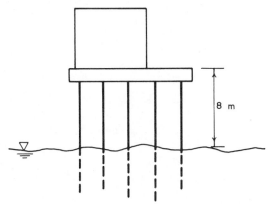

Figure E3.12

Therefore
$$P(H > 8 \text{ m in 10 years}) = 1 - (0.90)^{10} = 1 - 0.3487$$
$$= 0.6513$$

If it is assumed that, when subjected to waves exceeding the design height, there is a probability of 20% that the structure may be damaged, what is the probability of damage to the structure within 3 years?

This probability should take into consideration that there may be 0, 1, 2, or 3 exceedances in 3 years, assuming the likelihood of more than one such wave in a year is negligible. Furthermore, assume that structural damages from more than one exceedance are statistically independent. Then, according to the total probability theorem,

$$P \text{ (no damage in 3 years)} = 1.00(0.90)^3$$
$$+ 0.80[3(0.10)(0.90)^2]$$
$$+ (0.80)^2[3(0.10)^2(0.90)]$$
$$+ (0.80)^3(0.10)^3$$
$$= 0.9412$$

Therefore
$$P \text{ (damage in 3 years)} = 0.0588$$

Observe now that the probability of an event occurring within its return period $\bar{T}$ is

$$P \text{ (no occurrence in } \bar{T}) = (1 - p)^{\bar{T}}$$

where $p = 1/\bar{T}$. Expanding the above with the binomial theorem,

$$(1 - p)^{\bar{T}} = 1 - \bar{T}p + \frac{\bar{T}(\bar{T} - 1)}{2!} p^2 - \frac{\bar{T}(\bar{T} - 1)(\bar{T} - 2)}{3!} p^3 + \cdots$$

But for large $\bar{T}$ (and thus small p), it may be recognized that this is also approximately equal to $e^{-\bar{T}p}$. Hence, for large $\bar{T}$,

$$P \text{ (no occurrence in } \bar{T}) \simeq e^{-\bar{T}p} = e^{-1} = 0.368$$

Therefore
$$P \text{ (occurrence in } \bar{T}) \simeq 1 - 0.368 = 0.632 \text{ for large } \bar{T}$$

In other words, for a rare event (that is, large $\bar{T}$) the probability of the event's occurring within its return period is always 0.632. This result is a useful approximation even for return periods that are not very long; for instance, for $\bar{T} = 10$ (time units),

$$P \text{ (occurrence in } \bar{T}) = 1 - \left(1 - \frac{1}{10}\right)^{10} = 0.651$$

which shows that the error in the above approximation is less than 3%.

3.2.5.   The negative binomial distribution

We saw that the geometric distribution is the probability law governing the number of trials (or time units) until the first occurrence of an event in a Bernoulli sequence. The time until a subsequent occurrence of the same event is governed by the *negative binomial* distribution. That is, if T_k is the number of trials until the kth occurrence of the event in a series of Bernoulli trials, then

$$P(T_k = t) = \binom{t-1}{k-1} p^k q^{t-k} \qquad \text{for } t = k, k+1, \ldots \qquad (3.38)$$

$$= 0 \qquad \text{for } t < k$$

If the kth occurrence is realized at the tth trial, there must be exactly $(k-1)$ occurrences of the event in the prior $(t-1)$ trials and at the tth trial the event also occurs. Thus, from the binomial law,

$$P(T_k = t) = \binom{t-1}{k-1} p^{k-1} q^{t-k} p$$

yielding therefore Eq. 3.38.

EXAMPLE 3.11 *(continued)*

In the problem of Example 3.11, what is the probability that a second 50-year wind will occur exactly on the fifth year after completion of the structure?
From Eq. 3.38, the required probability is

$$P(T_2 = 5) = \binom{4}{1} (0.02)^2 (0.98)^3$$

$$= 0.0015$$

EXAMPLE 3.13

Suppose that a cable is composed of a number of independent wires (see Fig. E3.13). Occasionally the cable is subjected to high overloads; on such occasions the probability that one wire will fracture is 0.05. Assume that the failure of 2 or more wires during an overload is unlikely. If the cable must be replaced when 3 of the wires have failed, determine the probability that the cable can withstand at least 5 overload applications before being replaced.

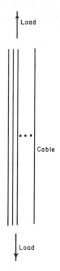

Figure E3.13

This means that the third failure must occur at or after the sixth overloading. Hence, using Eq. 3.38, the required probability is

$$P(T_3 \geq 6) = 1 - P(T_3 < 6)$$

$$= 1 - \binom{2}{2}(0.05)^3 - \binom{3}{2}(0.05)^3(0.95) - \binom{4}{2}(0.05)^3(0.95)^2$$

$$= 1 - 0.00116$$

$$= 0.99884$$

3.2.6.   The Poisson process and Poisson distribution

Many physical problems of interest to engineers involve the possible occurrences of events at any point in time and/or space. For example, fatigue cracks may occur anywhere along a continuous weld; earthquakes could strike at any time and anywhere over a seismically active region; and traffic accidents could happen at any time on a given highway. Conceivably, such space-time problems may be modeled also with the Bernoulli sequence, by dividing the time or space into small intervals, and assuming that an event will either occur or not occur (only two possibilities) within each interval, thus constituting a trial. However, if the event can occur at any instant (or at any point in space), it may occur more than once at a given time or space interval. In such cases, the occurrences of the event may be more appropriately modeled with a *Poisson sequence* or *Poisson process*.

Formally, the Poisson process is based on the following assumptions.

1.   An event can occur at random and at any time or any point in space.

2. The occurrence(s) of an event in a given time (or space) interval is independent of that in any other nonoverlapping intervals.
3. The probability of occurrence of an event in a small interval Δt is proportional to Δt, and can be given by $\nu \Delta t$, where ν is the mean rate of occurrence of the event (assumed to be constant); and the probability of two or more occurrences in Δt is negligible (of higher orders of Δt).

On the basis of these assumptions, the number of occurrences of an event in t is given by the *Poisson distribution*; that is, if X_t is the number of occurrences in time (or space) interval t, then

$$P(X_t = x) = \frac{(\nu t)^x}{x!} e^{-\nu t} \qquad x = 0, 1, 2, \ldots \qquad (3.39)$$

where ν is the *mean occurrence rate*; that is, the average number of occurrences of the event per unit time (or space) interval. It follows then that $E(X_t) = \nu t$; it can be shown that the variance of X_t is also νt.

A derivation of Eq. 3.39 based on the preceding assumptions is given in Appendix C.

The similarities and differences between the Bernoulli sequence and the Poisson process may be clarified with the following illustration. Suppose that, from a previous traffic count, an average of 60 cars per hour was observed to make left turns at an intersection. What is the probability that exactly 10 cars will be making left turns in a 10-minute interval?

An approximate solution would be to divide the hour into 120 30-second intervals; the probability of a left turn in any 30-second interval would be

$$p = \frac{60}{120} = 0.5$$

Then, assuming no more than one left turn is possible in a 30-second interval, the problem is reduced to the binomial probability of 10 events in 20 trials, in which the probability of an event in each trial is 0.5; thus

$$P \text{ (10 L.T. in 10 minutes)} = \binom{20}{10}(0.5)^{10}(0.5)^{20-10}$$

Physically, the solution is approximate because it is implicitly assumed that no more than one car would be making left turns during any 30-second interval; obviously, 2 or more left turns are actually possible.

The solution would be improved if a shorter time interval was chosen. For example, if 10-second intervals were used, then $p = 60/360 = 1/6$, and

$$P \text{ (10 L.T. in 10 minutes)} = \binom{60}{10}\left(\frac{1}{6}\right)^{10}\left(\frac{5}{6}\right)^{60-10}$$

Further improvements can be made by taking even shorter time intervals. In general, if the time t is divided into n equal intervals, then

$$P\ (x \text{ events in time } t) = \binom{n}{x}\left(\frac{\lambda}{n}\right)^{x}\left(1 - \frac{\lambda}{n}\right)^{n-x}$$

where λ is the *average* number of events in time t. If an event can occur at any time (as in the case of left-turn traffic), the process may tend to the case with $n \to \infty$; then

$$P\ (x \text{ events in } t)$$

$$= \lim_{n\to\infty}\binom{n}{x}\left(\frac{\lambda}{n}\right)^{x}\left(1 - \frac{\lambda}{n}\right)^{n-x}$$

$$= \lim_{n\to\infty}\left[\frac{n!}{x!(n-x)!}\left(\frac{\lambda}{n}\right)^{x}\left(1 - \frac{\lambda}{n}\right)^{n-x}\right]$$

$$= \lim_{n\to\infty}\left[\frac{n}{n}\cdot\frac{(n-1)}{n}\cdots\frac{(n-x+1)}{n}\cdot\frac{\lambda^{x}}{x!}\left(1 - \frac{\lambda}{n}\right)^{n}\left(1 - \frac{\lambda}{n}\right)^{-x}\right]$$

But

$$\lim_{n\to\infty}\left(1 - \frac{\lambda}{n}\right)^{n} = 1 - \lambda + \frac{\lambda^{2}}{2!} - \frac{\lambda^{3}}{3!} + \cdots = e^{-\lambda}$$

Therefore the limit yields

$$P\ (x \text{ events in } t) = \frac{\lambda^{x}}{x!}e^{-\lambda}$$

which is the Poisson distribution, Eq. 3.39, with $\lambda = \nu t$.

EXAMPLE 3.14

Historical records of rainstorms in a town indicate that on the average there had been 4 rainstorms per year over the last 20 years. Assuming that the occurrence of rainstorms is a Poisson process, what is the probability that there would be no rainstorms next year?

$$P(X_t = 0) = \frac{4^0}{0!}e^{-4}$$

$$= 0.018$$

The probability that exactly 4 rainstorms will occur in the next year is given by

$$P(X_t = 4) = \frac{4^4}{4!}e^{-4}$$

$$= 0.195$$

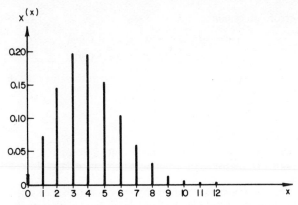

Figure E3.14 PMF of number of rainstorms in a year

This last result indicates that although the average yearly occurrences of rainstorms is 4, the probability of having exactly 4 rainstorms in a year is only about 20%. The probability of 2 or more rainstorms in the next year is

$$P(X_t \geq 2) = \sum_{x=2}^{\infty} \frac{4^x}{x!} e^{-4}$$
$$= 1 - \sum_{x=0}^{1} \frac{e^{-4} 4^x}{x!}$$
$$= 1 - 0.018 - 0.074$$
$$= 0.908$$

The PMF of the number of rainstorms in a year is shown graphically in Fig. E3.14.

EXAMPLE 3.15 (Design of left-turn bay)

For the purpose of designing a left-turn bay at a highway intersection, the left turns of vehicles may be modeled as a Poisson process. If the cycle time of the traffic light (for left turns) is 1 minute, and the design criterion requires a left-turn lane that will be sufficient 96% of the time (which is the criterion in California), what should the lane distance (in terms of car lengths) be to allow for an average of 100 left turns per hour?

Solution

Let k car lengths be the design length of the left-turn lane. The mean rate of left turns is $\nu = 100/60$ per minute. Therefore, during a 1-minute cycle of the traffic light, the probability of no more than k cars waiting for left turns must be at least 96%; thus

$$P(X_t \leq k) = \sum_{x=0}^{k} \frac{1}{x!} \left(\frac{100}{60}\right)^x e^{-100/60} = 0.96$$

By trial-and-error, we obtain

$$\text{if } k = 4, \quad P(X_t \leq 4) = 0.968$$
$$\text{if } k = 3, \quad P(X_t \leq 3) = 0.91$$

Therefore a left-turn bay of 4 car lengths is sufficient to satisfy the design requirements.

EXAMPLE 3.16

The street width at a school crosswalk is D ft, and a child crossing the street walks at a speed of 3.5 ft/sec. In other words, it takes a child $t = D/3.5$ sec to cross the street.

Suppose 60 free intervals (t seconds each) in an hour, on the average, are desired at this crossing; how much average traffic volume can be allowed at this crosswalk before crossing controls will be necessary? Assume that cars crossing the crosswalk constitute a Poisson process.

The number of t-sec intervals in an hour is $3600/t$, whereas in an interval of t sec the probability of no cars passing through the crosswalk (according to Eq. 3.39) is e^{-vt}, if v is the average vehicular traffic per second. Therefore the maximum average traffic volume that can be allowed is such that the mean number of free intervals equals 60; that is,

$$\left(\frac{3600}{t}\right) e^{-vt} = 60$$

or

$$\frac{3600 \times 3.5}{D} e^{-vD/3.5} = 60$$

From which

$$v = \frac{3.5}{D} \ln \frac{3600 \times 3.5}{60D}$$

For $D = 25$ ft,

$$v = \frac{3.5}{25} \ln \frac{3600 \times 3.5}{60 \times 25} = 0.298 \text{ vehicles/sec}$$
$$= 1073 \text{ vehicles/hr}$$

For other street widths D, the corresponding traffic volumes are as follows:

D(ft):	25	40	60	75
v(veh./hr):	1073	522	263	173

Therefore, for various street widths, the hourly traffic volumes given above are the maximum traffic flow that can be allowed before pedestrian crossing controls should be installed. This example points out how critical the street width is to the problem of pedestrian crossings, and indicates how important crossing controls for school children must be for heavily traveled highways. This also means that the wider streets involve much greater hazard to pedestrians. The above method has been adopted by the Joint Committee of the Institute of Traffic Engineers and the International Association of Chiefs of Police (Gerlough, 1955).

There are also problems in which both the Bernoulli and Poisson models are useful, as illustrated in the following.

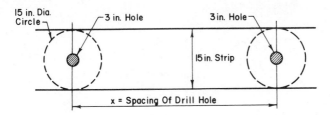

Figure E3.17

EXAMPLE 3.17

Suppose that the soil deposit in a given region contains 0.25 % boulders by volume. What is the probability that a 50-ft-deep 3-in.-diameter boring will encounter boulders? Assume hypothetically that boulders are 12-in.-diameter spheres.

Solution

Assume that boulders are randomly located in the soil mass, and the presence of boulders may be modeled by the Poisson process; the probability of n boulders in a soil volume V is, therefore,

$$P(N = n) = \frac{1}{n!}\left(\frac{0.0025\,V}{v}\right)^n e^{-0.0025V/v}$$

where v = volume of one boulder, given by

$$v = \frac{\pi}{6}(1)^3 = \frac{\pi}{6}\ \text{cu ft}$$

Thus

$$P(N = n) = \frac{1}{n!}\left(\frac{0.0025\,V}{\pi/6}\right)^n e^{-0.0025V/(\pi/6)}$$

$$= \frac{1}{n!}(0.00477\,V)^n e^{-0.00477V}$$

Next, suppose that a boulder will be encountered whenever the drill hole touches the circumference of the boulder. Then, if the borings are spaced x ft apart (see Fig. E3.17), we determine the probability of an encounter (or hit) per foot of boring depth as follows.

For a 3-in. drill hole, any boulder with its center inside the 15-in. circle as shown in Fig. E3.17 will be hit by the boring. Therefore, if there is one boulder in the 15-in. strip, the probability of hitting it per foot of boring depth is

$$P\ (\text{hit per ft boring}\mid 1\ \text{boulder}) = \frac{\text{area of 15-in. circle}}{\text{area of 15-in. strip}}$$

$$= \frac{(\pi/4)(15/12)^2}{(15/12)x} = \frac{0.982}{x}$$

However, there may be any number of boulders in the 15-in. strip. According to the Poisson model, the probability of n boulders in the strip of length x (with volume $V = (15/12)x$ per ft depth) is

$$P \text{ (}n \text{ boulders in strip)} = \frac{1}{n!}\left[0.00477\left(\frac{15}{12}x\right) \right]^{n} \exp\left[-0.00477\left(\frac{15}{12}x\right) \right]$$

$$= \frac{1}{n!}(0.00596x)^{n}e^{-0.00596x}$$

Also, if the probability of hitting any boulder within the strip is the same and statistically independent of other hits, we have

$$P \text{ (no hit per ft boring} \mid n \text{ boulders)} = \left(1 - \frac{0.982}{x}\right)^{n}$$

Then, applying the total probability theorem,

$$q \equiv P \text{ (no hit per ft boring)} = \sum_{n=0}^{\infty} \left(1 - \frac{0.982}{x}\right)^{n} \cdot \frac{1}{n!}(0.00596x)^{n}e^{-0.00596x}$$

$$= e^{-0.00596x} \cdot e^{0.00596x(1-0.982/x)}$$

$$= e^{-0.00596(0.982)}$$

and

$$p \equiv P \text{ (hit per ft boring)} = 1 - e^{-0.00596(0.982)} = 1 - e^{-0.00585}$$

$$\simeq 0.00585$$

Then, assuming the 50 ft of boring to be a Bernoulli sequence with $p = 0.00585$ per ft of boring, we obtain the required probability

$$P \text{ (hit in 50 ft boring)} = 1 - (1 - p)^{50} = 1 - (0.99415)^{50}$$

$$= 0.254$$

(See Problem 3.45 for an alternative approach to this problem.)

Before leaving the Bernoulli and Poisson models, we should point out for emphasis that in both processes the occurrences of an event between trials (in the case of the Bernoulli) and between intervals (in the case of the Poisson model) are *statistically independent*. More generally, the occurrence of a given event in one trial (or interval) may affect the occurrence or non-occurrence of the same event in subsequent trials (or intervals). In other words, the probability of occurrence of an event in a given trial may, in general, depend on what happened at earlier trials, and thus is a conditional probability. If this conditional probability depends only on the immediately preceding trial (or interval), the resulting model is called a Markov chain (or Markov process). The elements and applications of Markov chains are developed in Vol. II.

3.2.7. The exponential distribution   *geometric*

The *exponential distribution* (also known as the *negative exponential*) is related to the Poisson process as follows. If events occur according to a Poisson process, then the time T_1 till the first occurrence of the event has an exponential distribution. We observe that $(T_1 > t)$ means that no event occurs in time t; hence, according to Eq. 3.39,

$$P(T_1 > t) = P(X_t = 0) = e^{-\nu t}$$

T_1 is the *first occurrence time* in a Poisson process. However, since the occurrences of an event in nonoverlapping time intervals in a Poisson process are statistically independent, T_1 is also the *recurrence time* or the time between two consecutive occurrences of the event.

The distribution function of T_1, therefore, is

$$F_{T_1}(t) = P(T_1 \leq t) = 1 - e^{-\nu t} \tag{3.40a}$$

and its density function is

$$f_{T_1}(t) = \frac{dF}{dt} = \nu e^{-\nu t} \qquad t \geq 0 \tag{3.40b}$$

If ν is constant (independent of t), the mean value of T_1 is (see Example 3.5)

$$\mu_{T_1} = \frac{1}{\nu} \qquad \therefore \ \nu = \frac{1}{\mu_T} \approx \frac{1}{\bar{x}} \tag{3.41}$$

which means that the *mean recurrence time* or return period for a simple Poisson process is $1/\nu$. This should be compared with the return period of $1/p$ for the Bernoulli model. However, for events with small occurrence rate ν, $1/\nu \simeq 1/p$. To show this, we observe that in a Poisson process with mean occurrence rate ν, the probability of an event occurring in a unit time interval is $p = \nu e^{-\nu} = \nu(1 - \nu + \frac{1}{2}\nu^2 + \cdots)$; thus, for small ν, $p \simeq \nu$.

EXAMPLE 3.18

Historical records of earthquakes in San Francisco, California, show that during the period 1836–1961 (see Benjamin, 1968), there were 16 earthquakes of intensity VI or more. If the occurrence of such high-intensity earthquakes in this region is assumed to follow a Poisson process, what is the probability that such earthquakes will occur within the next 2 years?

$$\nu = \frac{16}{125} = 0.128 \text{ quake per year}$$

Then

$$P(T_1 \leq 2) = 1 - e^{-0.128(2)} = 0.226$$

The probability that no earthquake of this high intensity will occur in the next 10 years is

$$P(T_1 > 10) = e^{-10(0.128)} = 0.278$$

The return period of an intensity-VI earthquake in San Francisco, according to

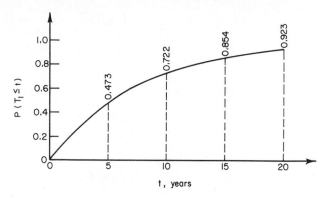

Figure E3.18 Probabilities of high-intensity earthquakes in San Francisco, California

Eq. 3.41, therefore, is

$$E(T_1) = \frac{1}{\nu} = \frac{1}{0.128} = 7.8 \text{ years}$$

meaning that an earthquake of at least intensity VI can be expected, *on the average*, once in every 7.8 years in San Francisco (assuming that the Poisson process is a reasonable model for the occurrence of high-intensity earthquakes in the area).

More generally, the probabilities of the occurrences of such earthquakes within a given time t is given by Eq. 3.40a; in the present case, this is

$$P(T_1 \le t) = 1 - e^{-0.128t}$$

which is portrayed graphically in Fig. E3.18.

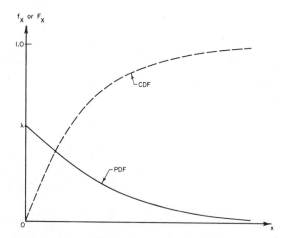

Figure 3.9 PDF and CDF of the exponential distribution

In particular, the probability of high-intensity earthquakes occurring within the return period of 7.8 years is

$$P(T_1 \leq 7.8) = 1 - e^{-0.128 \times 7.8}$$
$$= 1 - e^{-1.0} = 0.632$$

In fact, for a Poisson process, the probability of an event occurring (once or more) within its return period is always $1 - e^{-\nu \bar{T}} = 1 - e^{-1} = 0.632$. Compare this with the corresponding probability for large return period of the Bernoulli model (Section 3.2.3).

The exponential distribution is useful also as a general-purpose probability function. In general, its density function can be given as

$$f_X(x) = \lambda e^{-\lambda x} \qquad x \geq 0 \qquad\qquad (3.42a)$$
$$= 0 \qquad\qquad x < 0$$

where λ is a constant parameter. The corresponding distribution function is

$$F_X(x) = 1 - e^{-\lambda x} \qquad x \geq 0$$
$$= 0 \qquad\qquad x < 0 \qquad\qquad (3.42b)$$

Graphically, the PDF and CDF of the exponential function would appear as shown in Fig. 3.9.

Shifted exponential distribution. In Eq. 3.42 the density function starts at the origin $x = 0$. The PDF of the exponential distribution, however, can start at any value; the resulting distribution may be called the *shifted exponential distribution*, with the corresponding PDF and CDF

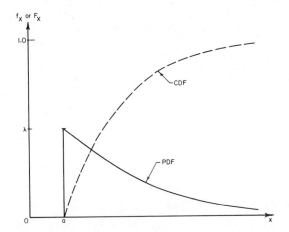

Figure 3.10 PDF and CDF of the shifted exponential distribution

given as follows:

$$f_X(x) = \lambda e^{-\lambda(x-a)} \quad x \geq a \tag{3.43a}$$
$$= 0 \quad x < a$$

and

$$F_X(x) = 1 - e^{-\lambda(x-a)} \quad x \geq a \tag{3.43b}$$
$$= 0 \quad x < a$$

Graphically, these functions would appear as shown in Fig. 3.10.

The exponential distribution may be derived also from other considerations; that is, other than as a consequence of the Poisson process, as described earlier. In particular, this distribution arises, in the *theory of reliability* (see Vol. II), as the model for the distribution of life or time-to-failure of systems under "chance" failure condition. In this connection, the parameter λ is related to the *mean life* or *mean time-to-failure* $E(T)$ as

$$E(T) = \frac{1}{\lambda}$$

See Example 3.5.

EXAMPLE 3.19

Suppose that four identical diesel engines are used as prime movers to generate backup electrical power for the emergency control system of a nuclear power plant. Assume that at least two units are required to supply the needed emergency power; in other words, at least two engines must start automatically during an emergency, otherwise this backup system will not be able to deliver the needed power. The operational life T of each engine has an exponential distribution, with a rated mean operational life of 15 years.

Determine the reliability of the emergency backup system for a period of two years; that is, what is the probability that at least two of the four engines will start automatically during any emergency within the first two years of the life of the system?

First, we observe that for each engine, the probability that there will be no failure to start in two years is

$$P(T > 2) = e^{-2/15} = 0.875$$

Then, denoting N as the number of reliable engines, the reliability of the backup system in two years is

$$P(N \geq 2) = \sum_{n=2}^{4} \binom{4}{n} (0.875)^n (0.125)^{4-n}$$
$$= 0.993$$

3.2.8. The gamma distribution

If the occurrences of an event constitute a Poisson process, then the time until the kth occurrence of the event is described by the *gamma probability distribution.* Let T_k denote the time till the kth event; then $(T_k \leq t)$ means that k or more events occur in time t. Hence, on the basis of Eq. 3.39, the distribution function of T_k is

$$F_{T_k}(t) = \sum_{x=k}^{\infty} P(X_t = x)$$

$$= 1 - \sum_{x=0}^{k-1} \frac{(\nu t)^x}{x!} e^{-\nu t} \tag{3.44a}$$

Equation 3.44a may be obtained also by observing that $(T_k > t)$ means there are at most $(k-1)$ events occurring within t.

The corresponding density function, therefore, is

$$f_{T_k}(t) = \frac{\nu(\nu t)^{k-1}}{(k-1)!} e^{-\nu t} \quad t \geq 0 \qquad \begin{matrix} k \approx \alpha \\ \nu \approx \beta \end{matrix} \tag{3.44b}$$

The gamma distribution (with integer k) is known also as the *Erlang distribution*. The mean time till the occurrence of the kth event is $E(T_k) = k/\nu$, with variance $\text{Var}(T_k) = k/\nu^2$.

EXAMPLE 3.20

Suppose that fatal accidents on a particular highway occur on the average about once every 6 months. If the occurrence of accidents on this road constitutes a Poisson process, the time till the first accident is given by the exponential law with $\nu = \frac{1}{6}$ accident per month; that is,

$$f_{T_1}(t) = \frac{1}{6}e^{-t/6}$$

The time till the second accident is described by the gamma distribution, or

$$f_{T_2}(t) = \frac{1}{6}\left(\frac{t}{6}\right)e^{-t/6}$$

whereas the time till the third accident would be

$$f_{T_3}(t) = \frac{1}{2}\cdot\frac{1}{6}\left(\frac{t}{6}\right)^2\cdot e^{-t/6}$$

The foregoing density functions are shown graphically in Fig. E3.20. The corresponding mean times are 6, 12, and 18 months, respectively.

It may be recognized that the exponential and gamma distributions are the continuous analogues, respectively, of the geometric and negative-binomial distributions, in the sense that the exponential and gamma distributions are related to the Poisson process in the same way that the

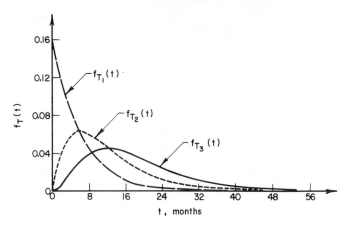

Figure E3.20   PDF for time till the 1st, 2nd and 3rd accidents on a highway

geometric and negative binomial distributions are related to the Bernoulli sequence.

The gamma distribution is useful also as a general-purpose probability distribution. For such purposes, however, it is usually given in a more general form.

We recall that the generalization of the factorial to noninteger numbers is the *gamma function*,

$$\Gamma(k) = \int_0^\infty x^{k-1} e^{-x}\, dx \qquad k > 0 \tag{3.45}$$

which we can show that integration-by-parts yields, for $k > 0$,

$$\Gamma(k) = (k-1)\,\Gamma(k-1)$$

Therefore Eq. 3.44*b* can be generalized for a random variable X by replacing $(k-1)!$ with the gamma function. Thus, in general, the gamma density function is

$$f_X(x) = \frac{\nu(\nu x)^{k-1}}{\Gamma(k)}\, e^{-\nu x} \qquad x \geq 0 \tag{3.46}$$

where ν and k are parameters. Its mean and variance remain k/ν and k/ν^2, respectively.

Calculation of probability involving the gamma distribution can be performed using tables of incomplete gamma function. Incomplete gamma functions are usually tabulated for the ratio (for example, Harter, 1963)

$$I(u, k) = \frac{\displaystyle\int_0^u y^{k-1}e^{-y}\,dy}{\Gamma(k)}$$

Then, if X has a gamma distribution we obtain for $a \geqslant 0$ and $b \geqslant 0$;

$$P\,(a < X \leq b) = \frac{\nu^k}{\Gamma(k)} \int_a^b x^{k-1}e^{-\nu x}\,dx$$

Letting $y = \nu x$, this integral becomes

$$P\,(a < X \leq b) = \frac{1}{\Gamma(k)}\left[\int_0^{\nu b} y^{k-1}e^{-y}\,dy - \int_0^{\nu a} y^{k-1}e^{-y}\,dy\right]$$

$$= I(\nu b, k) - I(\nu a, k)$$

In effect, therefore, the *incomplete gamma function ratio* is the CDF of the gamma distribution (with $\nu = 1$).

3.2.9. The hypergeometric distribution

The *hypergeometric distribution* arises when samples from a finite population (consisting of two types of elements, for example, "good" and "bad") are being examined. It is the basic distribution underlying many sampling plans used in connection with acceptance sampling and quality control (see Chapter 9).

Consider a lot of N items, m of which are defective and the remaining $(N - m)$ are good. If a sample of n items is taken (at random) from this lot, the probability of x defective items in the sample is given by the hypergeometric distribution

$$P(X = x) = \frac{\dbinom{m}{x}\dbinom{N - m}{n - x}}{\dbinom{N}{n}} \qquad x = 1, 2, \ldots, m \qquad (3.47)$$

SAMPLING W/OUT REPLACEMENT

In the lot, the number of samples of size n is $\binom{N}{n}$; among these, there are $\binom{m}{x}\binom{N-m}{n-x}$ samples with x defectives. Hence, assuming that the samples are equally likely to be chosen, we obtain Eq. 3.47.

EXAMPLE 3.21

A box contains 25 strain gages, and 4 of them are known to be defective gages. If 6 gages were used in an experiment, what is the probability that there was one defective gage in the experiment?
In this case, $N = 25$, $m = 4$, and $n = 6$. Hence the required probability is

$$P(X = 1) = \frac{\binom{4}{1}\binom{21}{5}}{\binom{25}{6}} = 0.46$$

The probability that none of the defective gages were used in the experiment is

$$P(X = 0) = \frac{\binom{21}{6}}{\binom{25}{6}} = 0.31$$

EXAMPLE 3.22

Suppose that 100 concrete cylinders are to be taken daily at a large construction project. To ensure quality, the acceptance criterion requires that 10 of these cylinders (chosen at random) must be tested and at least 9 of these must have a specified minimum crushing strength. What can we say about the acceptance criterion—is it too stringent?
Whether the acceptance criterion is too stringent, or not stringent enough, depends on whether it is difficult or easy for poor-quality material to go undetected. For example, if there is d percent of defective concrete, then on the basis of the specified acceptance criterion, the probability of the daily concrete mixes' being rejected is (denoting X as the number of defective cylinders)

$$P(X > 1) = 1 - P(X \le 1)$$

$$= 1 - \left[\frac{\binom{100(1-d)}{10}}{\binom{100}{10}} + \frac{\binom{100\,d}{1}\binom{100(1-d)}{9}}{\binom{100}{10}} \right]$$

For example, if $d = 5\%$,

$$P(\text{rejection}) = 1 - \left[\frac{\binom{95}{10}}{\binom{100}{10}} + \frac{\binom{5}{1}\binom{95}{9}}{\binom{100}{10}} \right]$$

$$= 1 - (0.5837 + 0.0034)$$

$$= 0.4129$$

whereas, if $d = 2\%$,

$$P(\text{rejection}) = 1 - \left[\frac{\binom{98}{10}}{\binom{100}{10}} + \frac{\binom{2}{1}\binom{98}{9}}{\binom{100}{10}}\right]$$

$$= 1 - (0.8091 + 0.1818)$$

$$= 0.0091$$

Therefore, according to these calculations, if 5% of the concrete is defective, it is likely (with 41% probability) to be discovered with the proposed acceptance criterion, whereas if there is only 2% defectives, the likelihood that the material will be rejected is almost nil (0.9% probability).

Hence, if the contract requires concrete with less than 2% defectives then the acceptance criterion is not stringent enough; on the other hand, if material with as much as 5% defectives is acceptable, then the proposed criterion may be satisfactory.

3.2.10. The beta distribution

A probability distribution appropriate for a random variable whose values are bounded, say between finite limits a and b, is the *beta distribution*. The density function of such a distribution is

$$f_X(x) = \frac{1}{B(q, r)} \frac{(x - a)^{q-1}(b - x)^{r-1}}{(b - a)^{q+r-1}} \qquad a \le x \le b \qquad (3.48)$$

$$= 0 \qquad\qquad\qquad \text{elsewhere}$$

in which q and r are parameters of the distribution, and $B(q, r)$ is the *beta function*

$$B(q, r) = \int_0^1 x^{q-1}(1 - x)^{r-1} \, dx \qquad (3.49)$$

which is related to the gamma function as follows:

$$B(q, r) = \frac{\Gamma(q)\,\Gamma(r)}{\Gamma(q + r)} \qquad (3.49a)$$

Depending on the parameters q and r, the density function of the beta distribution will have different shape. Figure 3.11 shows the density function between 2 and 12 with $q = 2.0$ and $r = 6.0$.

If the values of the variate are limited between 0 and 1.0 (that is, $a = 0$ and $b = 1.0$), Eq. 3.48 becomes

$$f_X(x) = \frac{1}{B(q, r)} x^{q-1}(1 - x)^{r-1} \qquad 0 \le x \le 1.0 \qquad (3.48a)$$

$$= 0 \qquad\qquad\qquad \text{elsewhere}$$

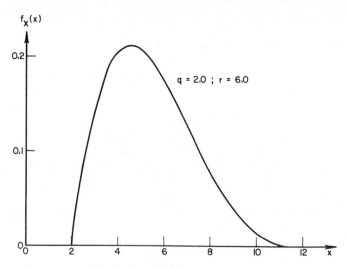

Figure 3.11   A beta distribution

which can be called the *standard beta distribution*. Figure 3.12 shows the standard beta density function with different values of q and r.

The probability associated with a beta distribution can be evaluated in terms of the *incomplete beta function*, which is defined as

$$B_x(q, r) = \int_0^x y^{q-1}(1 - y)^{r-1}\, dy \qquad 0 < x < 1.0 \qquad (3.50)$$

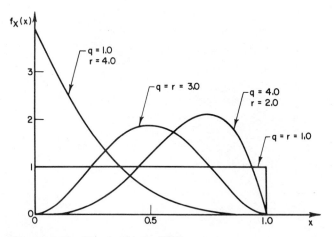

Figure 3.12   Standard beta PDF

For example, to evaluate the probability between $x = x_1$ and $x = x_2$, we have

$$P(x_1 < X \leq x_2) = \int_{x_1}^{x_2} \frac{1}{B(q, r)} \frac{(x - a)^{q-1}(b - x)^{r-1}}{(b - a)^{q+r-1}} dx$$

Let

$$y = \frac{x - a}{b - a}$$

so that

$$dy = \frac{dx}{b - a} \quad \text{and} \quad (1 - y) = \frac{b - x}{b - a}$$

With this change of variable, the preceding integral can be shown to be

$$P(x_1 < X \leq x_2)$$

$$= \frac{1}{B(q, r)} \left[\int_0^{(x_2-a)/(b-a)} y^{q-1}(1 - y)^{r-1} dy - \int_0^{(x_1-a)/(b-a)} y^{q-1}(1 - y)^{r-1} dy \right]$$

We recognize that each of the last two integrals is an incomplete beta function, $B_u(q, r)$ and $B_v(q, r)$, respectively, where $u = (x_2 - a)/(b - a)$ and $v = (x_1 - a)/(b - a)$. Thus the required probability is

$$P(x_1 < X \leq x_2) = \frac{1}{B(q, r)} [B_u(q, r) - B_v(q, r)] \qquad (3.51)$$

Values of the *incomplete beta function ratio* $[B_u(q, r)]/[B(q, r)]$ have been tabulated; for example, by Pearson (1934), and Pearson and Johnson (1968). Therefore, probabilities involving the beta distribution can be evaluated conveniently using tables of the incomplete beta function ratio. In fact, by virtue of Eq. 3.50, we also observe that the CDF of the standard beta distribution, Eq. 3.48a, with parameters q and r, is given by

$$\beta(x \mid q, r) = \frac{B_x(q, r)}{B(q, r)} \qquad (3.51a)$$

Effectively, therefore, tables of the *incomplete beta function ratio* are also the tables for the CDF of the standard beta distribution.

The mean and variance of the beta distribution, Eq. 3.48, are

$$\mu_X = a + \frac{q}{q + r} (b - a) \qquad (3.52)$$

$$\sigma_X^2 = \frac{qr}{(q + r)^2(q + r + 1)} (b - a)^2 \qquad (3.53)$$

whereas its mode is

$$\tilde{x} = a + \frac{1 - q}{2 - q - r} (b - a) \qquad (3.54)$$

and its coefficient of skewness

$$\theta = \frac{2(r - q)}{(q + r)(q + r + 2)\sigma_X} \qquad (3.55)$$

It can be observed that the skewness of the beta distribution is positive when $q < r$, and negative when $q > r$, whereas when $q = r$ the distribution is symmetrical ($\theta = 0$) about the mean value, as illustrated in Fig. 3.12. Therefore, with suitable choice of the parameters q and r, the beta distribution may be used to fit a wide variety of shapes of frequency diagrams.

EXAMPLE 3.23

The duration of an activity in a construction project has been estimated by the contractor to be as follows:

> minimum duration = 5 days
> maximum duration = 10 days
> expected duration = 7 days

The contractor also estimated the coefficient of variation of the duration to be 10%. Determine the beta distribution for the duration T of the activity.

It is obvious that a and b will be 5 and 10 days, respectively. By equating the expression for the mean value, we have

$$5 + \frac{q}{q + r} (10 - 5) = 7$$

giving $q = 2r/3$. Substituting this into the expression for the variance,

$$\frac{qr}{(q + r)^2(q + r + 1)} (10 - 5)^2 = (0.1 \times 7)^2$$

Thus obtaining $q = 3.26$ and $r = 4.89$. The appropriate beta distribution, therefore, has parameters $q = 3.26$ and $r = 4.89$.

The probability that this activity will be completed within 9 days is given by

$$P(T \leq 9) = \frac{B_u(3.26, 4.89)}{B(3.26, 4.89)}$$

with $u = (9 - 5)/(10 - 5) = 0.8$. From tables of the incomplete beta function ratio (Pearson and Johnson, 1968) we obtain (after suitable interpolation)

$$P(T \leq 9) = \frac{B_{0.8}(3.26, 4.89)}{B(3.26, 4.89)}$$

$$= 1 - \frac{B_{0.2}(4.89, 3.26)^*}{B(4.89, 3.26)}$$

$$= 1 - 0.008$$

$$= 0.992$$

3.2.11.   Other distributions

The probability distributions described thus far are among the most useful and important. However, these are not inclusive; for specific applications other distributions may prove to be more appropriate and useful, including the triangular and uniform distributions. Among other widely known distributions are the t-distribution, the chi-square (χ^2) distribution, the F-distribution, and the Pearson system (Elderton, 1953). The first three are important in statistical analysis; for example, the t-distribution is useful for determining the confidence interval of the population mean with unknown variance, whereas the chi-square distribution is useful in the interval estimation of the population variance (see Chapter 5).

Another group of probability distributions of special importance to engineering design is the distribution of extreme values. Extreme-value distributions are presented in Vol. II, with special reference to problems associated with extreme natural hazards.

3.3.   MULTIPLE RANDOM VARIABLES

The concept of a random variable and its probability distribution can be extended to two or more random variables. In order to identify numerically events that are the results of two or more physical processes, the events in a sample space may be mapped into two (or more) dimensions of the real space; implicitly this requires two or more random variables.

Consider, for example, the rainfall intensity at a gage station and the resulting runoff of a river; we may use a random variable X whose values x denote the values of the measured rainfall intensity (in inches), and another random variable Y whose values y are the possible runoffs in the river. Ac-

---

* Tables of the incomplete beta function ratio are usually given for $q \geq r$. For $q < r$, the ratio is

$$\frac{B_u(q, r)}{B(q, r)} = 1 - \frac{B_{1-u}(r, q)}{B(r, q)}$$

cordingly, $(X = x, Y = y)$ and $(X \leqslant x, Y \leqslant y)$ are joint events* defined by values of the random variables in the xy-space. Obviously, this notion can be extended to multiple random variables.

3.3.1. Joint and conditional probability distributions

Since values of X and Y represent events, there are probabilities associated with any pair of values x and y; the probabilities for all possible pairs of x and y may be described with the *joint distribution function* of the random variables X and Y, defined as

$$F_{X,Y}(x, y) = P(X \leqslant x, Y \leqslant y) \tag{3.56}$$

which is the cumulative probability of the joint occurrences of the events identified by $X \leqslant x$ and $Y \leqslant y$. In order to comply with the axioms of probability, the joint distribution function must satisfy the following:

(a) $F_{X,Y}(-\infty, -\infty) = 0;$ $F_{X,Y}(\infty, \infty) = 1.0$

(b) $F_{X,Y}(-\infty, y) = 0;$ $F_{X,Y}(\infty, y) = F_Y(y)$

   $F_{X,Y}(x, -\infty) = 0;$ $F_{X,Y}(x, \infty) = F_X(x)$

(c) $F_{X,Y}(x, y)$ is nonnegative, and a nondecreasing function of x and y

If the random variables X and Y are discrete, the probability distribution may also be described with the *joint probability mass function* (PMF), which is simply

$$p_{X,Y}(x, y) = P(X = x, Y = y) \tag{3.57}$$

Then the distribution function becomes

$$F_{X,Y}(x, y) = \sum_{\{x_i \leq x, \, y_j \leq y\}} p_{X,Y}(x_i, y_j) \tag{3.58}$$

which is simply the sum of probabilities associated with all point pairs (x_i, y_j) in the subset $\{x_i \leqslant x, y_j \leqslant y\}$.

The probability of $(X = x)$ may depend on the values of Y, or vice versa; accordingly, by virtue of Eq. 2.11, we have the *conditional probability mass function*

$$p_{X|Y}(x \mid y) \equiv P(X = x \mid Y = y) = \frac{p_{X,Y}(x, y)}{p_Y(y)} \tag{3.59}$$

* We will use the notation:

$$(X = x, Y = y) = [(X = x) \cap (Y = y)]$$

$$(X \leqslant x, Y \leqslant y) = [(X \leqslant x) \cap (Y \leqslant y)]$$

if $p_Y(y) \neq 0$. Similarly,

$$p_{Y|X}(y \mid x) = \frac{p_{X,Y}(x, y)}{p_X(x)} \tag{3.59a}$$

if $p_X(x) \neq 0$.

The PMF of the individual random variables may be obtained from the joint PMF; applying the theorem of total probability, Eq. 2.19, we obtain the *marginal* PMF of X as

$$p_X(x) = \sum_{\text{all } y_j} P(X = x \mid Y = y_j) \, P(Y = y_j)$$

$$= \sum_{\text{all } y_j} P(X = x, Y = y_j)$$

$$= \sum_{\text{all } y_j} p_{X,Y}(x, y_j) \tag{3.60}$$

By the same token,

$$p_Y(y) = \sum_{\text{all } x_i} p_{X,Y}(x_i, y) \tag{3.60a}$$

If the random variables X and Y are statistically independent (meaning that the events $X = x$ and $Y = y$ are statistically independent),

$$p_{X|Y}(x \mid y) = p_X(x) \qquad \text{and} \qquad p_{Y|X}(y \mid x) = p_Y(y)$$

Hence, Eq. 3.57 becomes

$$p_{X,Y}(x, y) = p_X(x) \, p_Y(y) \tag{3.61}$$

EXAMPLE 3.24

Suppose that, from a survey of construction labor, the work duration (in number of hours) per day and the average productivity (in terms of percent efficiency)

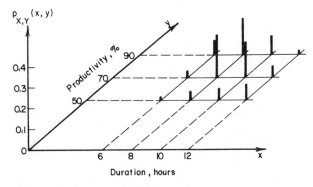

Figure E3.24a Joint PMF $p_{X,Y}(x, y)$

were recorded as shown below. For simplicity, the work duration is recorded as 6, 8, 10, and 12 hr, whereas the average productivity is categorized into 50%, 70%, and 90%. Data show the following results.

Duration and productivity (x, y)	No. of observations	Relative frequencies
6, 50	2	0.014
6, 70	5	0.036
6, 90	10	0.072
8, 50	5	0.036
8, 70	30	0.216
8, 90	25	0.180
10, 50	8	0.058
10, 70	25	0.180
10, 90	11	0.079
12, 50	10	0.072
12, 70	6	0.043
12, 90	2	0.014
	Total = 139	

These may be portrayed graphically as shown in. Fig. E3.24a.
The marginal PMF for X, the distribution of work duration, is

$$p_X(x) = \sum_{(y_j = 50, 70, 90)} p_{X,Y}(x, y_j)$$

and would appear as shown in Fig. E3.24b. For example, the ordinate at $X = 8$ is obtained as

$$p_X(8) = 0.036 + 0.216 + 0.180 = 0.432$$

Similarly, the marginal PMF for Y, representing the distribution of productivity, is shown in Fig. E3.24c.

If the work duration per day is 8 hr, the probability that the average productivity will be 90% is given by the conditional probability of Eq. 3.59a as

$$p_{Y|X}(90\% \mid 8) = \frac{p_{X,Y}(8, 90\%)}{p_X(8)}$$
$$= \frac{0.180}{0.432}$$
$$= 0.417$$

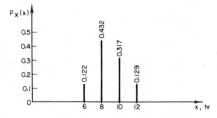

Figure E3.24b   Marginal PMF $p_X(x)$

In Fig. E3.24*d* is shown the conditional PMF for the average productivity of an 8-hr day, $p_{Y|X}(y \mid 8)$.

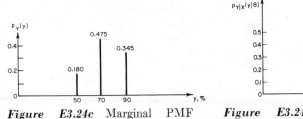

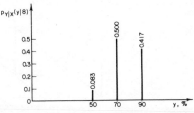

Figure E3.24c Marginal PMF $P_Y(y)$

Figure E3.24d Conditional PMF $p_{Y|X}(y|8)$

If the random variables X and Y are continuous, the probability distribution may also be described with the *joint probability density function* (PDF), which may be defined as

$$f_{X,Y}(x, y) \, dx \, dy = P(x < X \leqslant x + dx, y < Y \leqslant y + dy) \qquad (3.62)$$

Then

$$F_{X,Y}(x, y) = \int_{-\infty}^{x} \int_{-\infty}^{y} f_{X,Y}(u, v) \, dv \, du \qquad (3.63)$$

Conversely, if the partial derivatives exist,

$$f_{X,Y}(x, y) = \frac{\partial^2 F_{X,Y}(x, y)}{\partial x \, \partial y} \qquad (3.64)$$

Also,

$$P(a < X \leqslant b, c < Y \leqslant d) = \int_{a}^{b} \int_{c}^{d} f_{X,Y}(u, v) \, dv \, du \qquad (3.65)$$

which is the volume under the surface $f(x, y)$ as shown in Fig. 3.13.

Analogous to Eq. 3.59, the *conditional density function* of X given Y, is

$$f_{X|Y}(x \mid y) = \frac{f_{X,Y}(x, y)}{f_Y(y)} \qquad (3.66)$$

Therefore, in general,

$$f_{X,Y}(x, y) = f_{X|Y}(x \mid y) \, f_Y(y)$$

or $$\qquad (3.67)$$

$$f_{X,Y}(x, y) = f_{Y|X}(y \mid x) \, f_X(x)$$

However, if X and Y are statistically independent, that is, $f_{X|Y}(x \mid y) =$

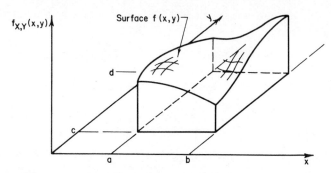

Figure 3.13 Joint PDF of X and Y

$f_X(x)$ and $f_{Y|X}(y \mid x) = f_Y(y)$, then

$$f_{X,Y}(x, y) = f_X(x) f_Y(y) \tag{3.68}$$

Applying the total probability theorem, we obtain the *marginal density functions*,

$$f_X(x) = \int_{-\infty}^{\infty} f_{X|Y}(x \mid y) f_Y(y) \, dy$$

$$= \int_{-\infty}^{\infty} f_{X,Y}(x, y) \, dy \tag{3.69}$$

and, similarly,

$$f_Y(y) = \int_{-\infty}^{\infty} f_{X,Y}(x, y) \, dx \tag{3.70}$$

The characteristics of a joint density function for two random variables X and Y, and the associated marginal densities, are portrayed in Fig. 3.14.

EXAMPLE 3.25

An example of a joint density function of two continuous random variables X and Y is the *bivariate normal* density function given by

$$f_{X,Y}(x, y) = \frac{1}{2\pi \sigma_X \sigma_Y \sqrt{1 - \rho^2}} \exp\left[\frac{-1}{2(1 - \rho^2)} \left\{ \left(\frac{x - \mu_X}{\sigma_X} \right)^2 \right. \right.$$

$$\left. \left. - 2\rho \left(\frac{x - \mu_X}{\sigma_X} \right) \left(\frac{y - \mu_Y}{\sigma_Y} \right) + \left(\frac{y - \mu_Y}{\sigma_Y} \right)^2 \right\} \right]$$

$$-\infty < x < \infty; \quad -\infty < y < \infty;$$

in which ρ is the *correlation coefficient* between X and Y (see Section 3.3.2). Such

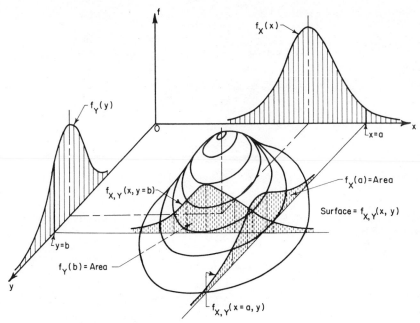

Figure 3.14 Joint and marginal PDF of continuous random variables X and Y

a function can be written also as

$$f_{X,Y}(x, y) = \frac{1}{\sqrt{2\pi}\sigma_X} \exp\left[-\frac{1}{2}\left(\frac{x - \mu_X}{\sigma_X}\right)^2\right] \cdot \frac{1}{\sqrt{2\pi}\sigma_Y\sqrt{1 - \rho^2}}$$
$$\exp\left[-\frac{1}{2}\left\{\frac{y - \mu_Y - \rho(\sigma_Y/\sigma_X)(x - \mu_X)}{\sigma_Y\sqrt{1 - \rho^2}}\right\}^2\right]$$

Then, in view of Eq. 3.67, we see that the conditional density function of Y given $X = x$ is

$$f_{Y|X}(y \mid x) = \frac{1}{\sqrt{2\pi}\sigma_Y\sqrt{1 - \rho^2}} \exp\left[-\frac{1}{2}\left\{\frac{y - \mu_Y - \rho(\sigma_Y/\sigma_X)(x - \mu_X)}{\sigma_Y\sqrt{1 - \rho^2}}\right\}^2\right]$$

whereas the marginal density function of X is

$$f_X(x) = \frac{1}{\sqrt{2\pi}\sigma_X} \exp\left[-\frac{1}{2}\left(\frac{x - \mu_X}{\sigma_X}\right)^2\right]$$

both of which are Gaussian. In particular, the conditional density function is normal with mean value

$$E(Y \mid X = x) = \mu_Y - \rho(\sigma_Y/\sigma_X)(x - \mu_X)$$

and variance

$$\text{Var}(Y \mid X = x) = \sigma_Y^2(1 - \rho^2)$$

Similarly, it can be shown that

$$f_{X|Y}(x\,|\,y) = \frac{1}{\sqrt{2\pi}\sigma_X\sqrt{1-\rho^2}}\exp\left[-\frac{1}{2}\left(\frac{x-\mu_X-\rho(\sigma_X/\sigma_Y)(y-\mu_Y)}{\sigma_X\sqrt{1-\rho^2}}\right)^2\right]$$

and

$$f_Y(y) = \frac{1}{\sqrt{2\pi}\sigma_Y}\exp\left[-\frac{1}{2}\left(\frac{y-\mu_Y}{\sigma_Y}\right)^2\right]$$

3.3.2. Covariance and correlation

The joint second moment of X and Y is

$$E(XY) = \int_{-\infty}^{\infty}\int_{-\infty}^{\infty} xy\, f_{X,Y}(x,y)\, dx\, dy \qquad (3.71)$$

and if X and Y are statistically independent, Eq. 3.71 becomes (by virtue of Eq. 3.68)

$$E(XY) = \int_{-\infty}^{\infty}\int_{-\infty}^{\infty} xy f_X(x) f_Y(y)\, dx\, dy$$

$$= \int_{-\infty}^{\infty} x f_X(x)\, dx \int_{-\infty}^{\infty} y f_Y(y)\, dy = E(X)\,E(Y) \qquad (3.71a)$$

The joint second moment about the means μ_X and μ_Y is the *covariance* of X and Y; that is,

$$\mathrm{Cov}(X,Y) = E[(X-\mu_X)(Y-\mu_Y)]$$

$$= E(XY) - E(X)\,E(Y) \qquad (3.72)$$

In view of Eq. 3.71a, $\mathrm{Cov}(X,Y) = 0$ if X and Y are statistically independent.

The physical significance of the covariance can be inferred from Eq. 3.72. If the $\mathrm{Cov}(X,Y)$ is *large and positive*, the values of X and Y tend to be both large or both small relative to their respective means, whereas if the $\mathrm{Cov}(X,Y)$ is *large and negative*, the values of X tend to be large when the values of Y are small, and vice versa, relative to their respective means; and if the $\mathrm{Cov}(X,Y)$ is small or zero, there is little or no (linear) relationship between the values of X and Y (or if a strong relationship exists, it is nonlinear).

Therefore, the $\mathrm{Cov}(X,Y)$ is a measure of the degree of (linear) interrelationship between the variates X and Y. For this purpose, however, it is preferable to use the normalized covariance or *correlation coefficient*, which is defined as

$$\rho = \frac{\mathrm{Cov}(X,Y)}{\sigma_X\sigma_Y} \qquad (3.73)$$

The values of ρ range between -1 and $+1$; that is,

$$-1 \le \rho \le +1 \tag{3.74}$$

which we can verify as follows.

According to Schwarz's inequality (Hardy, Littlewood, Polya, 1959),

$$\left[\int_{-\infty}^{\infty} \int_{-\infty}^{\infty} (x - \mu_X)(y - \mu_Y) f_{X,Y}(x, y) \, dx \, dy \right]^2$$

$$\le \int_{-\infty}^{\infty} \int_{-\infty}^{\infty} (x - \mu_X)^2 f_{X,Y}(x, y) \, dx \, dy \cdot \int_{-\infty}^{\infty} \int_{-\infty}^{\infty} (y - \mu_Y)^2 f_{X,Y}(x, y) \, dx \, dy$$

But the left-hand side is the $[\mathrm{Cov}(X, Y)]^2$, whereas

$$\int_{-\infty}^{\infty} \int_{-\infty}^{\infty} (x - \mu_X)^2 f_{X,Y}(x, y) \, dx \, dy = \int_{-\infty}^{\infty} (x - \mu_X)^2 f_X(x) \, dx = \sigma_X^2$$

and

$$\int_{-\infty}^{\infty} \int_{-\infty}^{\infty} (y - \mu_Y)^2 f_{X,Y}(x, y) \, dx \, dy = \int_{-\infty}^{\infty} (y - \mu_Y)^2 f_Y(y) \, dy = \sigma_Y^2$$

Hence we have

$$[\mathrm{Cov}(X, Y)]^2 \le \sigma_X^2 \sigma_Y^2 \tag{3.75}$$

or

$$\rho^2 \le 1.0 \tag{3.75a}$$

thus verifying Eq. 3.74.

When $\rho = \pm 1.0$, X and Y are linearly related as shown in Figs. 3.15a and 3.15b, respectively, whereas, when $\rho = 0$, values of X and Y may appear as in Fig. 3.15c. For intermediate values of ρ, values of X and Y would appear as in Fig. 3.15d—the "scatter" decreases as ρ increases. However, we also observe from Figs. 3.15e and 3.15f that when the relation between X and Y is nonlinear, $\rho = 0$ even when there is a perfect functional relationship between the variables.

Therefore the magnitude of the correlation coefficient ρ (between 0 and 1) is a measure of the degree of linear interrelationship between two variables.

It is also important to point out that although ρ is a measure of the degree of (linear) relationship between two variables, this does not necessarily imply a causal effect between the variables. Two variables X and Y may both depend on another variable (or variables), in which case there will be a strong correlation between the values of X and Y, but the values of one variable may have no direct effect on the values of the other. For example, the flood flow of a river and the productivity of a construction

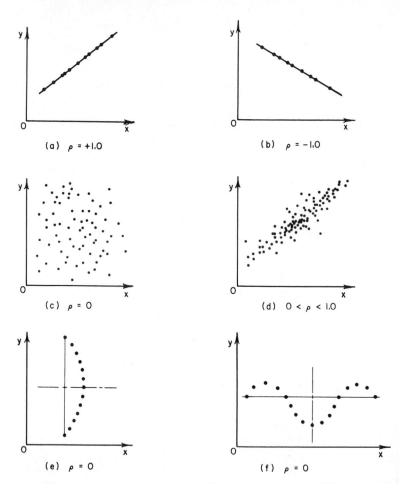

Figure 3.15   Significance of correlation coefficient ρ. $(a)\,\rho = +1.0.$ $(b)\,\rho = -1.0.$ $(c)\,\rho = 0.$ $(d)\,0 < \rho < 1.0.$ $(e)\,\rho = 0.$ $(f)\,\rho = 0.$

crew may be highly correlated because both depend on the weather condition; however, the flood flow may have no direct influence on the productivity of the construction crew, or vice versa. Consider also the following problem from mechanics.

EXAMPLE 3.26

A cantilever beam is subjected to two random loads S_1 and S_2 (Fig. E3.26), which are statistically independent with means and standard deviations μ_1, σ_1 and μ_2, σ_2 respectively.

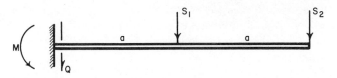

Figure E3.26

The shear force Q and bending moment M at the fixed support are

$$Q = S_1 + S_2$$

and

$$M = aS_1 + 2aS_2$$

which are also random variables with means and variances as follows (see Section 4.3.2)

$$\mu_Q = \mu_1 + \mu_2 \qquad \sigma_Q^2 = \sigma_1^2 + \sigma_2^2$$
$$\mu_M = a\mu_1 + 2a\mu_2 \qquad \sigma_M^2 = a^2(\sigma_1^2 + 4\sigma_2^2)$$

Although S_1 and S_2 are statistically independent, Q and M will be correlated; this correlation can be evaluated as follows:

$$E(QM) = E[(S_1 + S_2)(aS_1 + 2aS_2)]$$
$$= aE(S_1^2) + 3aE(S_1S_2) + 2aE(S_2^2)$$

but $E(S_1S_2) = E(S_1)E(S_2)$ by Eq. 3.71a, and $E(S_1^2) = \sigma_1^2 + \mu_1^2$, $E(S_2^2) = \sigma_2^2 + \mu_2^2$; thus

$$E(QM) = a(\sigma_1^2 + \mu_1^2) + 2a(\sigma_2^2 + \mu_2^2) + 3a\mu_1\mu_2$$
$$= a(\sigma_1^2 + 2\sigma_2^2) + \mu_Q\mu_M$$

Therefore

$$\text{Cov}(Q, M) = E(QM) - \mu_Q\mu_M$$
$$= a(\sigma_1^2 + 2\sigma_2^2)$$

and the corresponding correlation coefficient is

$$\rho_{QM} = \frac{\text{Cov}(Q, M)}{\sigma_Q \cdot \sigma_M} = \frac{\sigma_1^2 + 2\sigma_2^2}{\sqrt{(\sigma_1^2 + \sigma_2^2)(\sigma_1^2 + 4\sigma_2^2)}}$$

Hence, if $\sigma_2 = \sigma_1$,

$$\rho_{QM} = \frac{3}{\sqrt{10}} = 0.948$$

indicating strong correlation between the shear and moment at the support. This correlation arises because Q and M are functions of the same loads S_1 and S_2; however, there is no causal relation between Q and M.

3.3.3. Conditional mean and variance*

If there are two random variables, the mean and variance of one variable may depend on the value of the other variable; in such cases we have

* This section may be skipped over on first reading; the material is not needed for understanding the remaining chapters of the book.

conditional means and *conditional variances*. Indeed, it would be meaningful to speak of conditional moments of any order.

If X and Y are discrete random variables with joint PMF $p_{X,Y}(x, y)$, the conditional mean of X, given $Y = y$, is

$$\mu_{X|y} = E(X \mid Y = y) = \sum_{\text{all } x} x p_{X|Y}(x \mid y) \tag{3.76}$$

and if X and Y are statistically independent, that is, $p_{X|Y}(x \mid y) = p_X(x)$, then

$$E(X \mid Y = y) = E(X) \tag{3.77}$$

From Eqs. 3.59 and 3.60, we can write

$$E(X) = \sum_{\text{all } x} x p_X(x) = \sum_{\text{all } y} \sum_{\text{all } x} x p_{X,Y}(x, y)$$

$$= \sum_{\text{all } y} \sum_{\text{all } x} x p_{X|Y}(x \mid y) p_Y(y)$$

Thus, substituting Eq. 3.76,

$$E(X) = \sum_{\text{all } y} E(X \mid Y = y) \, p_Y(y) \tag{3.78}$$

If X and Y are continuous random variables, the conditional mean of X, given $Y = y$, becomes

$$\mu_{X|y} = \int_{-\infty}^{\infty} x f_{X|Y}(x \mid y) \, dx \tag{3.76a}$$

and the relationship in Eq. 3.78 becomes

$$\mu_X = \int_{-\infty}^{\infty} \mu_{X|y} f_Y(y) \, dy \tag{3.78a}$$

We should emphasize that whereas $E(X \mid Y = y)$ is a constant, $E(X \mid Y)$ is a random variable whose mean is

$$E_Y[E(X \mid Y)] = \sum_{\text{all } y} E(X \mid Y = y) p_Y(y)$$

$$= \sum_{\text{all } y} \sum_{\text{all } x} x p_{X|Y}(x \mid y) p_Y(y)$$

$$= \sum_{\text{all } x} x p_X(x) = E(X) \tag{3.79}$$

The subscript Y on E emphasizes that the expectation is with respect to Y. The conditional variance of X, given Y, is

$$\text{Var}(X \mid Y = y) = E[(X - \mu_{X|Y})^2 \mid Y = y] \tag{3.80}$$

Thus, for discrete X and Y,

$$\text{Var}(X \mid Y = y) = \sum_{\text{all } x} (x - \mu_{X|Y})^2 p_{X|Y}(x \mid y) \qquad (3.80a)$$

and for continuous X and Y,

$$\text{Var}(X \mid Y = y) = \int_{-\infty}^{\infty} (x - \mu_{X|Y})^2 f_{X|Y}(x \mid y) \, dx \qquad (3.80b)$$

The total (unconditional) variance can be expanded as follows:

$$\text{Var}(X) = E[(X - \mu_x)^2] = E_Y\{E[(X - \mu_{X|Y})^2 \mid Y]\}$$

The last equality follows from Eq. 3.79. This last term, however, is

$$E_Y\{E[(X - \mu_{X|Y})^2 \mid Y]\} = E_Y\{E[(X - \mu_x + \mu_x - \mu_{X|Y})^2 \mid Y]\}$$
$$= E_Y\{E[(X - \mu_x)^2 \mid Y]$$
$$+ 2E[(X - \mu_x)(\mu_x - \mu_{X|Y}) \mid Y]$$
$$+ [(\mu_{X|Y} - \mu_x)^2 \mid Y]\}$$

Recognizing that the second term is zero, and $E_Y(\mu_{X|Y}) = \mu_X$ according to Eq. 3.79, we have

$$\text{Var}(X) = E_Y[\text{Var}(X \mid Y)] + \text{Var}_Y[E(X \mid Y)] \qquad (3.81)$$

Equation 3.81 says that the total variance is equal to the *mean value of the conditional variance* plus the *variance of the conditional mean*.

3.4. CONCLUDING REMARKS

The principal concepts introduced in this chapter include the notions of a random variable and its associated probability distribution. Several of the more useful probability distribution functions and their properties are also described and developed. However, the list of distributions is incomplete; a number of other important distributions were omitted including the several extreme-value distributions that will be presented in Vol. II.

The complete description of a random variable would be accomplished by specifying its probability distribution (including the values of its parameters). However, a random variable may also be described approximately with its mean-value and variance (or standard deviation); physically, these *main descriptors* of a random variable represent its central value and measure of dispersion. For two (or more) random variables, the main descriptors must include also the covariance or correlation coefficient between the variables.

Thus far (and this will continue through Chapter 4), we have been dealing with idealized theoretical models. In particular, we have assumed, tacitly at least, that the probability distribution of a random variable, or its main descriptors, are known. In a real problem, of course, these must be estimated and inferred or derived on the basis of real-world data and conditions. The concepts and methods for these purposes are the subjects of Chapters 5 to 8.

PROBLEMS

Section 3.1

3.1 A contractor is submitting bids to 3 jobs, A, B, and C. The probabilities that he will win each of the three jobs are $P(A) = 0.5$, $P(B) = 0.8$, and $P(C) = 0.2$, respectively. Assume events A, B, C are statistically independent. Let X be the total number of jobs the contractor will win.
   (a) What are the possible values of X? Compute and plot the probability mass function (PMF) of the random variable X.
   (b) Plot the distribution function of X.
   (c) Determine $P(X \leq 2)$.   *Ans. 0.92.*
   (d) Determine $P(0 < X \leq 2)$.   *Ans. 0.84.*

3.2 The settlement of a structure has the probability density function shown in Fig. P3.2.
   (a) What is the probability that the settlement is less than 2 cm?
   (b) What is the probability that the settlement is between 2 and 4 cm?
   (c) If the settlement is observed to be more than 2 cm, what is the probability that it will be less than 4 cm?

3.3 The bearing capacity of the soil under a column-footing foundation is known to vary between 6 and 15 kips/sq ft. Its probability density within this range is given as

$$f_X(x) = \frac{1}{2.7}\left(1 - \frac{x}{15}\right) \qquad 6 \leq x \leq 15$$

$$= 0 \qquad \qquad \text{elsewhere}$$

If the column is designed to carry a load of 7.5 kips/sq ft, what is the probability of failure of the foundation?

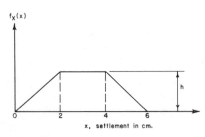

Figure P3.2

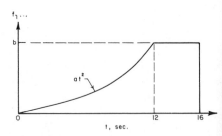

Figure P3.4

3.4 The time duration of a force acting on a structure has been found to be a random variable having the density function shown in Fig. P3.4.
  (a) Determine the appropriate values of a and b for the density function.
  (b) Calculate the mean and median for the variable T.
  (c) Calculate the probability that T will be equal to or greater than 6 sec, that is $P(T \geq 6)$.

3.5 A construction project consisted of building a major bridge across a river and a road linking it to a city (Fig. P3.5a). The contractual time for the entire project is 15 months.

The contractor knows that the construction of the road will require between 12 and 18 months, and the bridge could take between 10 and 20 months. The probability density functions of the respective completion times, however, are uniform for the road, and triangular for the bridge, as shown in Figs. P3.5b and c. Construction of the road and bridge can proceed simultaneously, and the completion of the bridge and the road are statistically independent.

Determine the probability of completing the project within the contractual time.

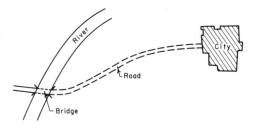

Figure P3.5a

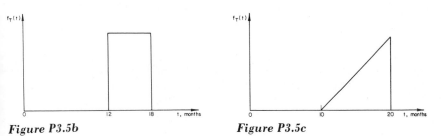

Figure P3.5b   **Figure P3.5c**

3.6 In order to repair the cracks that may exist in a 10-ft weld, a nondestructive testing (NDT) device is used first to detect the location of cracks. Because cracks may exist in various shapes and sizes, the probability that a crack will be detected by the NDT device is only 0.8. Assume that the events of each crack being detected are statistically independent.
  (a) If there are two cracks in the weld, what is the probability that they would not be detected?
  (b) The actual number of cracks N in the weld is not known. However, its

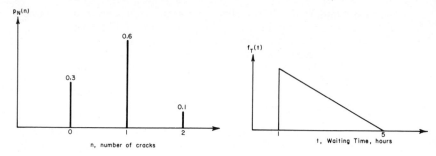

Figure P3.6

Figure P3.8 PDF of waiting time

PMF is given as in Fig. P3.6. What is the probability that the NDT device will fail to detect any crack in this weld?

(c) Determine the mean, variance, and coefficient of variation of N based on the PMF given in Fig. P3.6.

(d) If the device fails to detect any crack in the weld, what is the probability that the weld is flawless (that is, no crack at all)?

3.7 Suppose the duration (in months) of a construction job can be modeled as a continuous random variable T whose cumulative distribution function (CDF) is given by

$$F_T(t) = t^2 - 2t + 1 \qquad 1 \leq t \leq 2$$
$$= 0 \qquad\qquad\quad t < 1$$
$$= 1 \qquad\qquad\quad t > 2$$

(a) Determine the corresponding density function $f_T(t)$.

(b) Compute $P(T > 1.5)$.

3.8 The waiting time at airport A of city B has a density function shown in Fig. P3.8. The waiting time is measured from the time a traveler enters the terminal to the time when he is airborne.

The travel time from hotel C to the airport depends on the transportation mode and may be assumed to be 0.75, 1.00, and 1.25 hours corresponding to travel by rapid transit, taxi, and limousine, respectively. The probability of a traveler's taking each mode of transportation is as follows:

$$P \text{ (rapid transit)} = 0.3$$
$$P \text{ (taxi)} \qquad\quad = 0.4$$
$$P \text{ (limousine)} \quad = 0.3$$

(a) What is the probability that a traveler will be airborne in at most 3 hr after leaving hotel C? *Ans. 0.436.*

(b) Given that the traveler is airborne within 3 hr, what is the probability that he took the limousine? *Ans. 0.234.*

3.9 Two reservoirs are located upstream of a town; the water is held back by two dams A and B. Dam B is 40 m high. (See Fig. P3.9a.) During a strong-motion earthquake, dam A will suffer damage and water will flow downstream into the lower reservoir. Depending on the amount of water in the upper

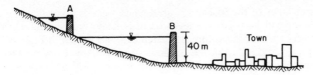

Figure P3.9a

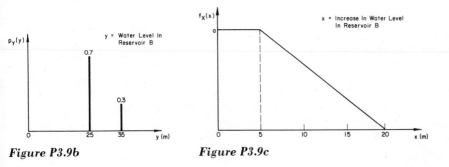

Figure P3.9b   **Figure P3.9c**

reservoir when such an earthquake occurs, the lower reservoir water may or may not overflow dam B. Suppose that the water level at reservoir B, during an earthquake, is either 25 m or 35 m, as shown in Fig. P3.9b; and the increase in the elevation of water level in B caused by the additional water from reservoir A is a continuous random variable with the probability density function given in Fig. P3.9c.

(a) Determine the value of a in Fig. P3.9c.

(b) What is the probability of overflow at B during a strong-motion earthquake?

(c) If there were no overflow at B during an earthquake, what is the probability that the original water level in reservoir B is 25 m?

3.10 A stretch of an intercity freeway has 3 one-way lanes and 2 convertible lanes. The capacity of the highway when the 3 lanes are used is 100 cars per minute. Its capacity when 5 lanes are used is 140 cars per minute.

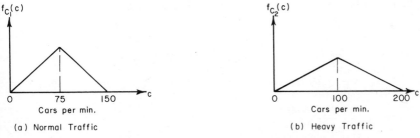

(a) Normal Traffic   (b) Heavy Traffic

Figure P3.10 PDF of traffic volume. (a) Normal traffic. (b) Heavy traffic

Three lanes of the freeway is used when there is normal traffic whereas all five lanes will be used whenever there is heavy traffic volume. The density function of the traffic volumes in each case are shown in Figs. P3.10a and b.

On a given day, if normal traffic is twice as likely as heavy traffic, what is the probability that the capacity of the freeway will be surpassed?

3.11 A traveler going from city A to city C must pass through city B (Fig. P3.11a). The quantities T_1 and T_2 are the times of travel from city A to city B and from city B to city C, in hours, respectively, which are statistically independent random variables. The probability mass functions of T_1 and T_2 are as shown in Figs. P3.11b and c. The time required to go through city B may be considered a deterministic quantity equal to 1 hr.

(a) Calculate the mean, the variance, the standard deviation, and the coefficient of variation of T_1.

(b) Determine the PMF of the total time of travel from city A to city C. Sketch your results graphically.

(c) What is the probability that the travel time from city A to city C will be at least 8 hr?

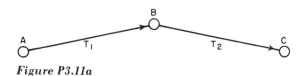

Figure P3.11a

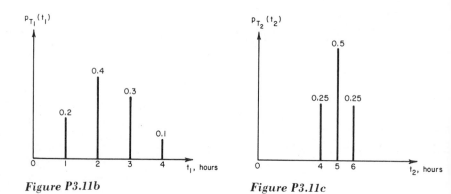

Figure P3.11b   *Figure P3.11c*

3.12 The hourly volume of traffic for a proposed highway is distributed as in Fig. P3.12.

(a) The traffic engineer may design the highway capacity equal to the following.

    (i) The mode of X.

    (ii) The mean of X.

    (iii) The median of X.

    (iv) $x_{.90}$, the 90-percentile value, which is defined as $F_X(x_{.90}) = 0.90$.

Determine the design capacity of the highway and the corresponding
probability of exceedance (that is, capacity is less than traffic volume)
for each of the four cases.
(b) Assume that the actual capacity of the highway after it is built is either
300 or 350 vehicles per hr with relative likelihoods of 1 to 4. What is the
probability that the capacity will be exceeded?

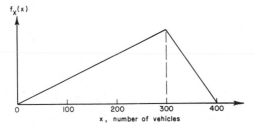

Figure P3.12 PDF of hourly traffic volume

3.13 The lateral resistance of a small building frame is random with the density
function

$$f_R(r) = \frac{3}{500}(r - 10)(20 - r) \qquad 10 \le r \le 20$$
$$= 0 \qquad\qquad\qquad\qquad \text{elsewhere}$$

(a) Plot the density function $f_R(r)$ and the cumulative distribution function
$F_R(r)$.
(b) Determine:
   (i) Mean value of R.
   (ii) Median of R.
   (iii) Mode of R.
   (iv) Standard deviation of R. *Ans.* $\sqrt{5}$.
   (v) Coefficient of variation of R. *Ans. 0.149.*
   (vi) Skewness coefficient. *Ans. 0.*

3.14 The delay time of a construction project is described with a random variable X.
Suppose that X is a discrete variate with probability mass function given in
Table P.3.14a. The penalty for late completion of the project depends on the

Table P3.14a. PMF of X			**Table P3.14b.** Penalty function	
x_i in days	$p_X(x_i)$		x_i (days)	$g(x_i)$ ($\$100,000$)
1	0.5		1	5
2	0.3		2	6
3	0.1		3	7
4	0.1		4	7

number of days of delay; that is, penalty $= g(x_i)$. The penalty function is given in Table P3.14*b* in units of $100,000.
(a) Calculate the mean penalty for this project.   *Ans. $570,000.*
(b) Calculate the standard deviation of the penalty.   *Ans. $78,000.*

Section 3.2

3.15   If the annual precipitation X in a city is a normal variate with a mean of 50 in. and a coefficient of variation of 0.2, determine the following.
(a) The standard deviation of X.
(b) $P(X < 30)$.
(c) $P(X > 60)$.
(d) $P(40 < X \le 55)$.
(e) Probability that X is within 5 in. from the mean annual precipitation.
(f) The value x_0 such that the probability of the annual precipitation exceeding x_0 is only 1/4 that of not exceeding x_0.

3.16   The present air traffic volume at an airport (number of landings and takeoffs) during the peak hour is a normal variate with a mean of 200 and a standard deviation of 60 airplanes (Fig. P3.16).
(a) If the present runway capacity (for landings and takeoffs) is 350 planes per hr, what is currently the daily probability of air traffic congestion? Assume there is one peak hour daily.   *Ans. 0.0062.*
(b) If no additional airports or expansion is built, what would be the probability of congestion 10 years hence? Assume that the mean traffic volume is increasing linearly at 10% of current volume per year, and the coefficient of variation remains the same.   *Ans. 0.662.*
(c) If the projected growth is correct, what airport capacity will be required 10 years from now to maintain the present service condition (that is, the same probability of congestion as now)?   *Ans. 700.*

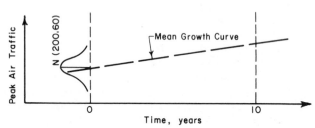

Figure P3.16

3.17   The moment capacity M for the cantilever beam shown in Fig. P3.17 is constant throughout the entire span. Because of uncertainties in material strength, M is assumed to be Gaussian with mean 50 kip-ft and coefficient of variation 20%. Failure occurs if the moment capacity is exceeded anywhere in the beam.
(a) If only a concentrated load 3 kips is applied at the free end, what is the probability that the beam will fail?   *Ans. 0.023.*
(b) If only a uniform load of 0.5 kips/ft is applied on the entire beam, what is the probability that the beam will fail?   *Ans. 0.006.*

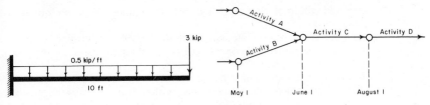

Figure P3.17   **Figure P3.18**

(c) In rare cases, the beam may be subjected to the combination of the concentrated load and the uniform load; what will be the reliability (probability of no failure) of the beam when this case occurs? *Ans. 0.308.*

(d) Suppose that the beam had survived under the concentrated load. What will be the probability that it will survive under the combined loads? *Ans. 0.316.*

(e) Suppose that a reliability level of 99.5% is desired, and the beam is subjected only to the uniform load w across the span. What will be the maximum allowable w? *Ans. 0.484 kip/ft.*

3.18 A portion of an activity network is shown in Fig. P3.18; an arrow indicates the starting and ending of an activity. Activity C can start only after completion of both activities A and B, whereas activity D can start only after completion of C. A, B, C, D are statistically independent activities.

The scheduled starting dates are as follows, and an activity cannot start earlier than its scheduled date. (For simplicity, assume all months have 30 days.)

$$\begin{array}{lll}
\text{Activities } A \text{ \& } B: & \text{May 1} \\
\text{Activity } C & : & \text{June 1} \\
\text{Activity } D & : & \text{August 1}
\end{array}$$

The times required to complete each activity are Gaussian random variables as follows.

$$\begin{array}{ll}
\text{Activity } A: & N(25 \text{ days, } 5 \text{ days}) \\
\text{Activity } B: & N(26 \text{ days, } 4 \text{ days}) \\
\text{Activity } C: & N(48 \text{ days, } 12 \text{ days}) \\
\text{Activity } D: & N(40 \text{ days, } 8 \text{ days})
\end{array}$$

Assume that both activities A and B started on schedule, that is, on May 1.

(a) Determine the probability that activity C will not start on schedule. *Ans. 0.292.*

(b) The availability of labor is such that unless C is started on schedule the necessary work force will be diverted to another project and thus will be unavailable for this activity for at least 90 days. What is the probability that activity D will start on schedule? *Ans. 0.596.*

3.19 A contractor estimates that the expected time for the completion of job A is 30 days. Because of uncertainties that exist in the labor market, materials supply, bad weather conditions, and so on, he is not sure that he will finish the job in exactly 30 days. However, he is 90% confident that the job will be completed within 40 days. Let X denote the number of days required to complete job A.

(a) Assume X to be a Gaussian random variable; determine μ and σ and also the probability that X will be less than 50, based on the given information. *Ans. 0.9948.*

(b) Recall that a Gaussian random variable ranges from $-\infty$ to $+\infty$. Thus X may take on negative values that are physically impossible. Determine the probability of such an occurrence. Based on this result, is the assumption of the normal distribution for X reasonable? *Ans. 0.00006.*

(c) Let us now assume that X has a log-normal distribution with the same expected value and variance as those in the normal distribution of part (a). Determine the parameters λ and ζ, and also the probability that X will be less than 50. Compare this with the result of part (a). *Ans. 0.9817.*

3.20 From records of repairs of construction equipments, it is found that the failure-free operation time (that is, time between breakdowns) of an equipment may be modeled with a log-normal variate, with a mean of 6 months and a standard deviation of 1.5 months. As the engineer in charge of maintaining the operational condition of a fleet of construction equipment, you wish to have at least a 90% probability that a piece of equipment will be operational at any time.

(a) How often should each piece of equipment be scheduled for maintenance? *Ans. 4.22 months.*

(b) If a particular piece of equipment is still in good operating condition at the time it is scheduled for maintenance, what is the probability that it can operate for at least another month without its regular maintenance? *Ans. 0.749.*

3.21 A system of storm sewers is proposed for a city. In order to evaluate the effectiveness of the sewer system in preventing flooding of the streets, the following information has been gathered. Figure P3.21a shows the probability mass function for the number of occurrences of rainstorm each year in the city. Figure P3.21b shows the distribution of the maximum runoff rate in each storm, which is log-normal with a median of 7 cfs (cubic feet/sec) and COV of 15%. From hydraulic analysis, the proposed sewer system is shown to be

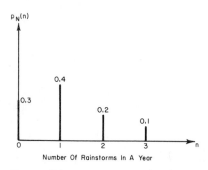

Figure P3.21a

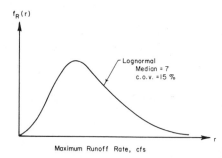

Figure P3.21b

adequate for any storm with runoff rate less than 8 cfs. Assume that the maximum runoff rates between storms are statistically independent.

(a) What is the mean and variance of the number of rainstorms in a year for the city?

(b) What is the probability of flooding during a rainstorm? *Ans. 0.187.*

(c) What is the probability of flooding in a year? *Ans. 0.189.*

3.22 The depth to which a pile can be driven without hitting the rock stratum is denoted as H (Fig. P3.22a). For a certain construction site, suppose that this depth has a log-normal distribution (Fig. P3.22b) with mean of 30 ft and COV of 20%. In order to provide satisfactory support, a pile should be embedded 1 ft into the rock stratum.

(a) What is the probability that a pile of length 40 ft will not anchor satisfactorily in rock? *Ans. 0.10.*

(b) Suppose a 40-ft pile has been driven 39 ft into the ground and rock has not yet been encountered. What is the probability that an additional 5 ft of pile welded to the original length will be adequate to anchor this pile satisfactorily in rock? *Ans. 0.71.*

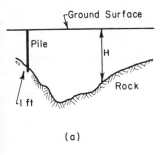

(a)

Figure P3.22a

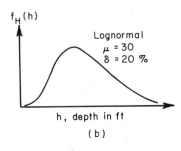

(b)

Figure P3.22b

3.23 A water distribution subsystem consists of pipes AB, BC, and AC as shown in Fig. P3.23. Because of differences in elevation and in hydraulic head loss in the pipes and associated uncertainties, the capacity of each pipe (which is defined as the maximum rate of flow) is given as follows, in cfs (cubic feet/sec):

AB: capacity is Gaussian with mean 5, COV 10%

BC: capacity is log-normal with median 5, COV 10%

AC: capacity equal to 8 or 9 with equal likelihood

(a) Determine the probability that the capacity of the branch ABC will exceed 4 cfs. *Ans. 0.963.*

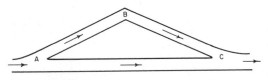

Figure P3.23

(b) Determine the probability that the total capacity of the subsystem shown above will exceed 13 cfs. *Ans. 0.607.* (*Hint.* Use conditional probability.)

3.24 A construction project is at present 30 days away from the scheduled completion date. Depending on the weather condition in the next month, the time required for the remaining construction will have log-normal distributions as follows:

Weather	Time required (days)	
Good	$\mu = 25,$	$\sigma = 4$
Bad	Median $= 30,$	$\sigma = 6$

Based on preliminary investigation, the weather in the next month would be equally likely to be good or bad.

(a) What is the probability that there will be a delay in the completion of the project? *Ans. 0.306.*

(b) A weather specialist is hired to obtain additional information on the weather condition for the next month. However, the specialist is not perfect in his prediction. In general, his predictions are correct 90% of the time, that is $P(PG \mid G) = 0.9$ and $P(PB \mid B) = 0.9$, where PG, PB denote the event that he predicts good and bad weather, respectively, and G, B denote the event that the weather is actually good and bad, respectively. Suppose that the specialist predicted good weather for the next month. What is the updated probability that there will be a delay in the completion of the project? *Ans. 0.150.*

3.25 A compacted subgrade is required to have a specified density of 110 pcf (pounds per cu ft). It will be acceptable if 4 out of 5 cored samples have at least the specified density.

(a) Assuming each sample has a probability of 0.80 of meeting the required density, what is the probability that the subgrade will be acceptable? *Ans. 0.737.*

(b) What should the probability of each sample be in order to achieve a 80% probability of an acceptable subgrade?

3.26 The following is the 20-year record of the annual maximum wind velocity V in town A (in kilometers per hour, kph).

Year	V (kph)	Year	V (kph)
1950	78.2	1960	78.4
1951	75.8	1961	76.4
1952	81.8	1962	72.9
1953	85.2	1963	76.0
1954	75.9	1964	79.3
1955	78.2	1965	77.4
1956	72.3	1966	77.1
1957	69.3	1967	80.8
1958	76.1	1968	70.6
1959	74.8	1969	73.5

(a) Based on this record, estimate the probability that V will exceed 80 kph in any given year.

(b) What is the probability that in the next 10 years there will be exactly 3 years with annual maximum wind velocity exceeding 80 kph?

(c) If a temporary structure is designed to resist a maximum wind velocity of 80 kph, what is the probability that this design wind velocity will be exceeded during the structure's lifetime of 3 years?

(d) How would the answer in part (c) change, if the design wind velocity is increased to 85 kph?

3.27 The sewers in a city are designed for a rainfall having a return period of 10 years.

(a) What is the probability that the sewers will be flooded for the first time in the third year after completion of construction?

(b) What is the probability of flooding in the first 3 years?

(c) What is the probability of flooding in 3 of the first 5 years?

(d) What is the probability of only one flood within 3 years?

3.28 A preliminary planning study on the design of a bridge over a river recommended a permissible probability of 30% of the bridge being inundated by flood in the next 25 years.

(a) If p denotes the probability that the design flood level for the bridge will be exceeded in 1 year, what should the value of p be to satisfy the design criterion given above ? [*Hint.* For small value of x, $(1 - x)^n \simeq 1 - nx$.]

(b) What is the return period of this design flood? *Ans. 83.4 years.*

3.29 Figure P3.29 shows a 40-ft soil stratum where boulders are randomly deposited. Piles are designed to be driven to rock. For simplicity, assume that the stratum can be divided into 4 independent layers of 10 ft each, that the probability of hitting a boulder within each 10-ft layer is 0.02, and that the probability of hitting 2 or more boulders within each layer is negligible.

(a) What is the probability that a pile will be successfully driven to rock without hitting any boulder?

(b) What is the probability that it will hit at most 1 boulder on its way to rock?

(c) What is the probability that a pile will hit the first boulder in layer C?

(d) Suppose the foundation of a small building requires a group of 4 such piles driven to rock. What is the probability that no boulders will be

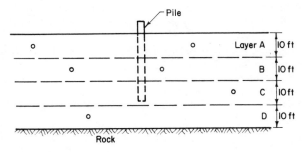

Figure P3.29

encountered in driving the piles? Assume that the pile-driving conditions between piles are statistically independent.

3.30 The useful life per mile of pavement (Fig. P3.30) is described as a log-normal variate with a median of 3 years and COV of 50%. Life means the usable time until repair is required. Assume that the lives between any 2 miles of pavement are statistically independent.

(a) What is the probability that a mile of pavement will require repair in a year?

(b) Suppose that the design life is specified to be the 5-percentile life $x_{.05}$ (that is, the pavement life will be less than the design life with probability 5%). Determine the design life.

(c) What is the probability that there will be no repairs required in the first year of a 4-mile stretch of pavement?

(d) What is the probability that 2 of the 4 miles will need repairs in the first year?

(e) What is the probability of repairs of the 4-mile stretch in the first 3 years of use?

(f) What is the probability that the first repair of the 4-mile stretch will occur in the second year? (Note that the condition in the second year is not independent of the first year.)   *Ans. 0.543.*

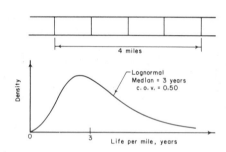

Figure P3.30

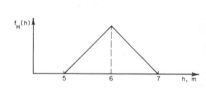

Figure P3.31

3.31 The maximum annual flood level of a river is denoted by H (in meters). Assume that the probability density of H is described by the triangular distribution shown in Fig. P3.31.

(a) Determine the flood height h_{20} which has a mean recurrence interval (return period) of 20 years.

(b) What is the probability that during the next 20 years the river height H will exceed h_{20} at least once?

(c) What is the probability that during the next 5 years the value of h_{20} will be exceeded exactly once?

(d) What is the probability that h_{20} will be exceeded at most twice during the next 5 years?

3.32 For the river in Problem 3.31, a control dam will be constructed according to

the following specification. The height of the dam will be so selected that in the next 3 years this height will be safe against floods with a probability of 94%.
   (a) Determine the required return period of the design flood. *Ans. 50 years.*
   (b) Determine the design height that will meet this requirement. *Ans. 6.8 m.*

3.33 For quality control purposes, 3 specimens in the form of 6-in.-diameter cylinders are taken at random from a batch of concrete, and each specimen is tested for its compressive strength. A specimen will pass the strength test if it survives an axial compressive load of 11 kips. From previous record, the contractor concludes that the histogram of crushing strength of similar concrete specimens can be satisfactorily modeled by a normal distribution with mean 14.68 kips and standard deviation 2.1 kips, that is, $N(14.68, 2.1)$.

   (a) What is the probability that a specimen picked at random will pass the test?

   (b) If the specification requires all 3 specimens to pass the test for the batch of concrete to be acceptable, what is the probability that a batch of concrete prepared by this contractor will be rejected?

   (c) The contractor prepares a batch of concrete each day. What is the probability that at most one batch of concrete will be rejected for a 2-day period?

   (d) Repeat part (b), if the specification is relaxed so that one failure out of the 3 specimens tested is allowed.

   (e) The contractor may use a better grade of concrete mix, and together with better workmanship and supervision, he can improve the mean crushing strength of concrete specimen to 16.5 kip, while reducing the coefficient of variation to 90% of its previous value. What is the probability for a batch to be acceptable now? Assume that the crushing strength of the concrete is a normal variate, and no failures are allowed in the 3 specimens tested. *Ans. 0.986.*

3.34 Three flood control dikes are built to prevent flooding of the low plain as shown in Fig. P3.34. The dikes are designed as follows.
   (i) Design flood of Dike I is the 20-year flood of river A.
   (ii) Design flood of Dike II is the 10-year flood of river A.
   (iii) Design flood of Dike III is the 25-year flood of river B.
Assume that the floods in rivers A and B are statistically independent; also, the failures of dikes I and II are statistically independent.

   (a) Within a year, determine the probability of flooding of the low plain caused by river A only. *Ans. 0.145.*

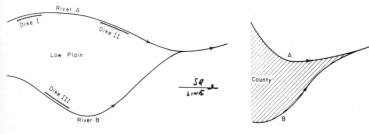

Figure P3.34   *Figure P3.35*

(b) What is the probability of flooding of the low plain area in a year?
Ans. 0.179.
(c) What is the probability of no flooding of the low plains in 4 consecutive years?   *Ans. 0.454.*

3.35 A county is bounded by streams A and B (Fig. P3.35). From flow record, the annual maximum flow in A may be modeled by a normal distribution with mean 1000 cfs and COV 20%, whereas that in B may be modeled by a lognormal distribution with mean 800 cfs and COV 20%. The capacities (defined as the maximum flow that can be carried without overflowing) of A and B are 1200 and 1000 cfs, respectively. Assume the stream flows in A and B are statistically independent.

(a) What is the probability that stream A will overflow in a year?
(b) What is the probability that stream B will overflow in a year?
(c) What is the probability that the county will be flooded in a year?
(d) What is the probability that the county will be free of floods in the next 3 years?
(e) If it is decided to reduce the probability of overflow in stream A to 5% a year by enlarging the stream bed at critical locations, what should be the new capacity of A?
(f) Suppose that, because of error in prediction, the capacity of stream B may not be 1000 cfs, and there is a 20% chance that the capacity may be 1100 cfs. In such a case, what is the probability that stream B will overflow in a year?

3.36 A cofferdam is to be built around a proposed bridge pier location so that construction of the pier may be carried out "dry" (see Fig. P3.36).

The height of the cofferdam should protect the site from overflow of wave water during the construction period with a reliability of 95%. The distribution of the monthly maximum wave height is Gaussian $N(5,2)$ ft above mean sea level.

(a) If the construction will take 4 months, what should be the design height of the cofferdam (above mean sea level)? Assume that monthly maximum wave heights are statistically independent.   *Ans. 9.46 ft.*
(b) If the time of construction can be shortened by 1 month with an additional cost of $600, and the cost of constructing the cofferdam is $2000 per ft (above mean sea level), should the contractor take this alternative? Assume that the same risk of overflow of wave water still applies.

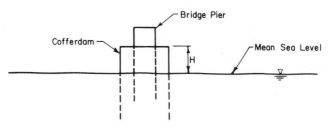

Figure P3.36

3.37 A contractor owns 5 trucks for use in his construction jobs. He decides to institute a new program of truck replacement, using the following procedure:
 (i) Any truck that has had more than 1 major breakdown on the job within a year will be evaluated to determine how many miles it gets per gallon of gas.
 (ii) Any truck given this special evaluation will be replaced if it gets less than 9 miles per gallon.
From prior experience, the contractor knows two facts with a high degree of confidence: (i) for each truck, the mean rate of major breakdowns is once every 0.8 year; and (ii) the gasoline consumption of trucks that have more than 1 major breakdown is a normal variate $N(10, 2.5)$ in miles per gallon.
 (a) What is the probability that a given truck will have more than 1 breakdown within a year?
 (b) What is the probability that a truck getting a special evaluation will fail to meet the miles-per-gallon test [see part (ii) above]?
 (c) What is the probability that a given truck will be replaced within a year?
 (d) What is the probability that the contractor will replace exactly 1 truck within a year?

3.38 On the average 2 damaging earthquakes occur in a certain country every 5 years. Assume the occurrence of earthquakes is a Poisson process in time. For this country, complete the following.
 (a) Determine the probability of getting 1 damaging earthquake in 3 years.
 (b) Determine the probability of no earthquakes in 3 years.
 (c) What is the probability of having at most 2 earthquakes in one year?
 (d) What is the probability of having at least 1 earthquake in 5 years?

3.39 (a) The occurrences of flood may be modeled by a Poisson process. If the mean occurrence rate of floods for a certain region A is once every 8 years, determine the probability of no floods in a 10-year period; of 1 flood; of more than 3 floods.
 (b) A structure is located in region A. The probability that it will be inundated, when a flood occurs, is 0.05. Compute the probability that the structure will survive if there are no floods; if there is 1 flood; if there are n floods. Assume statistical independence between floods.
 (c) Determine the probability that the structure will survive over the 10-year period. *Ans. 0.939.*

3.40 Traffic on a one-way street that leads to a toll bridge is to be studied. The volume of the traffic is found to be 120 vehicles per hr *on the average* and out of which $\frac{2}{3}$ are passenger cars and $\frac{1}{3}$ are trucks. The toll at the bridge is $0.50 per car and $2 per truck. Assume that the arrivals of vehicles constitute a Poisson process.
 (a) What is the probability that in a period of 1 minute, more than 3 vehicles will arrive at the toll bridge? *Ans. 0.1429.*
 (b) What is the *expected* total amount of toll collected at the bridge in a period of 3 hr?

3.41 Strikes among construction workers occur according to the Poisson process; on the average there is one strike every 3 years. The average duration of a strike is 15 days, and the corresponding standard deviation is 5 days.
 If it costs (in terms of losses) a contractor $10,000 per day of strike, answer the following.

(a) What would be the expected loss to the contractor during a strike?

(b) If the strike duration is a normal variate, what is the probability that the contractor may lose in excess of $20,000 during a strike?

(c) In a job that will take 2 years to complete, what would be the contractor's expected loss from possible strikes? (Remember that the occurrence of strikes is a Poisson process.)   *Ans.* $100,000.

3.42   The service stations along a highway are located according to a Poisson process with an average of 1 service station in 10 miles. Because of a gas shortage, there is a probability of 0.2 that a service station would not have gasoline available. Assume that the availabilities of gasoline at different service stations are statistically independent.

(a) What is the probability that there is at most 1 service station in the next 15 miles of highway?

(b) What is the probability that none of the next 3 stations have gasoline for sale?

(c) A driver on this highway notices that the fuel gauge in his car reads empty; from experience he knows that he can go another 15 miles. What is the probability that he will be stranded on the highway without gasoline?

3.43   Express rapid-transit trains run between two points (for example, between downtown terminal and airport). Suppose that the passengers arriving at the terminal and bound for the airport (Fig. P3.43) constitute a Poisson process with an average rate of 1.5 passengers per minute. If the capacity of the train is 100 passengers, how often should trains leave the terminal so that the probability of overcrowding is no more than 10%?

(a) *Formulate the problem* exactly.

(b) Determine an approximate solution by assuming that the number of airport-bound passengers is Gaussian with the same mean and standard deviation as the preceding Poisson distribution.

(c) If the trains depart from the terminal according to the schedule of part (b), what is the probability that in 5 consecutive departures 1 will be overcrowded? Assume statistical independence.

3.44   A large radio antenna system consisting of a dish mounted on a truss (see Fig. P3.44) is designed against wind load. Since damaging wind storms rarely occur, their occurrences may be modeled by a Poisson process. Local weather records show that during the past 50 years only 10 damaging wind storms have been reported. Assume that if damaging wind storm (or storms) occur in this period, the probabilities that the dish and the truss will be damaged in a storm are 0.2 and 0.05, respectively, and that damage to the

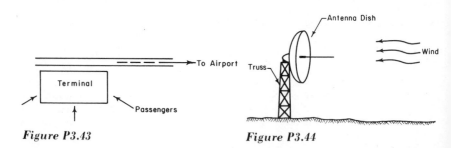

Figure P3.43   **Figure P3.44**

dish and truss are statistically independent. Determine the probabilities, during the next 10 years, for the following events.
(a) There will be more than 2 damaging wind storms.
(b) The antenna system will be damaged, assuming the occurrence of at most 2 damaging storms.
(c) The antenna system will be damaged.

3.45 The problem in Example 3.17 may be solved by assuming that whenever the center of a 12-in.-diameter boulder is inside the volume of a cylinder with 15 in. diameter and 50 ft depth, it will be hit by the 3-in. drill hole. On this basis and the assumption that the occurrence of boulders in the soil mass constitutes a Poisson process, develop the corresponding solution procedure for determining the probability of the 3-in. drill hole hitting boulders in a 50-ft depth boring.

3.46 Suppose that the hurricane record for the last 10 years at a certain coastal city in Texas is as follows.

Year	No. of hurricanes
1961	1
1962	0
1963	0
1964	2
1965	1
1966	0
1967	0
1968	2
1969	1
1970	1

The occurrence of hurricanes can be described by a Poisson process. The maximum wind speed of hurricanes usually shows considerable fluctuation. Suppose that those recorded at this city can be fitted satisfactorily by a log-normal distribution with mean = 100 ft/sec and standard deviation = 20 ft/sec.
(a) Based on the available data, find the probability that there will be at least 1 hurricane in this city in the next 2 years. *Ans. 0.798.*
(b) If a structure in this city is designed for a wind speed of 130 ft/sec, what is the probability that the structure will be damaged (design wind speed exceeded) by the next hurricane? *Ans. 0.08.*
(c) What is the probability that there will be at most 2 hurricanes in the next 2 years, and that no structure will be damaged during this period? *Ans. 0.718.*

3.47 Tornadoes may be divided into two types, namely I (strong) and II (weak). From 18 years of record in a city, the number of type I and type II tornadoes are 9 and 54, respectively. The occurrences of each type of tornado are assumed to be statistically independent and constitute a Poisson process.
(a) What is the probability that there will be exactly 2 tornadoes in the city next year?

(b) Assuming that exactly 2 tornadoes actually occurred, and 1 of the 2 is known to be of type I, what is the probability that the other is also type I?

3.48 Figure P3.48*a* shows a record of the earthquake occurrences in a county where a brick masonry tower is to be built to last for 20 years. The tower can withstand an earthquake whose magnitude is 5 or lower. However, if quakes with magnitude more than 5 (defined as damaging quake) occur, there is a likelihood that the tower may fail. The engineer estimated that the probability of failure of the tower depends on the number of damaging quakes occurring during its lifetime, which is described in Fig. P3.48*b*.

(a) What is the probability that the tower will be subjected to less than 3 damaging quakes during its lifetime? Assume earthquake occurrences may be modeled by a Poisson process.

(b) Determine the probability that the tower will not be destroyed by earthquakes within its useful life.

(c) Besides earthquakes, the tower may also be subjected to the attack of tornadoes whose occurrence may be modeled by a Poisson process with mean recurrence time of 200 years. If a tornado hits the tower, the tower will be destroyed. Assume that failures caused by earthquakes and tornadoes are statistically independent. What is the probability that the tower will fail by these natural hazards within its useful life?

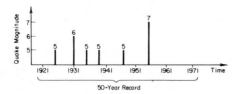

Figure P3.48a

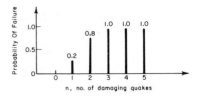

Figure P3.48b

3.49 A skyscraper is located in a region where earthquakes and strong winds may occur. From past record, the mean rate of occurrence of a large earthquake that may cause damage to the building is 1 in 50 years, whereas that for strong wind is 1 in 25 years. The occurrences of earthquake and strong wind may be modeled as independent Poisson processes. Assume that during a strong earthquake, the probability of damage to the building is 0.1, whereas the corresponding probability of damage under strong wind is 0.05. The damages caused by earthquake and wind may be assumed to be independent events.

(a) What is the probability that the skyscraper will be subjected to strong winds but not large earthquakes in a 10-year period? Also, determine the probability of the structure subjected to both large earthquakes and strong winds in the 10-year period?

(b) What is the probability that the building will be damaged in the 10-year period?

3.50 The daily water consumption of a city may be assumed to be a Gaussian random variable with a mean of 500,000 gal/day (gpd), and a standard deviation of 150,000 gpd. The daily water supply is either 600,000 or 750,000 gallons, with probabilities 0.7 and 0.3, respectively.

(a) What is the probability of water shortage in any given day?

(b) Assuming that the conditions between any consecutive days are statistically independent, what is the probability of shortage in any given week?

(c) On the average, how often would water shortage occur? If the occurrence of water shortage is a Poisson process, what would then be the probability of shortage in a week?

(d) If the city engineer wants the probability of shortage to be no more than 1 % in any given day, how much water supply is required?

3.51 Steel construction work on multistory buildings is a potentially hazardous occupation. A building contractor who is building a skyscraper at a steady pace finds that in spite of a strong emphasis on safety measures, he has been experiencing accidents among his large group of steel workers; on the average, about 1 accident occurs every 6 months.

(a) Assuming that the occurrence of a specific accident is not influenced by any previous accident, find the probability that there will be (exactly) 1 accident in the next 4 months.

(b) What is the probability of at least 1 accident in the next 4 months?

(c) What is the mean number of accidents that the contractor can expect in a year? What is the standard deviation for the number of accidents during a period of 1 year?

(d) If the contractor can go through a year without an accident among his steel construction workers, he will qualify for a safety award. What is the probability of his receiving this award next year?

(e) If the contractor's work is to continue at the same pace over the next 5 years, what is the probability that he will win the safety award twice during this 5-year period?

3.52 Two industrial plants are located along a stream (see Fig. P3.52). The solid and liquid wastes that are disposed from the plants into the stream are called effluents. In order to control the quality of the effluent from each plant, there is an effluent standard established for each plant. Assume that each day, the effluent of each plant may exceed this effluent standard with probability $p = 0.2$, during the actual operation. A good measure of the stream quality at A as a result of the pollution from these effluent wastes is given by the dissolved oxygen concentration (DO) at that location. Assume that the DO has a log-normal distribution with the following medians and COV (in mg/l).

Median	COV	
4.2	0.1	when both effluents do not exceed standard
2.1	0.15	when only 1 effluent exceeds standard
1.6	0.18	when both effluents exceed standard

(a) What is the probability that the DO concentration at A will be less than 2 mg/l in any given day?

(b) What is the probability that the DO concentration at A will be less than 2 mg/l in two consecutive days?

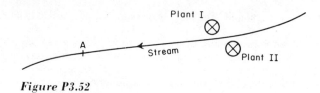

Figure P3.52

(c) It has been proposed as a stream standard that the probability of DO concentration at A falling below 2 mg/l in a day should not exceed 0.1. What should be the allowable maximum value of p (the probability of exceeding the effluent standard for each plant)?

3.53 The daily concentration of a certain pollutant in a stream has the exponential distribution shown in Fig. P3.53.

(a) If the mean daily concentration of the pollutant is $2 \text{ mg}/10^3$ liter, determine the constant c in the exponential distribution.

(b) Suppose that the problem of pollution will occur if the concentration of the pollutant exceeds $6 \text{ mg}/10^3$ liter. What is the probability of pollution problem resulting from this pollutant in a single day?

(c) What is the return period (in days) associated with this concentration level of $6 \text{ mg}/10^3$ liter? Assume that the concentration of the pollutant is statistically independent between days. *Ans. 20 days.*

(d) What is the probability that this pollutant will cause a pollution problem at most once in the next 3 days? *Ans. 0.993.*

(e) If instead of the exponential distribution, the daily pollutant concentration is Gaussian with the same mean and variance, what would be the probability of pollution in a day in this case? *Ans. 0.022.*

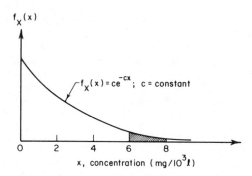

Figure P3.53

3.54 The interarrival times of vehicles on a road follows an exponential distribution with a mean of 15 sec. A gap of 20 sec is required for a car from a side street to cross the road or to join the traffic.

(a) What is the proportion of gaps that are less than 20 sec?

(b) What is the average (mean) interarrival time for all the gaps that are longer than 20 sec?

(c) In 1 hr, what is the expected total time occupied by gaps that are less than 20 sec? (*Hint.* What is the expected number of gaps that are less than 20 sec in 1 hr?)

3.55 The occurrences of tornadoes in a midwestern county may be modeled by a Poisson process with a mean occurrence rate of 2.5 tornadoes per year.
   (a) What is the probability that the recurrence time between tornadoes will be longer than 8 months?
   (b) Derive the distribution of the time till the occurrence of the second tornado. On the basis of this distribution, determine the probability that a second tornado will occur within a given year.

3.56 The time of operation of a construction equipment until breakdown follows an exponential distribution with a mean of 24 months. The present inspection program is scheduled at every 5 months.
   (a) What is the probability that an equipment will need repair at the first scheduled inspection date?
   (b) If an equipment has not broken down by the first scheduled inspection date, what is the probability that it will be operational beyond the next scheduled inspection date?
   (c) The company owns 5 pieces of a certain type of equipment; assuming that the service lives of equipments are statistically independent, determine the probability that at most 1 piece of equipment will need repair at the scheduled inspection date.
   (d) If it is desired to limit the probability of repair at each scheduled inspection date to not more than 10%, what should be the inspection interval? The conditions of part (c) remains valid.

3.57 The cost for the facilities to release and refill water for a navigation lock in a canal increases with decreasing time required for each cycle of operation. For purposes of design, it has been observed that the time of arrival of boats follows an exponential distribution with a mean interarrival time of 0.5 hr. Assume that the navigation lock is to be designed so that 80% of the incoming traffic can pass through the lock without waiting.
   (a) What should be the design time of each cycle of operation? *Ans. 0.11 hr.*
   (b) What is the probability that of 4 successive arrivals, none of them have to wait at the lock? *Ans. 0.41.*
   (c) Suppose that one boat leaves town A every 8 hr, and has to go through the lock to reach its destination. What is the probability that at least 1 of the boats leaving town A in a 24-hr day has to wait at the lock? *Ans. 0.488.*

3.58 A pipe carrying water is supported on short concrete piers that are spaced 20 ft apart as shown in Fig. P3.58a.The pipe is saddled on the piers as shown in Fig. P3.58b. When subjected to lateral earthquake motions, there is a horizontal inertia force that will tend to dislodge the pipe from its supports. The maximum lateral inertia force F at each pier may be estimated as

$$F = \frac{w}{g} \cdot a$$

where
   w = the weight of the pipe and water for a 20-ft section;
   g = acceleration of gravity = 32.2 ft/sec^2;
   a = maximum horizontal earthquake acceleration.

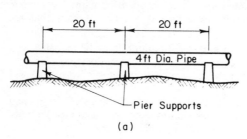

(a)

Figure P3.58a

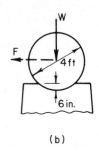

(b)

Figure P3.58b

The pipe has a diameter of 4 ft, so that the total weight per foot of pipe and contents is 800 lb per ft. Assume that the maximum acceleration during a strong-motion earthquake is a log-normal variate with a mean of $0.4\,g$ and a COV of 25%.

(a) What is the probability that during such an earthquake, the pipe will be dislodged from a pier support (by rolling out of the saddle)?

(b) If there are 5 piers supporting the pipe over a ravine, what is the probability that the pipe will not be dislodged anywhere? Assume the conditions between supports to be statistically independent.

(c) If the occurrence of strong-motion earthquakes is a Poisson process, and such earthquakes are expected (on the average) once every 3 years, what is the probability that the pipe may be dislodged from its supports over a period of 10 years?

3.59 Ten percent of the 200 tendons required to prestress a nuclear reactor structure have been corroded during the last year. Suppose that 10 tendons were selected at random and inspected for corrosion; what is the probability that none of the tendons inspected show signs of corrosion? What is the probability that there will be at least one corroded tendon among those inspected?

3.60 The fill in an earth embankment is compacted to a specified CBR (California Bearing Ratio). The entire embankment can be divided into 100 sections, of which 10 do not meet the required CBR.

(a) Suppose that 5 sections are selected at random and tested for their CBR, and acceptance here requires all 5 sections to meet the CBR limit. What is the probability that the compaction of the embankment will be accepted?

(b) If, instead of 5, 10 sections will be inspected and acceptance requires all 10 sections meeting the CBR limit. What is the probability of acceptance now?

Section 3.3

3.61 Both east and west bound rush-hour traffic on a toll bridge are counted at 10-sec intervals. The following table shows the number of observations for each combination of east and west bound traffic counts:

	Number of westbound vehicles				
	0	1	2	3	4
0	2	5	15	40	58
1	1	6	15	35	62
2	18	15	28	30	30
3	45	32	25	15	10
4	65	58	35	15	5

(Number of eastbound vehicles — row labels 0–4)

Total number of observations = 665

Let X = number of eastbound vehicles in a 10-sec interval.

Y = number of westbound vehicles in a 10-sec interval.

(a) Compute and plot the joint probability mass function of X and Y.
(b) Determine the marginal PMF of X.
(c) If there are 3 eastbound vehicles on the bridge in a 10-sec interval, determine the PMF of westbound vehicles in the same interval.
(d) In a 10-sec interval, what is the probability that 4 vehicles are going east if there are also 4 vehicles going west at the same time?
(e) Determine the covariance Cov (X, Y), and evaluate the corresponding correlation coefficient between X and Y.

3.62 The joint density function of the material and labor cost of a construction project is modeled as follows:

$$f_{X,Y}(x, y) = 2ye^{-y(2+x)} \qquad x, y \geq 0$$
$$= 0 \qquad \text{elsewhere}$$

where X = material cost in $100,000

Y = labor cost in $100,000

(a) What is the probability that the material and labor costs of the next construction project will be less than $100,000 and $200,000, respectively?
(b) Determine the marginal density function of material cost in a project.
(c) Determine the marginal density function of labor cost in a project.
(d) Are the material and labor costs in the construction project statistically independent? Why?
(e) If it is known that the cost of material in the project is $200,000, what is the probability that its labor cost will exceed $200,000?

4. Functions of Random Variables

4.1. INTRODUCTION

Engineering problems often involve the evaluation of functional relations between a dependent variable and one or more basic (independent) variables. If any of the basic variables are random, the dependent variable will likewise be random; its probability distribution, as well as its moments, will be functionally related to and may be derived from those of the basic random variables.

4.2. DERIVED PROBABILITY DISTRIBUTIONS

4.2.1. Function of single random variable

Consider first the function of a single random variable,

$$Y = g(X) \tag{4.1}$$

This means that when $Y = y$, $X = x = g^{-1}(y)$ where g^{-1} is the inverse function of g. [Assume for the moment that $g(x)$ is a monotonically increasing function of x with a unique inverse $g^{-1}(y)$.] Thus

$$P(Y = y) = P(X = x) = P[X = g^{-1}(y)]$$

That is, the PMF of Y is

$$p_Y(y) = p_X[g^{-1}(y)] \tag{4.2}$$

Also, it follows that

$$P(Y \le y) = P[X \le g^{-1}(y)]$$

Thus

$$F_Y(y) = F_X[g^{-1}(y)] \tag{4.3}$$

Hence, for discrete X,

$$F_Y(y) = \sum_{\text{all } x_i \le g^{-1}(y)} p_X(x_i) \tag{4.4}$$

whereas, for continuous X, Eq. 4.3 yields

$$F_Y(y) = \int_{\{x \le g^{-1}(y)\}} f_X(x)\, dx = \int_{-\infty}^{g^{-1}(y)} f_X(x)\, dx \tag{4.5}$$

In the latter case (that is, X continuous), we recall from calculus that by making a change of the variable of integration, Eq. 4.5 becomes

$$F_Y(y) = \int_{-\infty}^{g^{-1}(y)} f_X(x)\, dx = \int_{-\infty}^{y} f_X(g^{-1})\, \frac{dg^{-1}}{dy}\, dy$$

where $g^{-1} = g^{-1}(y)$. Therefore the density function of Y is

$$f_Y(y) = \frac{dF_Y(y)}{dy} = f_X(g^{-1})\, \frac{dg^{-1}}{dy}$$

This assumes that y increases with x. When y decreases with increasing x, $F_Y(y) = 1 - F_X(g^{-1})$; then

$$f_Y(y) = - f_X(g^{-1})\, \frac{dg^{-1}}{dy}$$

However, in this latter case (dg^{-1}/dy) is negative. Properly then the *derived density function* is

$$f_Y(y) = f_X(g^{-1})\, \left| \frac{dg^{-1}}{dy} \right| \tag{4.6}$$

EXAMPLE 4.1

Suppose that X is a normal variate with parameters μ and σ. Determine the density function of $Y = (X - \mu)/\sigma$.

The inverse function is $x = \sigma y + \mu$, and $\dfrac{dx}{dy} = \sigma$. Thus Eq. 4.6 yields

$$f_Y(y) = \frac{1}{\sqrt{2\pi}\, \sigma} \exp\left[\frac{-\frac{1}{2}(\sigma y + \mu - \mu)^2}{\sigma^2} \right] \cdot \sigma$$

$$= \frac{1}{\sqrt{2\pi}}\, e^{-y^2/2}$$

Therefore Y is a standard normal variate with density function $N(0, 1)$.

EXAMPLE 4.2

If X has a log-normal distribution with parameters λ and ζ, what is the distribution of $Y = \ln X$? In this case

$$f_X(x) = \frac{1}{\sqrt{2\pi}\, \zeta}\, \frac{1}{x} \exp\left[-\frac{1}{2}\left(\frac{\ln x - \lambda}{\zeta} \right)^2 \right]$$

and

$$x = e^y; \qquad \frac{dx}{dy} = e^y$$

Therefore, according to Eq. 4.6,

$$f_Y(y) = \frac{1}{\sqrt{2\pi}\,\zeta} \frac{1}{e^y} \exp\left[-\frac{1}{2}\left(\frac{y-\lambda}{\zeta}\right)^2 \right] \cdot e^y$$

$$= \frac{1}{\sqrt{2\pi}\,\zeta} \cdot \exp\left[-\frac{1}{2}\left(\frac{y-\lambda}{\zeta}\right)^2 \right]$$

Hence the distribution of Y is normal with mean value λ and standard deviation ζ; that is, $E(\ln X) = \lambda$, Var $(\ln X) = \zeta^2$.

We observe that the inverse function $g^{-1}(y)$ may not be single-valued; that is, there may be multiple values of x for a given value of y. In such cases, if $g^{-1}(y) = x_1, x_2, \cdots, x_k$, we have

$$(Y = y) = \bigcup_{i=1}^{k} (X = x_i)$$

Hence, for discrete X,

$$p_Y(y) = \sum_{i=1}^{k} p_X(x_i) \tag{4.7}$$

whereas, if X is continuous,

$$f_Y(y) = \sum_{i=1}^{k} f_X(g_i^{-1}) \left| \frac{dg_i^{-1}}{dy} \right| \tag{4.8}$$

in which $g_i^{-1} = x_i$, is the ith root of $g^{-1}(y)$.

EXAMPLE 4.3

The strain energy in a linearly elastic bar subjected to a force S is given by

$$U = \frac{L}{2AE} S^2$$

where
  L = length of the bar
  A = cross-sectional area of the bar,
  E = modulus of elasticity of the elastic material

Then, if S is a standard normal variate $N(0, 1)$, the density function of U is obtained on the basis of Eq. 4.8 as follows.
  Rewriting

$$U = cS^2$$

where $c = L/2AE$, we have

$$s = \pm\sqrt{\frac{u}{c}}$$

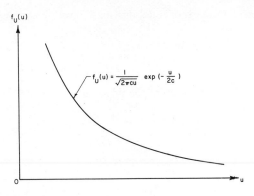

Figure E4.3

and thus

$$\frac{ds}{du} = \pm\frac{1}{2\sqrt{cu}}$$

or

$$\left|\frac{ds}{du}\right| = \frac{1}{2\sqrt{cu}}$$

Hence the density function of the strain energy U, in accordance with Eq. 4.8, is

$$f_U(u) = \left[f_S\left(\sqrt{\frac{u}{c}}\right) + f_S\left(-\sqrt{\frac{u}{c}}\right)\right]\frac{1}{2\sqrt{cu}}$$

$$= \frac{1}{\sqrt{2\pi cu}}\exp\left(-\frac{u}{2c}\right) \qquad u \geq 0$$

which is a *chi-square*-type distribution with one *degree of freedom* (see Eq. 5.40 of Chapter 5). Graphically, this distribution would appear as shown in Fig. E4.3.

EXAMPLE 4.4

The height of earth dams must allow sufficient freeboard above the maximum reservoir level to prevent waves from washing over the top. The determination of this height would include the consideration of wind tide and wave height.

The wind tide, in feet, above still-water level is

$$Z = \frac{F}{1400\,d}\,V^2$$

where

V = wind speed in *miles per hour*

F = fetch, or length of water surface over which the wind blows, in *feet*

d = average depth of lake along the fetch, in *feet*

If the wind speed has an exponential distribution with mean speed v_0; that is

$$f_V(v) = \frac{1}{v_0} e^{-v/v_0} \qquad v \geq 0$$
$$= 0 \qquad\qquad v < 0$$

then we determine the distribution of the tide Z as follows.

Denoting $a = F/1400\, d$, we have $Z = aV^2$; thus

$$v = \pm \sqrt{\frac{z}{a}}$$

and

$$\left| \frac{dv}{dz} \right| = \frac{1}{2\sqrt{az}}$$

Then, according to Eq. 4.8,

$$f_Z(z) = \left[f_V\left(\sqrt{\frac{z}{a}} \right) + f_V\left(-\sqrt{\frac{z}{a}} \right) \right] \frac{1}{2\sqrt{az}}$$

However, in this case since $f_V(x) = 0$ for $x < 0$, we have

$$f_Z(z) = \frac{1}{2\sqrt{az}} f_V\left(\sqrt{\frac{z}{a}} \right)$$
$$= \frac{1}{2v_0\sqrt{az}} \exp\left(-\frac{1}{v_0}\sqrt{\frac{z}{a}} \right) \qquad z \geq 0$$

4.2.2. Function of multiple random variables

Next, consider the function of two random variables X and Y,

$$Z = g(X, Y) \tag{4.9}$$

In this case, $(Z = z)$ refers to the same event as $[g(X, Y) = z]$; that is,

$$(Z = z) = [g(X, Y) = z] = \bigcup_{\{g(x_i,\, y_j)=z\}} (X = x_i, Y = y_j)$$

Hence, the PMF of Z is

$$p_Z(z) = \sum_{g(x_i,\, y_j)=z} p_{X,Y}(x_i, y_j) \tag{4.10}$$

and the corresponding CDF is

$$F_Z(z) = \sum_{g(x_i,\, y_j) \leq z} p_{X,Y}(x_i, y_j) \tag{4.11}$$

In particular, if $Z = X + Y$,

$$p_Z(z) = \sum_{x_i + y_j = z} p_{X,Y}(x_i, y_j) = \sum_{\text{all } x_i} p_{X,Y}(x_i, z - x_i) \tag{4.12}$$

Sum of random variables with Poisson distributions. Suppose that X and Y are statistically independent and have Poisson distributions with parameters ν and μ, respectively; that is,

$$p_X(x) = \frac{(\nu t)^x}{x!} e^{-\nu t}$$

$$p_Y(y) = \frac{(\mu t)^y}{y!} e^{-\mu t}$$

Then according to Eq. 4.12, the PMF of $Z = X + Y$ is

$$p_Z(z) = \sum_{\text{all } x} p_{X,Y}(x, z - x)$$

$$= \sum_{\text{all } x} \frac{(\nu t)^x (\mu t)^{z-x}}{x! (z - x)!} e^{-(\nu + \mu) t}$$

$$= e^{-(\nu + \mu) t} \cdot t^z \sum_{\text{all } x} \frac{\nu^x \mu^{z-x}}{x! (z - x)!}$$

But the sum is the binomial expansion of $(\nu + \mu)^z / z!$; thus

$$p_Z(z) = \frac{[(\nu + \mu) t]^z}{z!} e^{-(\nu + \mu) t}$$

which means that Z also has a Poisson distribution with parameter $(\nu + \mu)$. Generalizing this result, we infer that the sum of two or more independent Poisson processes is also a Poisson process; that is, if

$$Z = \sum_{i=1}^{n} X_i$$

where X_i has a Poisson PMF with parameter ν_i, the PMF of Z is also a Poisson distribution with parameter

$$\nu_Z = \sum_{i=1}^{n} \nu_i \qquad (4.13)$$

However, the difference of two Poisson processes is not a Poisson process; that is, it can be shown that the PMF of $Z = X\text{-}Y$ does not yield a Poisson distribution.

EXAMPLE 4.5

Suppose that a toll bridge serves three suburban residential districts A, B, C (see Fig. E4.5). It is estimated that during peak hours of the day, the average volumes of

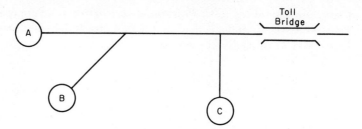

Figure E4.5

traffic from each of these three districts are, respectively, 2, 3, and 4 vehicles per minute. If the peak vehicular traffic from the respective districts is a Poisson process, the traffic crossing the toll bridge would also be a Poisson process with average crossing volume of 9 vehicles per minute.

If X and Y are continuous, Eq. 4.11 becomes

$$F_Z(z) = \underset{\{g(x,\,y)\,\le\,z\}}{\int \int} f_{X,Y}(x, y)\,dx\,dy$$

$$= \int_{-\infty}^{\infty} \int_{-\infty}^{g^{-1}} f_{X,Y}(x, y)\,dx\,dy \tag{4.14}$$

where $g^{-1} = g^{-1}(z, y)$. Changing the variable of integration from x to z, we have

$$F_Z(z) = \int_{-\infty}^{\infty} \int_{-\infty}^{z} f_{X,Y}(g^{-1}, y) \left| \frac{\partial g^{-1}}{\partial z} \right| dz\,dy$$

Thus the PDF of Z is

$$f_Z(z) = \int_{-\infty}^{\infty} f_{X,Y}(g^{-1}, y) \left| \frac{\partial g^{-1}}{\partial z} \right| dy \tag{4.15}$$

Alternatively, taking $g^{-1} = g^{-1}(x, z)$, we also have

$$f_Z(z) = \int_{-\infty}^{\infty} f_{X,Y}(x, g^{-1}) \left| \frac{\partial g^{-1}}{\partial z} \right| dx \tag{4.15a}$$

Specifically, if

$$Z = aX + bY$$

we have

$$x = \frac{z - by}{a} \quad \text{and} \quad \frac{\partial g^{-1}}{\partial z} = \frac{\partial x}{\partial z} = \frac{1}{a}$$

Then Eq. 4.15 would be

$$f_Z(z) = \int_{-\infty}^{\infty} \frac{1}{|a|} f_{X,Y}\left(\frac{z - by}{a}, y\right) dy \qquad (4.16)$$

and if X and Y are statistically independent,

$$f_Z(z) = \frac{1}{|a|} \int_{-\infty}^{\infty} f_X\left(\frac{z - by}{a}\right) f_Y(y) \, dy \qquad (4.16a)$$

or, based on Eq. 4.15a,

$$f_Z(z) = \frac{1}{|b|} \int_{-\infty}^{\infty} f_X(x) f_Y\left(\frac{z - ax}{b}\right) dx \qquad (4.16b)$$

EXAMPLE 4.6

Shown in Fig. E4.6 is an idealized model of a one-story building, with the total mass m concentrated at the roof level. When subjected to earthquake ground shaking, the building will vibrate about its original (at rest) position, inducing velocity components X and Y of the mass, with a resultant velocity $Z = \sqrt{X^2 + Y^2}$.

If X and Y are, respectively, standard normal variates, that is, with distribution $N(0, 1)$, determine the probability distribution of the resultant kinetic energy of the mass during an earthquake.

The resultant kinetic energy is

$$W = mZ^2 = m(X^2 + Y^2)$$

Let $U = mX^2$, and $V = mY^2$; then

$$W = U + V$$

From Example 4.3, we see that the distributions of U and V are, respectively, chi-square with one degree of freedom; that is,

$$f_U(u) = \frac{1}{\sqrt{2\pi mu}} e^{-u/2m} \qquad u \geq 0$$

$$f_V(v) = \frac{1}{\sqrt{2\pi mv}} e^{-v/2m} \qquad v \geq 0$$

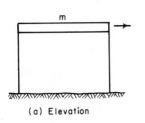

(a) Elevation

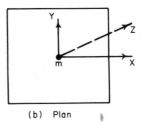

(b) Plan

Figure E4.6 (*a*) Elevation. (*b*) Plan

Then, according to Eq. 4.16*b* and observing that $v = w - u \geq 0$, we obtain the density function of the kinetic energy W as follows:

$$f_W(w) = \frac{1}{2\pi m} \int_0^w \frac{1}{\sqrt{u}} e^{-u/2m} \cdot \frac{1}{\sqrt{(w-u)}} e^{-(w-u)/2m} \, du$$

$$= \frac{1}{2\pi m} e^{-w/2m} \int_0^w u^{-1/2}(w-u)^{-1/2} \, du$$

Now let $r = u/w$; then $du = w \, dr$, and

$$f_W(w) = \frac{1}{2\pi m} e^{-w/2m} \int_0^1 r^{-1/2}(1-r)^{-1/2} \, dr$$

It can be observed that the above integral is the beta function $B(\frac{1}{2}, \frac{1}{2})$ of Eq. 3.49; furthermore, using Eq. 3.49*a* and observing that $\Gamma(\frac{1}{2}) = \sqrt{\pi}$ and $\Gamma(1) = 1.0$, we have

$$B(\tfrac{1}{2}, \tfrac{1}{2}) = \frac{\Gamma(\tfrac{1}{2})\Gamma(\tfrac{1}{2})}{\Gamma(1)} = \pi$$

Hence

$$f_W(w) = \frac{1}{2m} e^{-w/2m}$$

which is a chi-square-type distribution (see Eq. 5.40 of Chapter 5) with two degrees of freedom.

Sum (and difference) of independent normal variates. If X and Y are statistically independent normal variates with means and standard deviations μ_X, σ_X and μ_Y, σ_Y, respectively; the distribution of $Z = X + Y$, according to Eq. 4.16*a*, is

$$f_Z(z) = \frac{1}{2\pi\sigma_X\sigma_Y} \int_{-\infty}^{\infty} \exp\left[-\frac{1}{2}\left\{\left(\frac{z-y-\mu_X}{\sigma_X}\right)^2 + \left(\frac{y-\mu_Y}{\sigma_Y}\right)^2\right\}\right] dy$$

$$= \frac{1}{2\pi\sigma_X\sigma_Y} \exp\left[-\frac{1}{2}\left\{\left(\frac{\mu_Y}{\sigma_Y}\right)^2 + \left(\frac{z-\mu_X}{\sigma_X}\right)^2\right\}\right] \cdot$$

$$\cdot \int_{-\infty}^{\infty} \exp\left[-\frac{1}{2}\left(uy^2 - 2vy\right)\right] dy$$

where

$$u = \frac{1}{\sigma_X^2} + \frac{1}{\sigma_Y^2}$$

and

$$v = \frac{\mu_Y}{\sigma_Y^2} + \frac{z-\mu_X}{\sigma_X^2}$$

Completing the square for the last integrand above, and then substituting

$$w = y - \frac{v}{u}$$

the last integral above becomes

$$\int_{-\infty}^{\infty} \exp\left[-\tfrac{1}{2}(uy^2 - 2vy)\right] dy = e^{v^2/2u} \int_{-\infty}^{\infty} \exp\left(-\tfrac{1}{2}uw^2\right) dw$$

$$= \sqrt{\frac{2\pi}{u}} \exp\left(\frac{v^2}{2u}\right)$$

After some algebraic reduction, the final result for the density function of Z becomes

$$f_Z(z) = \frac{1}{\sqrt{2\pi(\sigma_X^2 + \sigma_Y^2)}} \exp\left[-\frac{1}{2}\left\{\frac{z - (\mu_X + \mu_Y)}{\sqrt{\sigma_X^2 + \sigma_Y^2}}\right\}^2\right]$$

which we recognize is also a normal density function with mean

$$\mu_Z = \mu_X + \mu_Y$$

and variance

$$\sigma_Z^2 = \sigma_X^2 + \sigma_Y^2$$

By the same procedure, it can be shown that $Z = X - Y$ is also Gaussian with mean $\mu_Z = \mu_X - \mu_Y$ and the same variance as above: $\sigma_Z^2 = \sigma_X^2 + \sigma_Y^2$. On the basis of these results, it can be shown inductively that if

$$Z = \sum_{i=1}^{n} a_i X_i$$

where a_i are constants, and X_i are statistically independent normal variates $N(\mu_{X_i}, \sigma_{X_i})$, then Z is also Gaussian with mean

$$\mu_Z = \sum_{i=1}^{n} a_i \mu_{X_i} \tag{4.17}$$

and variance

$$\sigma_Z^2 = \sum_{i=1}^{n} a_i^2 \sigma_{X_i}^2 \tag{4.18}$$

In other words, *any linear function* of normal variates is also a normal variate. The relationships of Eqs. 4.17 and 4.18, however, are not limited to normal variates. We shall observe later in Section 4.3.2 that these equations are, in fact, valid for linear functions of any statistically independent random variables regardless of their distributions.

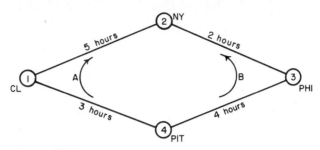

Figure E4.7

EXAMPLE 4.7

A trucking network links four cities, namely, (1) Cleveland, (2) New York, (3) Philadelphia, and (4) Pittsburgh. The expected travel time for each branch is indicated in Fig. E4.7, in hours. Assume that the travel times for each of the branches are independently Gaussian, with 20% coefficient of variation. Two trucks are dispatched at the same time from Pittsburgh to New York City, with truck A going via Cleveland and truck B via Philadelphia.

(a) What is the probability that truck A will arrive at the destination within 9 hr?

(b) What is the probability that truck A will arrive at the destination earlier than truck B?

Solution

(a) Let T_A be the total travel time for truck A. Hence

$$T_A = T_{41} + T_{12}$$

which is a sum of two independent normal random variables. The mean and variance of T_A are, respectively,

$$\mu_{T_A} = \mu_{41} + \mu_{12} = 3 + 5 = 8 \text{ hr}$$

and

$$\sigma_{T_A}^2 = \sigma_{41}^2 + \sigma_{12}^2 = (0.2 \times 3)^2 + (0.2 \times 5)^2$$
$$= 0.36 + 1 = 1.36 \text{ hr}^2$$

Therefore

$$P(T_A < 9) = \Phi\left(\frac{9-8}{\sqrt{1.36}}\right) = \Phi(0.858) = 0.805$$

(b) Let T_B be the total travel time for truck B. The event truck A will arrive at the destination earlier than truck B is $(T_A < T_B)$ or $(T_A - T_B < 0)$. It can be shown that T_B is normal with

$$\mu_{T_B} = \mu_{43} + \mu_{32} = 4 + 2 = 6 \text{ hr}$$
$$\sigma_{T_B}^2 = (0.2 \times 4)^2 + (0.2 \times 2)^2 = 0.64 + 0.16 = 0.8 \text{ hr}^2$$

If we let $Z = T_A - T_B$, Z is also normal with mean

$$\mu_Z = \mu_{T_A} - \mu_{T_B} = 8 - 6 = 2 \text{ hr}$$

and variance
$$\sigma_Z^2 = \sigma_{T_A}^2 + \sigma_{T_B}^2 = 1.36 + 0.8 = 2.16 \text{ hr}^2$$

Hence the required probability is

$$P(Z < 0) = \Phi\left(\frac{0 - 2}{\sqrt{2.16}}\right) = \Phi(-1.36) = 0.087$$

EXAMPLE 4.8

In considering the safety of a building, the total force acting on the columns of the building must be examined. This would include the effects of the dead load D (due to the weight of the structure), the live load L (due to human occupancy, movable furniture, and the like), and the wind load W.

Assume that the load effects on the individual columns are statistically independent Gaussian variates with

$$\mu_D = 4.2 \text{ kips} \qquad \sigma_D = 0.3 \text{ kips}$$
$$\mu_L = 6.5 \text{ kips} \qquad \sigma_L = 0.8 \text{ kips}$$
$$\mu_W = 3.4 \text{ kips} \qquad \sigma_W = 0.7 \text{ kips}$$

(a) Determine the mean and standard deviation of the total load acting on a column.

(b) If the strength of a column is also Gaussian with a mean equal to 1.5 times the total mean force, what is the probability of failure of the column? Assume that the coefficient of variation of the strength is 15% and that the strength and load effects are statistically independent.

Solution

(a) The combined load S is
$$S = D + L + W$$
which is also Gaussian with

$$\mu_S = \mu_D + \mu_L + \mu_W = 4.2 + 6.5 + 3.4 = 14.1 \text{ kip}$$
and
$$\sigma_S = \sqrt{\sigma_D^2 + \sigma_L^2 + \sigma_W^2} = \sqrt{(0.3)^2 + (0.8)^2 + (0.7)^2} = 1.1 \text{ kip}$$

(b) Failure of the column will occur when the strength R is less than the applied load S. Let X denote the difference $R - S$, namely,

$$X = R - S$$

Then $(X < 0)$ represents failure. Since R and S are independent Gaussian variates, X is also Gaussian with

$$\mu_X = \mu_R - \mu_S = 1.5 \times 14.1 - 14.1 = 7.05 \text{ kip}$$
$$\sigma_X = \sqrt{\sigma_R^2 + \sigma_S^2} = \sqrt{(\delta_R \mu_R)^2 + \sigma_S^2}$$
$$= \sqrt{(0.15 \times 1.5 \times 14.1)^2 + (1.1)^2}$$
$$= 3.36 \text{ kip}$$

Hence the probability of failure is

$$P(X < 0) = \Phi\left(\frac{0 - \mu_X}{\sigma_X}\right) = \Phi\left(\frac{-7.05}{3.36}\right) = \Phi(-2.1)$$
$$= 1 - 0.982 = 0.018$$

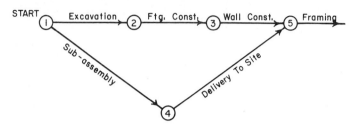

Figure E4.9   Construction activity network

EXAMPLE 4.9

The framing of a house may be done by subassembling the components in a plant and then delivering them to the site for framing. While this subassembly of components is being done, the preparation of the site, which includes the excavation through construction of the foundation walls, can proceed at the same time. These activities may be represented with the activity network shown in Fig. E4.9 and described in Table E4.9.

Table E4.9. Data of Example 4.9

| Activity | Description | Completion time (days) | |
		Mean	Std. dev.
1–2	Excavation	2	1
2–3	Construction of footings	1	$\frac{1}{2}$
3–5	Construction of foundation walls	3	1
1–4	Precutting and subassembly of components	5	1
4–5	Delivery of components to site	2	$\frac{1}{2}$

Assume that the completion time of each activity is a Gaussian random variable, with the respective means and standard deviations given in Table E4.9. Clearly, framing of the house cannot start until the foundation walls are completed and the components are delivered to the site. What is the probability that this will be at least 8 days after work started on the job? Completion times among the different activities may be assumed to be statistically independent.

Solution

Denote the durations of the activities listed above as X_1, X_2, X_3, X_4, and X_5, respectively. Let T_1 be the total time required to excavate and construct the footings and foundation walls, and T_2 be the corresponding time for assembly and delivery of the subcomponents. Then

$$T_1 = X_1 + X_2 + X_3$$
$$T_2 = X_4 + X_5$$

The required probability is

$$p = P(T_1 \geq 8 \cup T_2 \geq 8) = P(T_1 \geq 8) + P(T_2 \geq 8) - P(T_1 \geq 8) \cdot P(T_2 \geq 8)$$

According to Eqs. 4.17 and 4.18,

$$\mu_{T_1} = 2 + 1 + 3 = 6 \text{ days}$$
$$\sigma_{T_1} = \sqrt{1 + \tfrac{1}{4} + 1} = 1.5 \text{ days}$$

and

$$\mu_{T_2} = 5 + 2 = 7 \text{ days}$$
$$\sigma_{T_2} = \sqrt{1 + \tfrac{1}{4}} = 1.11 \text{ days}$$

Since T_1 and T_2 are also Gaussian, we have

$$P(T_1 \geq 8) = 1 - \Phi\left(\frac{8 - 6}{1.5}\right) = 1 - \Phi(1.33) = 0.0918$$

$$P(T_2 \geq 8) = 1 - \Phi\left(\frac{8 - 7}{1.11}\right) = 1 - \Phi(0.90) = 0.1841$$

Thence, the required probability is

$$p = 0.0918 + 0.1841 - (0.0918 \times 0.1841) = 0.26$$

Alternatively, the probability may be calculated by observing that

$$p = P(T_1 \geq 8 \cup T_2 \geq 8) = 1 - P(T_1 < 8) \cdot P(T_2 < 8)$$
$$= 1 - (0.9082)(0.8159) = 0.26$$

Products and quoteints of random variables. For the product of two random variables, say $Z = XY$, we have

$$X = \frac{Z}{Y}$$

$$\frac{dx}{dz} = \frac{1}{y}$$

Then Eq. 4.15 yields

$$f_Z(z) = \int_{-\infty}^{\infty} \left| \frac{1}{y} \right| f_{X,Y}\left(\frac{z}{y}, y\right) dy \qquad (4.19)$$

Similarly, for the quotient of two random variables, for example $Z = X/Y$, the density function of Z would be

$$f_Z(z) = \int_{-\infty}^{\infty} |y| f_{X,Y}(zy, y) \, dy \qquad (4.20)$$

In this regard, we observe that by virtue of the result for sums (and differences) of normal variates, it follows that the product and quotient of statistically independent log-normal variates is also a log-normal variate. Suppose

$$Z = \prod_{i=1}^{n} X_i$$

where the X_i's are statistically independent log-normal random variables with respective parameters λ_{X_i} and ζ_{X_i}. Then

$$\ln Z = \sum_{i=1}^{n} \ln X_i$$

Since each $\ln X_i$ is normal (see Example 4.2), it follows that $\ln Z$ is also normal with mean and variance, according to Eqs. 4.17 and 4.18, as follows:

$$\lambda_Z = E(\ln Z) = \sum_{i=1}^{n} \lambda_{X_i}$$

$$\zeta_X^2 = \text{Var}(\ln Z) = \sum_{i=1}^{n} \zeta_{X_i}^2$$

Hence Z is log-normal with the above parameters λ_Z and ζ_Z.

EXAMPLE 4.10

The settlement of a footing on sand may be estimated on the basis of the theory of elasticity as follows:

$$S = \frac{PBI}{M}$$

where

S = footing settlement, in feet

P = average applied bearing pressure in tons per square foot (tsf)

B = smallest footing dimension, in feet

I = influence factor dependent on footing geometry, depth of embedment, and depth to hard stratum

M = modulus of compressibility

Assume that P, B, I, and M are independent log-normal variates with parameters λ_P, λ_B, λ_I, λ_M and ζ_P, ζ_B, ζ_I, ζ_M, respectively. The following values are given for the design of a particular footing.

	Mean	Coefficient of variation
P (tsf)	1.0	0.10
B (ft)	6.0	0
I	0.6	0.10
M (tsf)	32.0	0.15

(a) Determine the mean settlement of the footing and its coefficient of variation.

(b) If the maximum allowable settlement is 2.5 in., what is the reliability against excessive settlement; that is, probability of no excessive settlement?

(c) If the variability in M can be decreased by investing for better information, say reducing the coefficient of variation to 5% at an expense of $100, would you spend this money? Assume that the exceedance of the maximum allowable settlement would involve a damage cost of $50,000.

Solution

(a)

$$\zeta_S{}^2 = (0.1)^2 + 0 + (0.1)^2 + (0.15)^2$$
$$= 0.01 + 0.01 + 0.0225 = 0.0425$$
$$\lambda_P = \ln(1.0) - \tfrac{1}{2}(0.1)^2 = 0 - 0.005 = -0.005$$
$$\lambda_B = \ln 6 = 1.792$$
$$\lambda_I = \ln(0.6) - \tfrac{1}{2}(0.1)^2 = -0.511 - 0.005 = -0.516$$
$$\lambda_M = \ln(32) - \frac{0.0225}{2} = 3.66 - 0.011 = 3.455$$

Hence

$$\lambda_S = -0.005 + 1.792 - 0.516 - 3.455 = -2.184$$

The mean settlement, therefore, is

$$\mu_S = \exp(\lambda_S + \tfrac{1}{2}\zeta_S{}^2) = \exp(-2.184 + 0.0212)$$
$$= \exp(-2.16) = 0.115 \text{ ft}$$

and the corresponding coefficient of variation is

$$\delta_S \simeq \zeta_S = \sqrt{0.0425} = 0.21$$

(b) Reliability $= P(S < 2.5 \text{ in.})$

$$= \Phi\left(\frac{\ln(2.5/12) - (-2.184)}{0.206}\right)$$
$$= \Phi(2.99) = 0.9986$$

(c) We have to first determine the reliability of the design if $\delta_M = 0.05$. In this case,

$$\lambda_M = 3.466 - \frac{0.0025}{2} = 3.65$$

and

$$\lambda_S = -2.194$$
$$\zeta_S{}^2 = 0.01 + 0.01 + 0.0025 = 0.0225$$
$$\zeta_S = 0.15$$

Hence the reliability $= \Phi\left(\dfrac{\ln(2.5/12) + 2.194}{0.15}\right)$

$$= \Phi\left(\frac{-1.572 + 2.194}{0.15}\right)$$
$$= \Phi(4.15) = 0.99998$$

Assume that the criterion for decision is based on minimizing the expected cost. The expected cost of the first design is

$$E(C_1) = C_0 + (1 - \text{reliability})(\text{cost of failure})$$
$$= C_0 + (1 - 0.9986)\,50{,}000$$
$$= C_0 + 70$$

where C_0 is the fixed initial cost of construction. Similarly, the expected cost of the second design is

$$E(C_2) = C_0 + 100 + (1 - 0.99998)\,50{,}000$$
$$= C_0 + 101$$

Therefore, on the basis of the expected costs, the decision would be that money should not be spent to gather more information on M.

EXAMPLE 4.11

A 15-ft-long 4-by-12-in. prismatic cantilever wood beam is carrying a uniformly distributed load w (see Fig. E4.11), with a mean load intensity of $\bar{w} = 180$ lb/ft and a COV of $\delta_w = 15\%$. The material is structural-grade California redwood with a rated average yield strength (parallel to grain under bending) of $\bar{\sigma}_y = 4000$ psi and COV $\delta_{\sigma_y} = 20\%$.

Prescribe log-normal distributions for w and σ_y.

(a) Determine the probability that the maximum extreme fiber stress in the beam will exceed the tensile yield strength of the wood.

The bending moment at any section of the beam is

$$M = \frac{wL^2}{2}$$

Since w is a log-normal variate, M will also be log-normal (see Example 4.2).

For a rectangular cross-section, the extreme fiber stress at any section of the beam may be given by

$$\sigma = \frac{6M}{bh^2}$$

where b and h are the width and depth, respectively, of the rectangular beam. It follows, therefore, that σ is also a log-normal variate.

In the present case, the maximum bending moment occurs at the support, with a

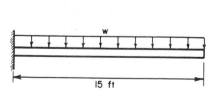
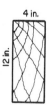

Figure E4.11

mean value of
$$\bar{M} = \tfrac{1}{2} \times 180 \times 15^2 \times 12 = 243{,}000 \text{ in.-lb}$$

Hence the mean maximum extreme fiber stress in the beam is
$$\bar{\sigma} = \frac{6 \times 243{,}000}{4(12)^2} = 2531 \text{ psi}$$

and COV
$$\delta_\sigma = 15\%$$

The required probability is
$$p_E = P(\sigma > \sigma_y) = P\left(\frac{\sigma}{\sigma_y} > 1.0\right) = P\left(\ln \frac{\sigma}{\sigma_y} > 0\right)$$

and since σ and σ_y are log-normal variates, $\ln(\sigma/\sigma_y)$ is Gaussian with mean
$$\lambda = (\ln \bar{\sigma} - \tfrac{1}{2}\delta_\sigma^2) - (\ln \bar{\sigma}_y - \tfrac{1}{2}\delta_{\sigma_y}^2)$$
$$= \ln \frac{\bar{\sigma}}{\bar{\sigma}_y} - \frac{1}{2}(\delta_\sigma^2 - \delta_{\sigma_y}^2)$$
$$= \ln \frac{2531}{4000} - \frac{1}{2}[(0.15)^2 - (0.20)^2] = -0.45$$

and standard deviation
$$\zeta \simeq \sqrt{\delta_\sigma^2 + \delta_{\sigma_y}^2} = \sqrt{(0.15)^2 + (0.20)^2} = 0.25$$

Thence,
$$p_e = 1 - \Phi\left(\frac{0 + 0.45}{0.25}\right) = 1 - \Phi(1.80)$$
$$= 0.036$$

(b) Suppose that in order to ensure an adequate level of safety, the probability of overstressing the wood beyond its yield strength is not permitted to exceed 0.001. Redesign the beam section, keeping the beam width the same at 4 in. (that is, determine h).

The limiting condition is
$$P\left(\ln \frac{\sigma}{\sigma_y} > 0\right) = 0.001$$

or
$$P\left(\ln \frac{6M}{bh^2 \sigma_y} > 0\right) = 0.001$$

But
$$\ln \frac{6M}{bh^2 \sigma_y}$$

is also Gaussian with
$$\lambda = \ln \frac{6(243{,}000)}{4(4000)h^2} - \frac{1}{2}[(0.15)^2 - (0.20)^2]$$
$$= \ln \frac{91.13}{h^2} - 0.01$$

and
$$\zeta \simeq 0.25$$

Therefore the required limiting condition becomes

$$1 - \Phi\left[\frac{0 - \ln(91.13/h^2) + 0.01}{0.25}\right] = 0.001$$

from which we obtain

$$h^2 = 91.13 \exp[0.25\,\Phi^{-1}(0.999) - 0.01]$$
$$= 91.13 \exp(0.25 \times 3.09 - 0.01)$$
$$= 195.35$$

Thus

$$h = 13.98 \text{ in.}$$

(c) *Determination of allowable design stress.* We observe that the beam size may be determined (or designed) using a mean allowable stress $\bar{\mathfrak{s}}_a$, as follows.

If we limit the maximum load-induced stress to a specified allowable stress $\mathfrak{s}_a$, we should have

$$\frac{6M}{bh^2} \leq \mathfrak{s}_a$$

Therefore, for a given width b, the minimum required depth h is

$$h^2 = \frac{6M}{b\mathfrak{s}_a}$$

Clearly the beam would be of larger cross-section and thus safer if we use lower values of $\mathfrak{s}_a$. However, if $\mathfrak{s}_a$ is too low, the design would be unnecessarily conservative and thus wasteful. In order to ensure an adequate level of safety without being overly conservative, the allowable stress $\mathfrak{s}_a$ may be determined on the basis of a specified tolerable probability p_E. For this purpose, we observe that, for small $\delta_{\mathfrak{s}}$ and $\delta_{\mathfrak{s}_y}$,

$$\lambda \simeq \ln\frac{\bar{\mathfrak{s}}}{\bar{\mathfrak{s}}_y}$$

and thus

$$p_E = 1 - \Phi\left(\frac{-\lambda}{\zeta}\right)$$
$$= 1 - \Phi\left(\frac{-\ln(\bar{\mathfrak{s}}/\bar{\mathfrak{s}}_y)}{\sqrt{\delta_{\mathfrak{s}}^2 + \delta_{\mathfrak{s}_y}^2}}\right)$$

from which

$$\ln\frac{\bar{\mathfrak{s}}_y}{\bar{\mathfrak{s}}} = \Phi^{-1}(1 - p_E)\sqrt{\delta_{\mathfrak{s}}^2 + \delta_{\mathfrak{s}_y}^2}$$

Thus, limiting the applied stress to $\bar{\mathfrak{s}} = \bar{\mathfrak{s}}_a$, we have

$$\frac{\bar{\mathfrak{s}}_y}{\bar{\mathfrak{s}}_a} = \exp[\Phi^{-1}(1 - p_E)\sqrt{\delta_{\mathfrak{s}}^2 + \delta_{\mathfrak{s}_y}^2}]$$

The ratio $\bar{\mathfrak{s}}_y/\bar{\mathfrak{s}}_a$ is called the mean "safety factor"; denoting this as γ_0, we obtain the mean allowable stress

$$\bar{\mathfrak{s}}_a = \frac{\bar{\mathfrak{s}}_y}{\gamma_0}$$

where

$$\gamma_0 = \exp\left[\Phi^{-1}(1 - p_E)\sqrt{\delta_{\tilde{\sigma}}^2 + \delta_{\sigma_y}^2}\right]$$

Applying this to the problem in (b) above, where $p_E = 0.001$, we obtain

$$\gamma_0 = \exp\left[\Phi^{-1}(0.999)(0.25)\right] = e^{0.25 \times 3.09} = 2.16$$

Thus

$$\bar{\sigma}_a = \frac{4000}{2.16} = 1852 \text{ psi}$$

Hence, for $b = 4$ in.,

$$h^2 = \frac{6\bar{M}}{b\bar{\sigma}_a} = \frac{6(243,000)}{4(1852)} = 196.81$$

and

$$h = 14.03 \text{ in.}$$

(this value of h is not exactly the same as that of part b because of the approximation introduced above for λ)

The central limit theorem. One of the most significant theorems in probability theory is that pertaining to the limiting distribution of a sum of random variables known as the *central limit theorem*. Stated loosely, the theorem says that the sum of a large number of individual random components, none of which is dominant, tends to the normal distribution as the number of components (regardless of their initial distributions) increases without limit. Therefore, if a physical process is the result of the totality of a large number of individual effects, then according to the central limit theorem the process would tend to be Gaussian; that is, the sum of the individual effects would tend to have a Gaussian distribution.

The proof of the central limit theorem is beyond our scope of interest; however, the essence of the proof may be demonstrated with the following example. Suppose that

$$S = \frac{1}{\sqrt{n}} \sum_{i=1}^{n} X_i$$

where the X_i's are statistically independent and identically distributed with PMF,

$$P(X_i = 1) = \tfrac{1}{2}$$

$$P(X_i = -1) = \tfrac{1}{2}$$

and $P(X_i = x) = 0$ otherwise. The factor $1/\sqrt{n}$ is necessary to retain a finite variance for S as $n \to \infty$.

Then the probability distribution of S, for increasing values of n ($n = 2$, 5, 10, 20), would be as shown in Fig. 4.1 (with the probabilities at specified values of S spread over the appropriate intervals).

By virtue of the central limit theorem, the product of a large number of independent factors (none of which dominates the product) will tend to the

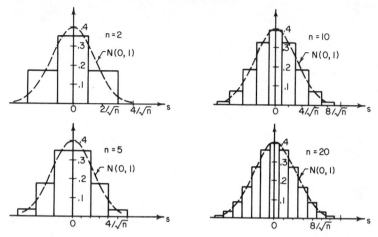

Figure 4.1   Demonstration of the central limit theorem. (Area of each rectangle is probability of centered value)

log-normal distribution. That is, regardless of the distributions of X_i, the product

$$P = c \prod_{i=1}^{n} X_i$$

will approach a log-normal distribution as $n \to \infty$.

Generalization   The method described above for a function of two variables can be generalized to derive the distribution of a function of n random variables. Briefly, if

$$Z = g(X_1, X_2, \cdots, X_n) \tag{4.21}$$

then, generalizing Eq. 4.14, we have

$$F_Z(z) = \int \cdots \int_{\{g(x_1,\ldots,x_n) \leq z\}} f_{X_1,\ldots,X_n}(x_1, \cdots, x_n) \, dx_1 \cdots dx_n$$

$$= \int_{-\infty}^{\infty} \cdots \int_{-\infty}^{\infty} \int_{-\infty}^{g^{-1}} f_{X_1,\ldots,X_n}(x_1, \cdots, x_n) \, dx_1 \cdots dx_n \tag{4.22}$$

where $g^{-1} = g^{-1}(z, x_2, \cdots, x_n)$. Changing the variable of integration from x_1 to z, we have

$$F_Z(z) = \int_{-\infty}^{\infty} \cdots \int_{-\infty}^{\infty} \int_{-\infty}^{z} f_{X_1,\ldots,X_n}(g^{-1}, x_2, \cdots, x_n) \cdot$$

$$\cdot \left| \frac{\partial g^{-1}}{\partial z} \right| dz \, dx_2 \cdots dx_n$$

Therefore

$$f_Z(z) = \int_{-\infty}^{\infty} \cdots \int_{-\infty}^{\infty} f_{X_1,\ldots,X_n}(g^{-1}, x_2, \cdots, x_n) \left| \frac{\partial g^{-1}}{\partial z} \right| dx_2 \cdots dx_n \quad (4.23)$$

4.3. MOMENTS OF FUNCTIONS OF RANDOM VARIABLES

4.3.1. Introduction

According to Section 4.2, the probability distribution of a function of random variables can, theoretically, be derived from the probability distributions of the basic random variables; however, such derivations are generally difficult, especially when the function is nonlinear. In such circumstances, the moments—particularly the mean and variance—of the function may be the only practically obtainable information. In many instances, this may be sufficient for practical purposes even if the correct probability distribution must be left undetermined. Such moments are functionally related to the moments of the individual basic variates, and therefore may be derived as functions of the moments of the basic variates.

Mathematical expectation.   The mathematical expectation of a function of several random variables can be obtained as a generalization of Eq. 3.8; thus, for a function of n variables, $Z = g(X_1, X_2, \cdots, X_n)$, its mathematical expectation is

$$E(Z) = E[g(X_1, X_2, \cdots, X_n)]$$

$$= \int_{-\infty}^{\infty} \cdots \int_{-\infty}^{\infty} g(x_1, x_2, \cdots, x_n) \, f_{X_1, X_2, \ldots, X_n}(x_1, \cdots, x_n) \cdot$$

$$\cdot \, dx_1 \, dx_2 \cdots dx_n \quad (4.24)$$

In the following, we shall use Eq. 4.24 as the basis for deriving the (first and second) moments of linear functions of random variables; the results will be the basis for the first-order approximate moments of nonlinear functions.

4.3.2.   Mean and variance of a linear function

Consider the moments of linear functions. First of all, suppose that

$$Y = aX + b$$

where a and b are constants. Then according to Eq. 3.8, the mean value of Y is the mathematical expectation of $aX + b$, or

$$E(Y) = E(aX + b) = \int_{-\infty}^{\infty} (ax + b) f_X(x) \, dx$$

$$= a \int_{-\infty}^{\infty} x f_X(x) \, dx + b \int_{-\infty}^{\infty} f_X(x) \, dx$$

$$= aE(X) + b \tag{4.25}$$

whereas the variance of Y is

$$\text{Var}(Y) = E[(Y - \mu_Y)^2]$$

$$= E[(aX + b - a\mu_X - b)^2]$$

$$= a^2 \int_{-\infty}^{\infty} (x - \mu_X)^2 f_X(x) \, dx$$

$$= a^2 \, \text{Var}(X) \tag{4.26}$$

Furthermore, if $Y = a_1 X_1 + a_2 X_2$, where a_1 and a_2 are constants, then according to Eq. 4.24,

$$E(Y) = \int_{-\infty}^{\infty} \int_{-\infty}^{\infty} (a_1 x_1 + a_2 x_2) f_{X_1, X_2}(x_1, x_2) \, dx_1 \, dx_2$$

$$= a_1 \int_{-\infty}^{\infty} x_1 f_{X_1}(x_1) \, dx_1 + a_2 \int_{-\infty}^{\infty} x_2 f_{X_2}(x_2) \, dx_2$$

The integrals are, respectively, $E(X_1)$ and $E(X_2)$; hence

$$E(Y) = a_1 E(X_1) + a_2 E(X_2) \tag{4.27}$$

That is, the *expected value of a sum is the sum of the expected values*. The corresponding variance is

$$\text{Var}(Y) = E[(a_1 X_1 + a_2 X_2) - (a_1 \mu_{X_1} + a_2 \mu_{X_2})]^2$$
$$= E[a_1(X_1 - \mu_{X_1}) + a_2 (X_2 - \mu_{X_2})]^2$$
$$= E[a_1^2 (X_1 - \mu_{X_1})^2 + 2 a_1 a_2 (X_1 - \mu_{X_1})(X_2 - \mu_{X_2})$$
$$+ a_2^2 (X_2 - \mu_{X_2})^2]$$

Recognizing that the expected values of the first and third terms are variances, whereas that of the middle term is a covariance, Eq. 3.72, we obtain

$$\text{Var}(Y) = a_1^2 \, \text{Var}(X_1) + a_2^2 \, \text{Var}(X_2) + 2a_1 a_2 \, \text{Cov}(X_1, X_2) \tag{4.28}$$

If $Y = a_1X_1 - a_2X_2$, the results would be

$$E(Y) = a_1 E(X_1) - a_2 E(X_2) \qquad (4.29)$$

and

$$\text{Var}(Y) = a_1{}^2 \text{Var}(X_1) + a_2 \text{Var}(X_2) - 2 a_1a_2 \text{Cov}(X_1, X_2) \qquad (4.30)$$

If X_1 and X_2 are statistically independent, $\text{Cov}(X_1, X_2) = 0$ and Eqs. 4.28 and 4.30 reduce to

$$\text{Var}(Y) = a_1{}^2 \text{Var}(X_1) + a_2{}^2 \text{Var}(X_2) \qquad (4.31)$$

More generally, if

$$Y = \sum_{i=1}^{n} a_iX_i$$

where a_i are constants, we have, on extending the results of Eqs. 4.27 through 4.30,

$$E(Y) = \sum_{i=1}^{n} a_i E(X_i) \qquad (4.32)$$

and

$$\text{Var}(Y) = \sum_{i=1}^{n} a_i{}^2 \text{Var}(X_i) + \sum_{i \neq j}^{n}\sum^{n} a_ia_j \text{Cov}(X_i, X_j) \qquad (4.33)$$

$$= \sum_{i=1}^{n} a_i{}^2\sigma_{X_i}^2 + \sum_{i \neq j}^{n}\sum^{n} a_ia_j\rho_{ij}\sigma_{X_i}\sigma_{X_j} \qquad (4.33a)$$

where ρ_{ij} is the *correlation coefficient* between X_i and X_j. Moreover, if Z is another linear function of the X_i's, that is,

$$Z = \sum_{i=1}^{n} b_iX_i$$

then the covariance between Y and Z can be shown to be

$$\text{Cov}(Y, Z) = \sum_{i=1}^{n} a_ib_i \text{Var}(X_i) + \sum_{i \neq j}^{n}\sum^{n} a_ib_j \text{Cov}(X_i, X_j) \qquad (4.34)$$

$$= \sum_{i=1}^{n} a_ib_i\sigma_{X_i}^2 + \sum_{i \neq j}^{n}\sum^{n} a_ib_j\rho_{ij}\sigma_{X_i}\sigma_{X_j} \qquad (4.34a)$$

EXAMPLE 4.12

The lengths of two rods will be determined by two measurements with an unbiased instrument that makes random error with mean 0 and standard deviation σ in each measurement. Compute the variance in the estimation of the lengths T_1 and T_2 by the following methods:

(a) The two rods are measured separately.

(b) The sum and difference of the lengths of the two rods are measured instead of the individual lengths.

(a) Let M_1 and M_2 denote the measurements obtained for the two rods, then

$$T_1 = M_1 + \varepsilon_1$$

and

$$T_2 = M_2 + \varepsilon_2$$

where ε_1 and ε_2 are the errors involved in the measurements. Then the variance in the estimation of T_1 is

$$\text{Var}\,(T_1) = \text{Var}\,(M_1 + \varepsilon_1) = \text{Var}\,(M_1) + \text{Var}\,(\varepsilon_1) = 0 + \sigma^2 = \sigma^2$$

Similarly, $\text{Var}\,(T_2) = \text{Var}\,(M_2 + \varepsilon_2) = \sigma^2$.

(b) Let M_3 denote the measured combined length of the two rods, and M_4 denote the measured difference between the lengths of the two rods; then

$$T_1 + T_2 = M_3 + \varepsilon_3$$

and

$$T_1 - T_2 = M_4 + \varepsilon_4$$

Solving these two equations simultaneously, we have

$$T_1 = \frac{M_3 + M_4}{2} + \frac{\varepsilon_3 + \varepsilon_4}{2}$$

and

$$T_2 = \frac{M_3 - M_4}{2} + \frac{\varepsilon_3 - \varepsilon_4}{2}$$

Assuming that the errors ε_3 and ε_4 are statistically independent, the variance in the estimation of T_1 is, therefore,

$$\text{Var}\,(T_1) = \text{Var}\left(\frac{M_3 + M_4}{2} + \frac{\varepsilon_3 + \varepsilon_4}{2}\right)$$

$$= \text{Var}\left(\frac{M_3 + M_4}{2}\right) + \text{Var}\left(\frac{\varepsilon_3 + \varepsilon_4}{2}\right)$$

$$= \frac{1}{4}\,[\text{Var}\,(\varepsilon_3) + \text{Var}\,(\varepsilon_4)]$$

$$= \frac{1}{4} \times 2\sigma^2 = \frac{\sigma^2}{2}$$

Similarly,

$$\text{Var}\,(T_2) = \text{Var}\left(\frac{M_3 - M_4}{2}\right) + \text{Var}\left(\frac{\varepsilon_3 - \varepsilon_4}{2}\right)$$

$$= \frac{1}{4}\,[\text{Var}\,(\varepsilon_3) + \text{Var}\,(\varepsilon_4)]$$

$$= \frac{\sigma^2}{2}$$

Therefore we see that the second method of measuring the lengths of the two rods is better, since the variances in the estimation of the true lengths T_1 and T_2 are smaller.

EXAMPLE 4.13

The total vertical load on the ground-floor columns of an n-story building would be the sum of the individual contributions from each of the n floors; thus

$$Y = \sum_{i=1}^{n} X_i$$

where X_i is the load on the column from the ith floor. Assuming the mean and variance to be the same for all floors (this appears to be the case from actual load surveys [Mitchell and Woodgate, 1970]), the mean load on the column is

$$\mu_Y = n\mu_X$$

and the variance, from Eq. 4.33a, is

$$\mathrm{Var}(Y) = n\,\mathrm{Var}(X) + \mathrm{Var}(X) \sum_{\substack{i \neq j}}^{n} \sum^{n} \rho_{ij}$$

where ρ_{ij} is the correlation between the loads on the ith and jth floors.

(a) If the loads on any two floors are assumed to be statistically independent, that is, $\rho_{ij} = 0$, then $\mathrm{Var}(Y) = n\,\mathrm{Var}(X)$, and the standard deviation would be

$$\sigma_Y = \sqrt{n}\,\sigma_X$$

and COV

$$\delta_Y = \frac{\sqrt{n}\,\sigma_X}{n\,\mu_X} = \frac{\delta_X}{\sqrt{n}}$$

The design load is usually specified to be on the high side; suppose that this is taken at k standard deviations above the mean. The design load for n floors then is

$$Y^* = \mu_Y + k\sigma_Y$$

$$= \mu_Y\left(1 + k\frac{\delta_X}{\sqrt{n}}\right)$$

$$= n\left(1 + k\frac{\delta_X}{\sqrt{n}}\right)\mu_X$$

This means that the total design load on the ground-floor column increases with the number of stories; however, on the average, the load from each floor is

$$\frac{Y^*}{n} = \left(1 + k\frac{\delta_X}{\sqrt{n}}\right)\mu_X$$

which means that the contribution from the individual floors decreases with the number of floors in the building.

The reduction factor that specifies the load contribution from each floor, which can be defined as $r = Y^*/nX^*$, therefore, becomes

$$r = \frac{n\mu_X\left(1 + k\dfrac{\delta_X}{\sqrt{n}}\right)}{n\mu_X(1 + k\delta_X)} = \left(\frac{1 + k\dfrac{\delta_X}{\sqrt{n}}}{1 + k\delta_X}\right)$$

(b) However, if the correlation between any two floors is the same and positive (that is, $\rho_{ij} = \rho$, a positive constant for any i and j), then the variance of Y becomes

$$\text{Var}(Y) = n\,\text{Var}(X) + \text{Var}(X)[n(n-1)\rho]$$
$$= \text{Var}(X)\,[n + n(n-1)\rho]$$

and

$$\delta_Y = \delta_X\sqrt{\frac{1 + (n-1)\rho}{n}}$$

In this case, the reduction factor would be

$$r = \frac{Y^*}{nX^*} = \frac{1 + k\delta_X\sqrt{\dfrac{1 + (n-1)\rho}{n}}}{1 + k\delta_X}$$

which is larger than the corresponding factor obtained earlier assuming statistical independence. Any correlation between the loads on different floors may be expected to be positive; on this basis, therefore, the assumption of statistical independence would yield results on the unsafe side.

4.3.3.   Product of independent variates

If n random variables $X_1, X_2, \ldots, X_n$ are statistically independent, the mean value of their product

$$Z = X_1 X_2 \ldots X_n$$

is

$$E(Z) = \int_{-\infty}^{\infty} \cdots \int_{-\infty}^{\infty} x_1 \cdots x_n f_{X_1}(x_1) \cdots f_{X_n}(x_n)\,dx_1 \cdots dx_n$$

$$= \int_{-\infty}^{\infty} x_1 f_{X_1}(x_1)\,dx_1 \int_{-\infty}^{\infty} x_2 f_{X_2}(x_2)\,dx_2 \cdots \int_{-\infty}^{\infty} x_n f_{X_n}(x_n)\,dx_n$$

$$= E(X_1)\,E(X_2) \cdots E(X_n)$$

Therefore

$$\mu_Z = \mu_{X_1}\mu_{X_2}\ldots\mu_{X_n} \tag{4.35}$$

Similarly, we can show that

$$E(Z^2) = E(X_1^2)\,E(X_2^2) \cdots E(X_n^2)$$

Hence the variance of the product of independent variates is

$$\sigma_Z^2 = E(X_1^2)\,E(X_2^2) \cdots E(X_n^2) - (\mu_{X_1}\mu_{X_2}\ldots\mu_{X_n})^2 \tag{4.36}$$

4.3.4.   Mean and variance of a general function

For a general function of a random variable X, that is

$$Y = g(X)$$

the exact moments of Y may be obtained as the mathematical expectation of $g(X)$; according to Eq. 3.8, the mean and variance would be

$$E(Y) = \int_{-\infty}^{\infty} g(x)\, f_X(x)\, dx$$

and

$$\text{Var}(Y) = \int_{-\infty}^{\infty} [g(x) - E(Y)]^2 f_X(x)\, dx$$

Obviously, to obtain the mean and variance of the function Y with the above relations, information on $f_X(x)$ is needed. In many applications, however, the density function $f_X(x)$ may not be known; information may be limited to the mean and variance of the original variate X. Furthermore, even when $f_X(x)$ is known, the integrations indicated above may be difficult to perform. For these reasons, approximate mean and variance of the function Y would be practically useful and may be obtained as follows.

Expand $g(X)$ in a Taylor series about the mean value μ_X; thus

$$Y = g(\mu_X) + (X - \mu_X)\frac{dg}{dX} + \frac{1}{2}(X - \mu_X)^2 \frac{d^2g}{dX^2} + \cdots \quad (4.37)$$

where the derivatives are evaluated at μ_X.

If the series is truncated at the linear terms, we obtain the *first-order* approximate mean and variance of Y:

$$E(Y) \simeq g(\mu_X) \quad (4.38)$$

and

$$\text{Var}(Y) \simeq \text{Var}(X - \mu_X)\left(\frac{dg}{dX}\right)^2$$

$$\simeq \text{Var}(X)\left(\frac{dg}{dX}\right)^2 \quad (4.39)$$

We can observe that if the function $g(X)$ is approximately linear for the entire range of values of X, Eqs. 4.38 and 4.39 should yield good approximations of the exact moments (Hald, 1952). Moreover, when the variance of X is small relative to $g(\mu_X)$, the above approximations should be adequate for many practical purposes even when the function is nonlinear.

The above first-order approximations may be successively improved by including the higher-order terms in the Taylor series; for example, if the second-order term in Eq. 4.37 is included, the *second-order* approximations are accordingly

$$E(Y) \simeq g(\mu_X) + \tfrac{1}{2}\text{Var}(X)\frac{d^2g}{dX^2} \quad (4.40)$$

and

$$\text{Var}(Y) \simeq \text{Var}(X) \left(\frac{dg}{dX}\right)^2 - \frac{1}{4}\text{Var}^2(X)\left(\frac{d^2g}{dX^2}\right)^2$$

$$+ E(X - \mu_X)^3 \frac{dg}{dX}\frac{d^2g}{dX^2} + \frac{1}{4}E(X-\mu_X)^4\left(\frac{d^2g}{dX^2}\right)^2 \quad (4.41)$$

Such improvements, of course, would involve the higher moments of the original variate in the evaluation of $\text{Var}(Y)$, as shown in Eq. 4.41, in which the third and fourth central moments of X are involved.

For practical purposes, of course, we may use Eq. 4.40 for $E(Y)$ and Eq. 4.39 for $\text{Var}(Y)$; in this way, we can take advantage of an improved mean value for Y, without involving more than the mean and variance of X.

EXAMPLE 4.14

The maximum impact pressures of ocean waves on coastal structures may be determined by

$$p_{\max} = 2.7\frac{\rho K U^2}{D}$$

where ρ = density of water; K = length of hypothetical piston; D = thickness of air cushion; and U = horizontal velocity of the advancing wave.

Suppose that the mean crest velocity is 4.5 ft/sec with a COV of 20%. The density of sea water is about 1.96 slugs/cu ft, and the ratio $K/D = 35$. Determine the mean and standard deviation of the peak impact pressure.

According to Eqs. 4.38 and 4.39, we obtain

$$E(p_{\max}) \simeq 2.7(1.96)(35)(4.5)^2 = 3750.70 \text{ psf}$$
$$= 26.05 \text{ psi}$$

and

$$\text{Var}(p_{\max}) \simeq \left(2.7\rho\frac{K}{D}\right)^2(2\bar{U})^2 \text{Var}(U)$$
$$= (2.7 \times 1.96 \times 35 \times 2 \times 4.5)^2(0.20 \times 4.5)^2$$

Thus

$$\sigma_{p_{\max}} = 2.7 \times 1.96 \times 35 \times 2 \times 0.2 \times (4.5)^2$$
$$= 1500.3 \text{ psf} = 10.42 \text{ psi}$$

Thus the COV of the maximum wave pressure is 0.40, which is twice that of the wave velocity.

If Y is a function of several random variables, that is

$$Y = g(X_1, X_2, \ldots, X_n)$$

we obtain the approximate mean and variance of Y similarly as follows. Expand the function $g(X_1, X_2, \ldots, X_n)$ in a Taylor series about the mean values $\mu_{X_1}, \mu_{X_2}, \ldots, \mu_{X_n}$; in this case, we have

$$Y = g(\mu_{X_1}, \mu_{X_2}, \ldots, \mu_{X_n}) + \sum_{i=1}^{n} (X_i - \mu_{X_i}) \frac{\partial g}{\partial X_i}$$

$$+ \frac{1}{2} \sum_{i=1}^{n} \sum_{j=1}^{n} (X_i - \mu_{X_i})(X_j - \mu_{X_j}) \frac{\partial^2 g}{\partial X_i \partial X_j} + \cdots \qquad (4.42)$$

where the derivatives are evaluated at $\mu_{X_1}, \mu_{X_2}, \ldots, \mu_{X_n}$.

Truncating the series at the linear terms, and by virtue of Eqs. 4.32 and 4.33, we obtain the first-order approximate mean and variance of Y as follows:

$$E(Y) \simeq g(\mu_{X_1}, \mu_{X_2}, \ldots, \mu_{X_n}) \qquad (4.43)$$

which says that the mean of the function is equal (approximately) to the function of the means; and

$$\mathrm{Var}(Y) \simeq \sum_{i=1}^{n} c_i^2 \, \mathrm{Var}(X_i) + \sum_{i \neq j}^{n} \sum^{n} c_i c_j \, \mathrm{Cov}(X_i, X_j) \qquad (4.44)$$

where c_i and c_j are the values of the partial derivatives $\partial g/\partial X_i$ and $\partial g/\partial X_j$, respectively, evaluated at $\mu_{X_1}, \mu_{X_2}, \ldots, \mu_{X_n}$. Observe that if X_i and X_j are uncorrelated (or statistically independent) for all i and j, then Eq. 4.44 reduces to

$$\mathrm{Var}(Y) \simeq \sum_{i=1}^{n} c_i^2 \, \mathrm{Var}(X_i) \qquad (4.44a)$$

Again, the conditions for the applicability of the above approximations are the same as those stated earlier for the single-variable case. The above approximate mean and variance may also be improved by including the higher-order terms of the Taylor series expansion of $g(X_1, X_2, \ldots, X_n)$. In particular, from Eq. 4.42, the second-order approximate mean of Y would be

$$E(Y) \simeq g(\mu_{X_1}, \mu_{X_2}, \ldots, \mu_{X_n})$$

$$+ \frac{1}{2} \sum_{i=1}^{n} \sum_{j=1}^{n} \left(\frac{\partial^2 g}{\partial X_i \partial X_j} \right) \mathrm{Cov}(X_i, X_j) \qquad (4.43a)$$

where the derivatives are evaluated at $\mu_{X_1}, \mu_{X_2}, \ldots, \mu_{X_n}$. Again, if X_i and X_j are uncorrelated, Eq. 4.43a becomes

$$E(Y) \simeq g(\mu_{X_1}, \mu_{X_2}, \ldots, \mu_{X_n}) + \frac{1}{2} \sum_{i=1}^{n} \left(\frac{\partial^2 g}{\partial X_i^2} \right) \mathrm{Var}(X_i) \qquad (4.43b)$$

Equation 4.44 is the basis of error propagation analysis in measurement theory (Jordan, Eggert, and Kneissel, 1961; Richardus, 1966). However, it is also a useful approximation for many other engineering problems (Cornell, 1969; Ang, 1973); in particular, it is the basis for the general analysis of uncertainty as presented in Vol. II.

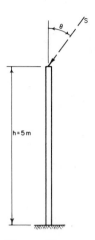

Figure E4.15

EXAMPLE 4.15

Consider a 5-meter-high column supporting a load S, which is inclined at an angle θ from the vertical as shown in Fig. E4.15. Here S and θ are random variables with respective means and standard deviations.

$$\bar{S} = 100 \text{ Newtons}, \qquad \sigma_S = 20 \text{ Newtons}$$
$$\bar{\theta} = 30°(0.524 \text{ rad}), \qquad \sigma_\theta = 5°(0.087 \text{ rad})$$

Determine the mean value and standard deviation of the maximum bending moment on the column induced by the inclined load. Assume that S and θ are statistically independent.

The maximum bending moment occurs at the fixed base of the column, which is (see Fig. E4.15)

$$M = hS \sin \theta$$

Therefore, on the basis of Eq. 4.43, the first-order approximate mean bending moment is

$$\bar{M} \simeq h\bar{S} \sin \bar{\theta}$$
$$= 5(100) \sin 30°$$
$$= 250 \text{ Nm (Newton-meter)}$$

The corresponding variance, according to Eq. 4.44a, is

$$\sigma_M{}^2 \simeq \sigma_S{}^2(h \sin \bar{\theta})^2 + \sigma_\theta{}^2(h\bar{S} \cos \bar{\theta})^2$$
$$= (20)^2(5 \sin 30°)^2 + (0.087)^2(5 \times 100 \cos 30°)^2$$
$$= 2500 + 1420$$
$$= 3920 \text{ (Nm)}^2$$

yielding a standard deviation

$$\sigma_M = 63 \text{ Nm}$$

The accuracy of the estimated mean bending moment may be improved by using the second-order approximation of Eq. 4.43a; thus

$$\bar{M} \simeq h\bar{S}\sin\bar{\theta} + \frac{1}{2}\left(\frac{\partial^2 M}{\partial S^2}\right)\sigma_S^2 + \frac{1}{2}\left(\frac{\partial^2 M}{\partial \theta^2}\right)\sigma_\theta^2 + \left(\frac{\partial^2 M}{\partial S\, \partial\theta}\right)\text{Cov}\,(S,\,\theta)$$

$$= 250 - \tfrac{1}{2}(\bar{S}h\sin\bar{\theta})\,\sigma_\theta^2$$

$$= 250 - \tfrac{1}{2}(100 \times 5\sin 30°)(0.087)^2$$

$$= 249 \text{ Nm}$$

This shows, therefore, that the first-order approximation is quite accurate for this case.

EXAMPLE 4.16

A 2-span bridge across a 400-ft-wide river is to be built with a center pier about 200 ft from one bank of the river. To locate the center position of the pier, a base line B is established along one bank as shown in Fig. E4.16 and the pier position is determined by intersecting the lines of sight from stations a and b, with θ_2 fixed at 90°.

Suppose that the pier is to be located 200 ft from the base line, which has a measured mean length $\bar{B} = 300$ ft and a standard deviation $\sigma_{\bar{B}} = 1$ in.

If the angle θ_1 measured from station b has $\bar{\theta}_1 = 33°40'$ and $\sigma_{\bar{\theta}_1} = 2'$, what are the mean and standard deviation of the measured distance D to the pier location?

$$D = \bar{B}\tan\bar{\theta}_1$$

Thus

$$D = 300 \tan 33.667°$$

$$= 199.82 \text{ ft}$$

$$\sigma_D^2 \simeq (\tan\bar{\theta}_1)^2\sigma_{\bar{B}}^2 + (\bar{B}\sec^2\bar{\theta}_1)^2\sigma_{\bar{\theta}_1}^2$$

$$= 0.444(\tfrac{1}{12})^2 + (300 \times 1.444)^2(5.818 \times 10^{-4})^2$$

$$= 0.0666 \text{ ft}^2$$

or

$$\sigma_D \simeq 0.2581 \text{ ft} = 3.10 \text{ in.}$$

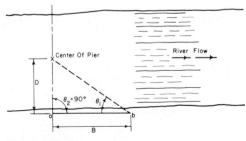

Figure E4.16

EXAMPLE 4.17

The capital cost (in $1000) of a combined municipal activated sludge plant may be estimated as follows:

$$C_c = 583Q^{0.84} + (110 + 37Q)\frac{S_o}{200}$$

$$+ (77 + 23Q)\left(\frac{S_s}{200} - 1\right)$$

in which Q is the flow rate in million gallons per day (mgd); S_o is the biological concentration of influent BOD (biological oxygen demand) in milligram per liter (mg/l); and S_s is the concentration of suspended solids (in mg/l).

Suppose that a waste water treatment plant is needed for the following conditions:

mean flow rate, $\bar{Q} = 5$ mgd
mean BOD concentration, $\bar{S}_o = 600$ mg/l
mean concentration of suspended solids, $\bar{S}_s = 200$ mg/l
with coefficients of variation 30 %, 20 %, and 15 %, respectively

Determine the average capital cost of the plant, and corresponding standard deviation.

$$\bar{C}_c \simeq 583(5)^{0.84} + (110 + 37 \times 5)\left(\frac{600}{200}\right) + (77 + 23 \times 5)\left(\frac{200}{200} - 1\right)$$

$$= \$3,138,219$$

$$\sigma_{C_c}^2 \simeq \left[583(0.84)\bar{Q}^{-0.16} + \frac{37\bar{S}_o}{200}\right]^2 \sigma_Q^2$$

$$+ \left(\frac{110 + 37\bar{Q}}{200}\right)^2 \sigma_{S_o}^2 + \left(\frac{77 + 23\bar{Q}}{200}\right)^2 \sigma_{S_s}^2$$

$$= \left[583(0.84)(5)^{-0.16} + \frac{37 \times 600}{200}\right]^2 (1.5)^2 + \left(\frac{110 + 37 \times 5}{200}\right)^2 (120)^2$$

$$+ \left(\frac{77 + 23 \times 5}{200}\right)^2 (30)^2$$

$$= 539,213 + 31,329 + 829$$

$$= 571,371$$

Therefore, $\sigma_{C_c} \simeq \$756,000$ and the COV is $\delta_{C_c} = 0.24$.

4.4. CONCLUDING REMARKS

In this chapter, we saw that the probabilistic characteristics of a function of random variables may be derived from those of the basic constituent variables. These include, in particular, the probability distribution and the main descriptors (mean and variance) of the function. The derivation of the distribution, however, may be complicated mathematically, especially for nonlinear functions of multiple variables. Therefore, even though the required distribution may (theoretically) be derived, they are often impractical to use, except for special cases (for instance, linear functions of independent normal variates). In view of this, it is often necessary, in many applications, to describe the function in terms only of its mean and variance. Even then, the mean and variance of linear functions are amenable to exact evaluation; however, for a general nonlinear function, (first-order) approximations must often be resorted to. In this chapter, we have introduced and developed the elements for such first-order analysis; these concepts will form the basis for the formal analysis of uncertainty covered in Vol. II.

PROBLEMS

Section 4.2

4.1 The force in the cable of the truss shown in Fig. P4.1, when subjected to a load W, is given by

$$F_{ac} = \frac{\sqrt{h^2 + l^2}}{h} W$$

(a) If the load W is a normal variate $N(\mu_W, \sigma_W)$, derive the density function of the force F_{ac}.

(b) If $\mu_W = 20$ metric tons, $\sigma_W = 5$ metric tons, and $h = \frac{1}{2}l$, what is the probability that the force F_{ac} will exceed 30 tons?   *Ans. 0.0934.*

4.2 A dike is proposed to be built to protect a coastal area from ocean waves (see Fig. P4.2). Assume that the wave height H is related to the wind velocity by the equation

$$H = 0.2V$$

where H is in meters and V is velocity in kilometers per hour (kph). The annual maximum wind velocity is assumed to have a log-normal distribution with a mean of 80 kph and a coefficient of variation of 15 %.

(a) Determine the probability distribution of the annual maximum wave height and its parameters.

(b) If the dike is designed for a 20-year wave height, what is the design height of the dike?

(c) With this design, what is the probability that the dike will be topped by waves within the first three years?

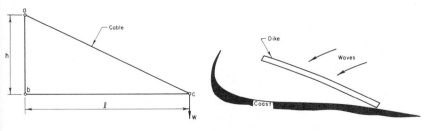

Figure P4.1   *Figure P4.2*

4.3 In Example 4.14, the maximum wave pressure on structures is given as

$$p_{max} = 2.7 \frac{\rho K}{D} U^2$$

where U is the horizontal velocity of the advancing wave.

(a) If U has a log-normal distribution with parameters λ_U and ζ_U, derive the distribution of p_{max} using Eq. 4.8.

(b) Using the data given in Example 4.14, determine the probability that the maximum impact pressure will exceed 40 psi.   *Ans. 0.121.*

4.4 The hydraulic head loss h_L in a pipe due to friction may be given by the Darcy-Weisbach equation

$$h_L = f \frac{L}{2gD} V^2$$

where L and D are, respectively, the length and diameter of the pipe; f is the friction factor; and V is the velocity of flow in the pipe. If V has an exponential distribution with a mean velocity v_0, derive the density function for the head loss h_L.

4.5 From the statistics collected for towns and cities in Illinois the average consumption of water, in gallons per capita per day, is found to increase with the size of population P as follows:

$$X = 19.5 \ln \frac{P}{40} - 17 \qquad \text{for} \quad P > 1000$$

Suppose that the population in 1974 for a certain developing town can be described by a log-normal distribution with a mean of 10,000 and a COV of 5%. It is expected that the median of the population will grow at 10% (of the 1974 population) per year, while the COV will remain roughly constant (see Fig. P4.5).

(a) Assume that the distribution of population is log-normal at any future time; determine the distribution of X, the average per capita water consumption in 1984.

(b) Determine approximately the mean and variance of D, the average total daily demand of water in 1984.   *Ans. 2.08 × 10⁶; 1.53 × 10¹⁰.*

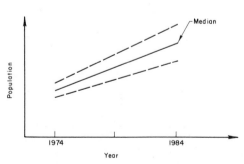

Figure P4.5

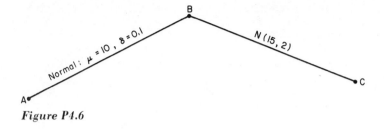

Figure P4.6

4.6 The time of travel between cities A and B is a normal variate with a mean value of 10 hr and a coefficient of variation of 0.10 (see Fig. P4.6). The time of travel between cities B and C is also normal with a mean and standard deviation of 15 and 2 hr, respectively. Assume that these two travel times are statistically independent.
    (a) Determine the density function of the travel time between cities A and C going through B if there is exactly 2 hr of waiting in city B.
    (b) What is the probability that the time of travel between A and C will exceed 30 hr; will be less than 20 hr?

4.7 A simple structure consisting of a cantilever beam AB and a cable BC is used to carry a load S (see Fig. P4.7). The magnitude of the load varies daily, and its monthly maximum has been observed to be Gaussian with a mean of 25,000 kg, and a coefficient of variation of 30%.
    (a) If the cable BC and beam AB are designed to withstand a 10-month maximum load (that is, a maximum load with a return period of 10 months) with factors of safety of 1.25 and 1.40, respectively, what are the probabilities of failure of the cable and of the beam?

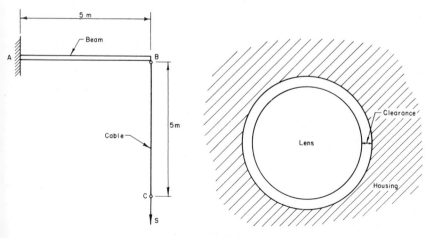

Figure P4.7   **Figure P4.9**

(b) Assuming statistical independence between the failures of the beam and cable, what is the probability of failure of the structure (that is, that it will be unable to carry the load)?

(c) If (instead of part [a]) the strength of the cable were random $N(50,000$ kg; 10,000 kg), what would be its failure probability under the load S?

4.8 The occurrences of hurricanes in a Texas county is described by a Poisson process. Suppose that 32 hurricanes have occurred in the last 50 years; 28 of the 32 hurricanes occurred in the hurricane season (August 1 to November 30).

(a) For this Texas county, estimate the mean rate of occurrence of hurricanes (i) per year; (ii) per month in the hurricane season; (iii) per month in the nonhurricane season.

(b) A temporary offshore structure is to be located off the coast of this county and it is expected that the structure will operate for 19 months between April 1 and October 31 of the following year. What is the mean number of hurricanes that will occur in this period of time?

(c) What is the probability that this structure will be hit by hurricanes during its period of operation?

(d) Suppose that whenever a hurricane occurs, the owner of the structure will incur a loss of $10,000, which includes repairs for damage, loss of revenue, and so on. What is the owner's *expected total loss* from hurricanes? The total loss T (in dollars) is given by

$$T = 10,000 \, N$$

where N is the number of hurricanes during the period of operation, which is assumed to be a Poisson random variable.

(e) What is the probability that the total loss T will *exceed* $10,000?

4.9 To insure proper mounting of a lens in its housing in an aerial camera, a clearance of not less than 0.10 cm and not greater than 0.35 cm is to be allowed. The clearance is the difference between the radius of the housing and the radius of the lens (see Fig. P4.9).

A lens was produced in a grinder whose past records indicate that the radii of such lenses can be regarded as a normal variate with mean of 20.00 cm and a coefficient of variation of 1%.

A housing was manufactured in a machine whose past records indicate that the radii of such housing can be regarded as a normal variate with mean of 20.20 cm and a coefficient of variation of 2%. What is the probability that the specified clearance will be met for this pair of lens and housing? *Ans. 0.216.*

4.10 The safety of a proposed design for the slope shown in Fig. P4.10 is to be analyzed. Suppose that the circular arc AB (with center at 0) represents the potential failure surface and that the wedge of soil contained within the arc will slide if the clockwise moment about point 0 due to the weight of the soil W exceeds the counterclockwise moment provided by the frictional forces F_1 and F_2. The following information is given:

	Mean (kips)	Standard deviation (kips)
W	400	60
F_1	100	30
F_2	300	60

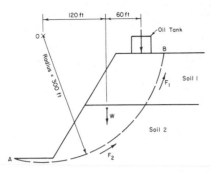

Figure P4.10

(a) Let M_R = total resisting (counterclockwise) moment. Determine $E(M_R)$, $Var(M_R)$.

(b) What is the probability that sliding along the arc AB will occur? Assume that W, F_1, F_2 are statistically independent normal random variables.

(c) An oil tank is proposed to be located as shown in Fig. P4.10. If the maximum permissible probability of sliding failure is 0.01, how heavy can the oil tank be?

4.11 The water supply to a city comes from two sources—namely, from the reservoir and from pumping underground water, as shown in Fig. P4.11. For the next 3 months, the amounts of water available from each source are independently Gaussian $N(30, 3)$ and $N(15, 4)$, respectively, in million gallons. Suppose that the demand in the next 3 months can be described by the probability mass function given in Fig. P4.11.

(a) Determine the probability that there will be insufficient supply of water in the next 3 months.

(b) Repeat part (a) if the demand is also Gaussian with the same mean and variance as those of Fig. P4.11.

4.12 The traffic on a bridge may be described by a Poisson process with mean

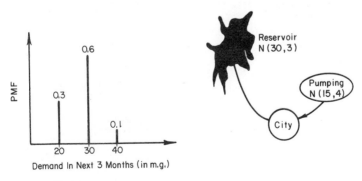

Figure P4.11

arrival rate of 18 vehicles per minute. The vehicles may be divided into two types: trucks and passenger cars. The weight of a truck is $N(15, 5)$ when empty and $N(30, 7)$ when loaded, whereas that of a passenger car is $N(2, 1)$. All units are in tons. Trucks make up only 20% of the total traffic and half the trucks are loaded. The weights of the vehicles are statistically independent.

(a) What is the probability that a vehicle observed at random will not exceed 5 tons? *Ans. 0.801.*

(b) What is the probability that 3 vehicles in a row will each exceed 5 tons?

(c) How would the probability in part (b) change if it is known that 2 vehicles are passenger cars and the remaining one is a truck?

(d) If there are 3 passenger cars and one empty truck on the bridge, what is the probability that the total vehicle load on the bridge will exceed 30 tons? *Ans. 0.0446.*

(e) What is the probability that there will be exactly one truck but no passenger cars arriving within a 10-second interval? *Ans. 0.03.*

4.13 The pole shown in Fig. P4.13 is acted upon by two loads P_1 and P_2 so that the bending moment at the bottom of the pole is

$$M_A = 30P_1 - 20P_2$$

Here, P_1 and P_2 are independent Gaussian random variables with the following parameters.

Load	Mean (kips)	Standard deviation (kips)
P_1	50	5
P_2	20	3

(a) Determine the mean and standard deviation of the moment M_A at the base of the pole.

(b) If the moment-resisting capacity at the bottom of the pole is M_R, a Gaussian random variable with a mean of 1750 ft-kips and a standard deviation of 150 ft-kips, what is the probability that the pole will fail under the loads P_1 and P_2?

(c) Five such poles are arranged in line to support a bank of critical electrical equipment. Adequate support of the equipment requires at least 3 adjacent poles. What is the probability of survival of the system?

(d) In contrast to part (c), what is the probability that exactly 3 poles will fail (regardless of the positions of the poles)?

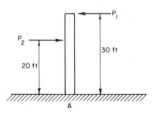

Figure P4.13

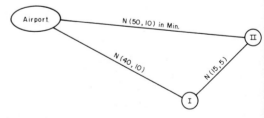

Figure P4.15

4.14 A traffic survey on the various modes of transportation between New York and Boston shows that the percentage of total trips by air, rail, bus, and car are 15.3, 10.6, 9.4, and 64.7 percent, respectively. The distributions of the trip times in each mode are approximately Gaussian with means 0.9, 4.5, 4.8, 4.5 hr, and coefficients of variation 0.15, 0.1, 0.15, 0.2, respectively.
  (a) What proportion of trips between New York and Boston can be completed in 4 hr?
  (b) What is the probability that transportation by bus is faster than by car between these two cities?

4.15 The feasibility of an airport location is to be evaluated. Among many other criteria, one of them is to minimize the travel time from the city to the airport. For simplicity, assume that the city may be subdivided into 2 regions, I and II, each with independent Gaussian travel times to the airport as indicated in Fig. P4.15 in minutes. The ratio of air passengers originating from the two regions is 7 to 3.
  (a) What percentage of the passengers will take more than 1 hr to get to the airport? *Ans. 0.0635*
  (b) A limousine service departs from the airport and picks up or unloads passengers at I and II, consecutively, before returning to the airport. Assume that the travel time for the limousine between stops is also Gaussian as shown in Fig. P4.15. What is the probability that the limousine will complete a round trip within 2 hr? Is this equal to the probability of making 2 rounds within 4 hours? Justify your answer. Assume that the travel times between rounds are also statistically independent.
  (c) A passenger is waiting at the airport for his friend so that they may leave together at a 9 A.M. flight. At 8:50 A.M., he still has not seen his friend. He becomes impatient and calls his friend's home at region II. If his friend had left home at 7:50 A.M., what is the probability that his friend will arrive at the airport in the next 10 minutes? *Ans. 0.857.*

4.16 The existing sewer network shown in Fig. P4.16 consists of pipes AC, BC, CD, ED, DF. The mean inflows from A, B, E are 30, 10, 20 cfs, respectively. Suppose that the flow capacity of pipe DF is 70 cfs, and all the other pipes can adequately handle their respective flows. Assume that the inflows are statistically independent normal variates with 10% COV.
  (a) What is the probability that the capacity of pipe DF will be exceeded in the existing network?

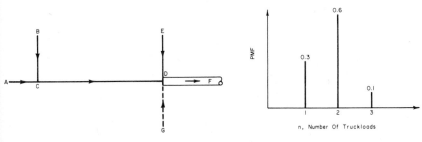

Figure P4.16　　　　　　　　　*Figure P4.17*

(b) Suppose that the sewer from a newly developed area is proposed to be hooked up to the present system at D, and this additional inflow is also Gaussian with a mean of 30 cfs and COV 10%. How should pipe DF be expanded (that is, what should be the new capacity of pipe DF) so that the risk of flow exceedance remains the same as that in the present system?

4.17 The number of truck loads of solid waste arriving at a waste treatment plant in the next hour is random with the PMF given in Fig. P4.17.
The time required for processing each truckload of solid waste is Gaussian ($N10$, 2) in minutes.

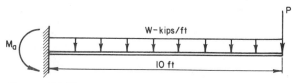

Figure P4.18

What is the probability that the total time needed for processing the solid waste arriving in the next hour will be less than 25 minutes? Assume that the processing times are statistically independent.   *Ans. 0.884.*

4.18 The cantilever beam shown in Fig. P4.18 is subjected to a random concentrated load P and a random distributed load W.
Assume

$$P \text{ is } N(5, 1), \text{ in kips}$$
$$W \text{ is } N(1, 0.2), \text{ in kips/ft}$$

(a) Determine the mean and variance of the applied bending moment $M_a = 50W + 10P$. Assume that $\rho_{W,P} = 0.5$ (that is, the loads are correlated).
(b) The resisting moment of the beam M_r, which is statistically independent of the applied moment M_a, is also Gaussian $N(200, 50)$ in ft-kips. Determine the probability of failure of the beam, $P(M_r < M_a)$ assuming that M_a is Gaussian.

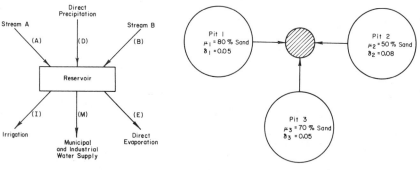

Figure P4.19   **Figure P4.20**

4.19 There are three sources of inflow into the reservoir, namely, streams A and B and direct precipitation D, as shown in Fig. P4.19. Each of these three sources depends on the total rainfall R in the watershed surrounding the reservoir. The following are relationships between A, B, D, and R:

$$A = 0.2R + 0.3$$
$$B = 0.15R + 0.4$$
$$D = 0.03R$$

All are in million gallons (mg). The rainfall R for the next 3-month period is assumed to have a normal distribution $N(15, 2)$ in inches. The outflows from the reservoir consist of irrigation I, municipal and industrial use M, and loss through direct evaporation E. During the next 3 months, each of these three outflows is a normal random variable with

$$I = N(1.5, 0.3)$$
$$M = N(1, 0.1) \quad \text{all in mg}$$
$$E = N(2.5, 0.4)$$

Let T denote the total inflow into the reservoir for the next 3-month period. Assume that T, I, M, E are statistically independent.
  (a) Are A, B, D statistically independent? Why?
  (b) Determine $E(T)$, $Var(T)$. What is the distribution of T?
  (c) Assume that the present storage of the reservoir is 30 mg. Let S be the storage of the reservoir 3 months from now. Determine: $E(S)$, $Var(S)$. What is the distribution of S?
  (d) What is the probability that there will be an increase in the reservoir storage 3 months from now?

4.20 A concrete mixing plant obtains sand and gravel mixtures from 3 gravel pits. The mean percentages of sand by weight in each pit are 80, 50, and 70, respectively, and the coefficients of variation of the percentage of sand are 0.05, 0.08, and 0.05, respectively (see Fig. P4.20). Assume that gravel makes up the remaining percentage by weight. Two, three, and five units of sand-gravel mixture are delivered, respectively, from the three pits and are mixed together. What is the probability that in the resultant mixture, the ratio of sand to gravel by weight does not exceed 2.5 to 1 and also does not fall below 1.5 to 1? Note that these two limits may represent the tolerable sand-to-gravel ratio for acceptable concrete aggregate. Assume that the contents of the pits are statistically independent. *Ans. 0.987.*

4.21 A catch basin is used to control flooding of a region. Aside from serving the immediate neighborhood, it also receives the storm water from another district through a storm sewer.
Suppose that the catch basin has a storage capacity of 50 in. of water; also, any water in the basin is drained at the average rate of 2.5 in. per minute with a COV of 20%.
The rate of inflow from the two sources of drainage water during a rainstorm is as follows.

	Mean rate (in. per minute)	Standard deviation (in. per minute)
Immediate neighborhood:	2	1
Distant district:	1.5	0.5

Assume that all variates are independent and normal, and that the rates of inflow and outflow are constant with time.

(a) Determine the mean and standard deviation of the rate of filling (per minute) of the catch basin.

(b) Determine the probability of flooding (basin capacity exceeded) in 30 minutes of rain (assume that the basin is dry before it rains).

(c) The probability of flooding may be decreased by increasing the capacity of the catch basin. If it is decided to decrease this probability to no more than 10% during a 30-minute rain, what should the catch basin capacity be?

4.22 A plain concrete column is subjected to an axial load W that is a log-normal variate with mean $\bar{W} = 3000$ kN (kilo Newton) and COV $\delta_W = 0.20$ (see Fig. P4.22).

The mean crushing strength of the concrete is $\bar{\sigma}_c = 35,000$ kN/m² (kilo Newtons/square meter) with COV $\delta_{\sigma_c} = 0.20$. Assume uniform compressive stress over the cross-sectional area of the column, so that the applied stress is

$$\sigma = \frac{W}{A}$$

where A = cross-sectional area of column.

(a) What is the density function of the applied stress?

(b) Determine the probability of crushing of a 0.40 m × 0.40 m column. Assume a convenient probability distribution for σ_c.

(c) If a failure (crushing) probability of 10^{-3} is permitted, determine the required cross-sectional area of the column.

(d) Derive the expression for the allowable design stress corresponding to a permissible risk or failure probability p_F (see Example 4.11).

4.23 Figure P4.23 shows a schematic procedure of the treatment system for the waste from a factory before it is dumped into a nearby river. Here X denotes the concentration of a pollutant feeding into the treatment system, and Y denotes the concentration of the same pollutant leaving the system. Suppose that for a normal day, X has a log-normal distribution with median 4 mg/l and the COV is 20%. Because of the erratic nature of biological and chemical reactions, the efficiency of the treatment system is unpredictable. Hence the fraction of the influent pollutant remaining untreated, denoted by F, is also a random variable. Assume F is also a log-normal variate with a median of 0.15 and COV of 10%. Assume X and F are statistically independent.

(a) Determine the distribution of Y and the values of its parameters. Note that

$$Y = FX$$

(b) Suppose that the maximum concentration of the pollutant permitted to

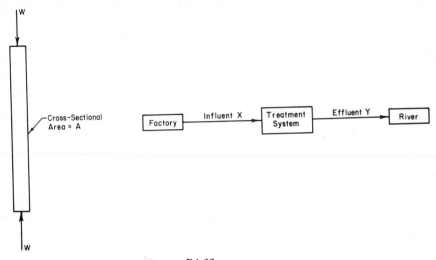

Figure P4.22   *Figure P4.23*

be dumped into the river is specified to be 1 mg/l. What is the probability that this specified standard will be exceeded on a normal day?

(c) On some working days, owing to heavy production in the factory, the influent X will have a median of 5 mg/l instead. Assume that the distribution of X is still log-normal with the same COV and that the efficiency of the treatment system does not change statistically. Suppose that such a heavy work day happens only 10% of the time. Then, on a given day selected at random, what is the probability that the specified standard of 1 mg/l for Y will be exceeded?

4.24 You are taking a plane from O'Hare to Kennedy Airport. Being conscious of the congested conditions at O'Hare, you would like to find out your chance of delay. Based on available data, the delay time (beyond the scheduled departure time) at O'Hare is an exponential random variable; its mean value depends on the weather condition as shown in Fig. P4.24.

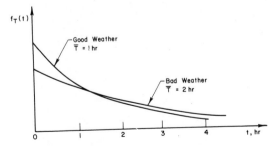

Figure P4.24   Waiting time at O'Hare

The relative likelihood of good and bad weather conditions at O'Hare is about 3 to 1.

(a) What is the probability that your delay at O'Hare will be at least 1.5 hr beyond the scheduled departure time? *Ans. 0.285.*

(b) The delay in landing at Kennedy Airport on a good weather day also has an exponential distribution with a mean of 1/2 hr. What is the probability that the total delay in arrival would be more than 2 hr if the weather is good at both O'Hare and Kennedy airports?
Assume that there is no delay in flight. (*Hint.* Derive the density function for the sum of two exponential variates using Eq. 4.16*a*.) *Ans. 0.252.*

4.25 A construction project consists of two major phases, namely, the construction of the foundation and the superstructure. Let T_1 and T_2 denote the respective durations of each phase. Assume that T_1 and T_2 are independent exponential random variables with $E(T_1) = 2$ months and $E(T_2) = 3$ months. Assume also that the superstructure phase will start only after the foundation phase has been completed.

(a) What is the probability that the project will be completed within 6 months?

(b) If T_1 and T_2 have the same exponential distribution with $E(T_1) = E(T_2) = \lambda$, show that the project duration has a gamma distribution.

4.26 The mean number of arrivals at an airport during rush hour is 20 planes per hour whereas the mean number of departures is 30 planes per hour. Suppose that the arrivals and departures can each be described by a respective Poisson process. The number of passengers in each arrival or departure has a mean of 100 and COV of 40%.

(a) What is the probability that there will be a total of two arrivals and/or departures in a 6-minute period?

(b) Suppose that in the last hour there have been 25 arrivals.

(i) What is the mean and variance of the total number of arriving passengers in the last hour?

(ii) What is the probability that the total number of arriving passengers exceeded 3000 in the last hour? State and justify any assumption you may make.

4.27 The total distance between A and B is composed of the sum of 124 independent measurements. See Fig. P4.27. The random error E in each measurement is uniformly distributed between ± 1 in. If the total distance AB is approximately 2 miles and the lengths of the segments are approximately equal, compute the following.

(a) The mean and variance of the error in each measurement.

(b) The probability that the error in each measurement is not more than 0.01% of the actual length of the segment.

(c) The mean and variance of the total error in the distance AB.

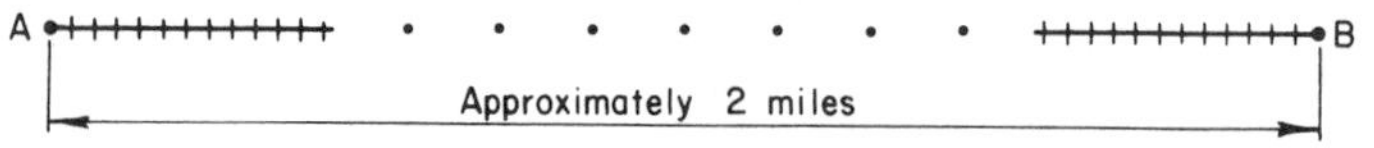

Figure P4.27

(d) The probability that the total error is not more than 0.01 % of the actual distance AB.

Section 4.3

4.28 A cylindrical volume has an average measured outside diameter of $D_o = 5$ m and an average inside diameter of $D_i = 3$ m, and a height of $H = 10$ m (see Fig. P4.28). If the COV of these measurements are, respectively, 2%, 1%, and 1%, what are the mean and variance of the volume, if D_o and D_i are perfectly correlated (that is, $\rho_{D_o, D_i} = 1.0$) whereas these are statistically independent of H?

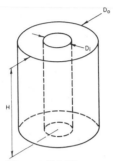

Figure P4.28

4.29 Small flaws (cracks) in metals grow when subjected to cyclic stresses. The rate of crack growth (per load cycle) may be given by

$$\frac{dA}{dn} = C(\Delta K)^m$$

where

$$\Delta K = \Delta S \sqrt{\pi A}$$
$$A = \text{existing crack size}$$
$$\Delta S = \text{applied stress increment}$$

and C and m are constants. If

$$C = 0.5 \times 10^{-5} \qquad m = 2$$

and

$$\bar{A} = 0.1 \text{ in.} \quad \text{with } \delta_A = 20\%$$
$$\overline{\Delta S} = 50 \text{ ksi} \quad \text{with } \delta_{\Delta S} = 30\%$$

determine the mean and COV of the crack growth rate per load cycle. Assume that A and ΔS are statistically independent. *Ans. 0.00392; 0.633.*

4.30 The range R of a projectile is given by the following:

$$R = \frac{v_0^2}{g} \sin 2\phi$$

where g is the gravitational acceleration, v_0 is the initial velocity of the

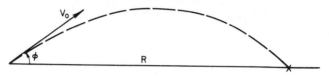

Figure P4.30

projectile, and ϕ is its direction from the horizontal (see Fig. P4.30). If

$$\bar{\phi} = 30° \quad \text{and} \quad \delta_\phi = 5\%$$
$$\bar{v}_0 = 500 \text{ ft/sec} \quad \text{and} \quad \sigma_{v_0} = 50 \text{ ft/sec}$$

determine the first-order mean and standard deviation of the range R. Assume that $g = 32.2 \text{ ft/sec}^2$; ϕ and v_0 are statistically independent. Evaluate also the second-order mean range. *Ans. 6723.6; 1359.8; 6781.6 ft.*

4.31 The number of airplanes arriving over Chicago O'Hare Airport during the peak hour from various major cities in the United States are listed below.

City	Average number of arrivals	Standard deviation
New York	5	2
Miami	3	1
Los Angeles	4	2
Washington, D.C.	4	1
San Francisco	4	2
Dallas	2	0
Seattle	3	1

Suppose (hypothetically) that the holding time T (in minutes) at O'Hare Airport is a function of the number of arrivals from the above cities; specifically,

$$T = 4\sqrt{N_A}, \text{ in minutes}$$

where N_A is the total number of arrivals (during the peak hour) from the cities listed above.

Assuming a log-normal distribution for T, determine the probability that the holding time will exceed 25 minutes. Assume that arrivals from different cities are statistically independent.

4.32 In a study of noise pollution, the noise level at C transmitted from two noise sources as shown in Fig. P4.32 is analyzed. Suppose that the intensities of the noise originating from A and B are statistically independent and denoted as I_A and I_B, with mean value 1000 and 2000 units, respectively, and the coefficient of variation is 10% for both I_A and I_B. Since the noise intensity decreases with distance from the source, the following equation has been suggested:

$$I(x) = \frac{I}{(x + 1)^2}$$

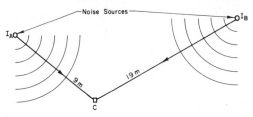

Figure P4.32

where

$$I = \text{intensity generated at a source}$$
$$I(x) = \text{intensity at distance } x \text{ from the source}$$

(a) Let I_C be the noise intensity at C, which is the sum of the two intensities transmitted from A and B. Determine $E(I_C)$, $Var\ (I_C)$.

(b) A common measure of noise intensity is in terms of decibels. Suppose that the number of decibels D is expressed as a function of intensity I as

$$D = 40 \ln 2I$$

Determine the approximate mean and variance of D_C, that is, the number of decibels at C.

4.33 The velocity of uniform flow, in feet per second, in an open channel is given by the Manning equation

$$V = \frac{1.49}{n} R^{2/3} S^{1/2}$$

where

$$S = \text{slope of the energy line}$$
$$R = \text{hydraulic radius, in feet}$$
$$n = \text{roughness coefficient of the channel}$$

Consider a rectangular open channel with concrete surface ($\bar{n} = 0.013$); R is estimated to be 2 ft (average value) and the average slope S is 1%. Because the determinations of R, S, and n are not very precise, the uncertainties associated with these values, expressed in terms of COV, are as follows: $\delta_R = 0.05$, $\delta_S = 0.10$, and $\delta_n = 0.30$.
Determine the first-order mean value of the flow velocity V, and the underlying uncertainty in terms of COV. Evaluate also the second-order approximation for the mean flow velocity.

4.34 The settlement of a column footing, shown in Fig. P4.34a,

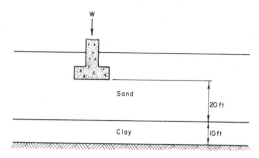

Figure P4.34a

is composed of two components—the settlements of the sand and clay strata. The flexibilities (that is, inches settlement per foot of strata per ton of applied load) of the two strata, denoted F_S and F_C, are independent normal variates N (0.001, 0.0002) and N (0.008, 0.002), respectively. The total column load is W, which may be assumed to be statistically independent of F_S and F_C.

(a) If $W = 20$ tons, what is the probability that the total settlement will exceed 3 in.? *Ans. 0.007.*

(b) Suppose the load W is also a random variable with the PMF given in Fig. P4.34b.

Figure P4.34b

With this PMF of W, determine the mean and variance of the total settlement by first-order approximation. In this case, what would be the probability that the settlement will exceed 3 in.? *Ans. 2.2; 0.51; 0.058.*

5. Estimating Parameters From Observational Data

5.1. THE ROLE OF STATISTICAL INFERENCE IN ENGINEERING

We have seen in the previous chapters that once we know (or assume) the distribution function of a random variable and the values of its parameters, the probabilities associated with events defined by values of the random variable can be computed. The calculated probability is clearly a function of the values of the parameters, as well as of the assumed form of distribution. Naturally, questions pertaining to the determination of the parameters, such as the mean value μ and variance σ^2, and the choice of specific distributions are of interest.

Answers to these questions often require observational data. For example, in determining the maximum wind speed for the design of a tall building, past records of measured wind velocities at or near the building site are pertinent and important; similarly, in designing a left-turn lane at an existing highway crossing, a traffic count of left turns at the intersection may be required. Based on these observations, information about the probability distribution may be inferred, and its parameters estimated statistically.

In many geographic regions, data on natural processes, such as rainfall intensities, flood levels, wind velocity, earthquake frequencies and magnitudes, traffic volumes, pollutant concentrations, ocean wave heights and forces, have been and continue to be collected and reported in published records. Field and laboratory data on the variabilities of concrete strength, yield strength of steel, fatigue lives of materials, shear strength of soils, efficiency of construction crews and equipment, measurement errors in surveying, and many others, continue to be collected. These statistical data provide the information from which the probability model and the corresponding parameters required in engineering design may be developed or evaluated.

The techniques of deriving probabilistic information and of estimating parameter values from observational data are embodied in the methods of

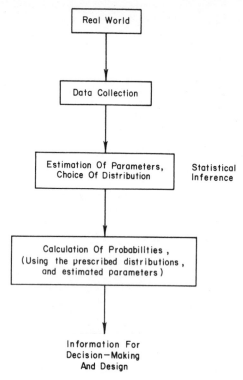

Figure 5.1   Role of statistical inference in decision-making process

statistical inference, in which information obtained from sampled data is used to make generalizations about the populations from which the samples were obtained. Inferential methods of statistics, therefore, provide a link between the real world and the (idealized) probability models assumed or prescribed in a probabilistic analysis. The role of statistical inference in the decision-making process is schematically shown in Fig. 5.1. This chapter is devoted to the estimation of statistical parameters; the subject of determining probability distribution is covered in Chapter 6.

Although there are other inferential methods of statistics, only those that are most basic and of wide applications in engineering are discussed here. These include principally the methods of estimation (point and interval estimations) in this chapter, determination of probability distributions in Chapter 6, and regression and correlation analyses in Chapter 7. Chapter 8 presents the Bayesian approach to the estimation problem. The more esoteric topics of statistics such as *design of experiments* and *analysis of variance* are not covered. Moreover, we shall not dwell on such theoretical questions as the unbiasedness, efficiency, consistency, and sufficiency of an estimator; only the concepts underlying the above methods are developed, and their significance to engineering problems are emphasized and illustrated.

5.1.1. Inherent variability and estimation error

It may be emphasized that even when the distribution function and its parameters of a random variable are known, we still cannot predict with certainty the occurrence (or nonoccurrence) of specific events. At best, we can say that an event will occur with an associated probability. The underlying uncertainty, in this case, is due to the inherent randomness of the natural phenomenon. However, uncertainty arises also from the inaccuracies in the estimation of the parameters and in the choice of the distribution. For example, when available data are limited, the estimated mean and variance may not be accurate and the distribution function determined on the basis of available data may not be the most appropriate. Such errors, therefore, would contribute additional uncertainty.

Uncertainties associated with errors of parameter estimation can be reduced by increasing the amount of data, whereas the uncertainty associated with the inherent variability may remain unchanged or may even increase with additional data.

More generally, errors would also include inaccuracies of modeling and prediction. For example, when an idealized mathematical equation is used to evaluate an engineering system, or its response to specified input, the imperfection of the mathematical model gives rise also to further uncertainty. Such imperfections may be due to factors whose effects were not explicitly reflected in the model, or to gross idealizations necessary for mathematical tractability and urgency of engineering solutions.

In general, therefore, we shall consider uncertainty to be the result of (inherent or natural) variability as well as of (prediction) error. A general model for the systematic assessment and analysis of uncertainty is developed in Vol. II.

5.2. CLASSICAL APPROACH TO ESTIMATION OF PARAMETERS

Classical estimation of parameters is divided into *point* and *interval* estimation. Point estimation is concerned with the calculation of a single number, from a set of observational data, to represent the parameter of the underlying population; interval estimation goes further to establish a statement of confidence in the estimated quantity, resulting in the determination of an interval indicating the range wherein the population parameter may be located (with the associated confidence).

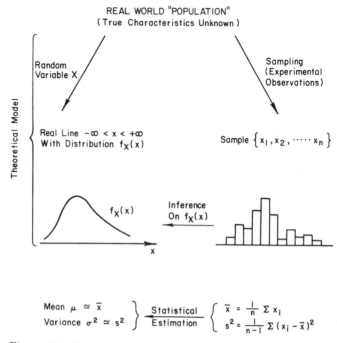

Figure 5.2 Role of sampling in statistical inference

5.2.1. Random sampling and point estimation

As alluded to earlier, the parameters of a probability model may be evaluated or estimated only on the basis of a set of observational data obtained from the "population"; such a data set represents a *sample* of the population, and thus the value of a parameter calculated on the basis of the sample values is necessarily an *estimator* of the parameter. In other words, the exact values of the population parameters are generally unknown; the best that can be done is to make estimates of these values by *sampling* the population. As indicated in Fig. 5.2, the real-world population may be modeled by a random variable X, with probability distribution $f_X(x)$ and associated parameters, for example μ and σ in the case of the normal distribution. The form of $f_X(x)$ may be derived on the basis of physical considerations, or determined empirically, as described in Chapter 6. Invariably, however, the parameters such as μ and σ must be estimated from sampled observational data.

Given a set of sample values, there are different methods for estimating the parameters; among these are the *method of moments* and *method of maximum likelihood*. Regardless of the method, estimation of parameters is necessarily based on a set of sample values, say $x_1, \cdots, x_n$, called a *sample of size n* from the population X. Usually such a sample is assumed to constitute a *random sample*; this means that the successive sample values are independent and the underlying population (or the distribution of X) remains the same from one sample value to another.

We should point out that there are certain properties that are desirable of a point estimator: the properties of *unbiasedness, consistency, efficiency,* and *sufficiency*. If the expected value of an estimator is equal to the parameter, the estimator is said to be *unbiased*. Unbiasedness, therefore, implies that on the average the value of the estimator will be equal to the parameter; however, nothing is said about whether the individual values of the estimator is close to the parameter itself. On the other hand, the property of *consistency* implies that as $n \to \infty$, the estimator approaches the value of the parameter. Consistency, therefore, is an asymptotic property—practically, it means that the error in the estimator decreases as the sample size increases. *Efficiency* refers to the variance of the estimator; if everything else is equal, an estimator θ_1 is said to be more efficient than another θ_2, if θ_1 has a smaller variance than that of θ_2. Finally, an estimator is said to be *sufficient* if it utilizes all the information in a sample that is pertinent to the estimation of the parameter.

In practice, however, it is impractical to require all or many of these properties; seldom, in fact, is there an estimator that possesses all the above properties. In the following sequel, we may refer to one or more of these properties in connection with specific estimators.

The method of moments. In Chapter 3, we saw that the mean and variance are the main descriptors of a random variable. These are related to the parameters of the distribution, as shown in Table 5.1 for a number of common distributions. For example, in the case of a normal random variable, the parameters μ and σ^2 of the distribution are also the mean and variance of the variate, whereas, in the case of the gamma distribution, the parameter s/ν and k are related to the mean and variance as follows:

$$E(X) = k/\nu$$
$$\mathrm{Var}(X) = k/\nu^2$$

On the basis of the relationships between the moments of a random variable and the parameters of the corresponding distribution, such as those shown in Table 5.1, it follows that the parameters of a distribution may be determined by first estimating the mean and variance (and higher moments, if necessary) of the random variable. This, in essence, is the basis of the *method of moments*.

Intuitively, it seems that the *sample moments* may be used as estimates

Table 5.1. Common Distributions and Their Parameters

Distribution	Probability density function (PDF) or mass function (PMF)	Parameters	Relation to Mean and Variance
Binomial	$p_X(x) = \binom{n}{x} p^x(1-p)^{n-x}$ $x = 0, 1, 2, \ldots, n$	p	$E(X) = np$ $\text{Var}(X) = np(1-p)$
Geometric	$p_X(x) = p(1-p)^{x-1}$ $x = 0, 1, 2, \ldots$	p	$E(X) = 1/p$ $\text{Var}(X) = (1-p)/p^2$
Poisson	$p_X(x) = \dfrac{(\nu t)^x}{x!} e^{-\nu t}$ $x = 0, 1, 2, \ldots$	ν	$E(X) = \nu t$ $\text{Var}(X) = \nu t$
Exponential	$f_X(x) = \dfrac{1}{\lambda} e^{-x/\lambda} \quad x \geq 0$	λ	$E(X) = \lambda$ $\text{Var}(X) = \lambda^2$
Gamma	$f_X(x) = \dfrac{\nu(\nu x)^{k-1} e^{-\nu x}}{\Gamma(k)} \quad x \geq 0$	ν, k	$E(X) = k/\nu$ $\text{Var}(X) = k/\nu^2$
Normal (Gaussian)	$f_X(x) = \dfrac{1}{\sqrt{2\pi}\,\sigma} \exp\left[-\dfrac{1}{2}\left(\dfrac{x-\mu}{\sigma}\right)^2\right]$ $-\infty < x < \infty$	μ, σ	$E(X) = \mu$ $\text{Var}(X) = \sigma^2$

Lognormal

$$f_X(x) = \frac{1}{\sqrt{2\pi}\,\zeta x} \exp\left[-\frac{1}{2}\left(\frac{\ln x - \lambda}{\zeta} \right)^2 \right]$$

$$x \geq 0$$

λ, ζ

$$E(X) = \exp\left(\lambda + \tfrac{1}{2}\zeta^2 \right)$$

$$\mathrm{Var}(X) = E^2(X)\left[e^{\zeta^2} - 1 \right]$$

Rayleigh

$$f_X(x) = \frac{x}{\alpha^2} \exp\left[-\frac{1}{2}\left(\frac{x}{\alpha} \right)^2 \right] \qquad x \geq 0$$

α

$$E(X) = \sqrt{\frac{\pi}{2}}\,\alpha$$

$$\mathrm{Var}(X) = \left(2 - \frac{\pi}{2} \right)\alpha^2$$

Uniform

$$f_X(x) = \frac{1}{b-a} \qquad a < x < b$$

a, b

$$E(X) = (a+b)/2$$

$$\mathrm{Var}(X) = \tfrac{1}{12}(b-a)^2$$

Triangular

$$f_X(x) = \frac{2}{b-a}\left(\frac{x-a}{u-a} \right) \qquad a \leq x \leq u$$

$$= \frac{2}{b-a}\left(\frac{b-x}{b-u} \right) \qquad u \leq x \leq b$$

a, b, u

$$E(X) = \tfrac{1}{3}(a+b+u)$$

$$\mathrm{Var}(X) = \frac{1}{18}\left(a^2 + b^2 + c^2 \right.$$
$$\left. - ab - au - bu \right)$$

Beta

$$f_X(x) = \frac{1}{B(q,r)}\frac{(x-a)^{q-1}(b-x)^{r-1}}{(b-a)^{q+r-1}}$$

$$a \leq x \leq b$$

a, b, q, r

$$E(X) = a + \frac{q}{q+r}(b-a)$$

$$\mathrm{Var}(X) = \frac{qr}{(q+r)^2(q+r+1)}(b-a)^2$$

225

of the corresponding moments of the random variable. In this regard, just as the mean and variance are the (weighted) averages of X and $(X - \mu)^2$, the *sample mean* and *sample variance* can be defined as the respective averages of a sample of size n, namely $x_1, \cdots, x_n$, as follows.

$$\bar{x} = \frac{1}{n} \sum_{i=1}^{n} x_i \qquad (5.1)$$

and

$$s^2 = \frac{1}{n} \sum_{i=1}^{n} (x_i - \bar{x})^2 \qquad (5.2)$$

Accordingly, $\bar{x}$ and s^2 are the *point estimates* of the population mean μ and population variance σ^2, respectively.

After the mean and variance of the random variable (or higher moments, if required) have been estimated, the parameters of its probability distribution can then be determined; for example, through the relationships given in Table 5.1 (see Example 5.2).

It should be pointed out that Eq. 5.2 is a "biased" estimate (Freund, 1962) for the variance. This bias can be removed by dividing the sum of squares with $(n - 1)$ instead of n (see Eq. 5.32); thus the unbiased sample variance,

$$s^2 = \frac{1}{(n - 1)} \sum_{i=1}^{n} (x_i - \bar{x})^2 \qquad (5.3)$$

is preferred over Eq. 5.2. Of course, for large n, there is little difference between the two estimates of Eqs. 5.2 and 5.3.

By expanding the squared terms, it can be shown that Eq. 5.3 may also be expressed as

$$s^2 = \frac{1}{n - 1} \left(\sum_{i=1}^{n} x_i^2 - n\bar{x}^2 \right) \qquad (5.3a)$$

EXAMPLE 5.1

Consider the data on the crushing strength of concrete for 25 specimens listed in Table E5.1. To determine the values of its mean and variance μ and σ^2, we apply Eqs. 5.1 and 5.3a, obtaining

$$\bar{x} = \frac{1}{25} \sum_{i=1}^{25} x_i = 5.6 \text{ ksi}$$

and

$$s^2 = \frac{1}{24} \left[\sum_{i=1}^{25} x_i^2 - 25(5.6)^2 \right] = 0.44 \text{ (ksi)}^2$$

On the basis of the data, the mean and variance of the crushing strength of the concrete are 5.6 ksi and 0.44 (ksi)^2, respectively.

Table E5.1. Computation of Mean and Variance for Concrete Crushing Strength of Example 5.1.

Specimen number	x_i	x_i^2
1	5.6	31.36
2	5.3	28.09
3	4.0	16.00
4	4.4	19.36
5	5.5	30.25
6	5.7	32.49
7	6.0	36.00
8	5.6	31.36
9	7.1	50.41
10	4.7	22.09
11	5.5	30.25
12	5.9	34.81
13	6.4	40.96
14	5.8	33.64
15	6.7	44.89
16	5.4	29.16
17	5.0	25.00
18	5.8	33.64
19	6.2	38.44
20	5.6	31.36
21	5.7	32.49
22	5.9	34.81
23	5.4	29.16
24	5.1	26.01
25	5.7	32.49
	$\Sigma = 140.00$	$\Sigma = 794.52$

$$\bar{x} = \frac{140}{25} = 5.60$$

$$s^2 = \frac{1}{24} [794.52 - 25(5.60)^2] = 0.44$$

The calculations for $\bar{x}$ and s^2 can be performed conveniently in tabular form as illustrated in Table E5.1.

EXAMPLE 5.2

Data for fatigue life of 75 S–T aluminum yield the histogram of Fig. 1.5. It is suggested that a log-normal distribution will fit the shape of the histogram well. Estimate the parameters λ and ζ of the log-normal distribution.

The sample mean and sample variance are computed to be

$$\bar{x} = 26.75 \text{ million cycles}$$
$$s^2 = 360.0 \text{ (million cycles)}^2$$

According to the relationships in Table 5.1, the mean and variance of a lognormal distribution are given by

$$E(X) = \exp\left(\lambda + \tfrac{1}{2}\zeta^2\right)$$
$$\text{Var}(X) = E^2(X)(e^{\zeta^2} - 1)$$

Hence, the estimates of the parameters λ and ζ, denoted as $\hat{\lambda}$ and $\hat{\zeta}$, are obtained as the solutions to the following equations.

$$\exp\left(\hat{\lambda} + \tfrac{1}{2}\hat{\zeta}^2\right) = \bar{x} = 26.75$$
$$(26.75)^2(e^{\hat{\zeta}^2} - 1) = s^2 = 360.0$$

Thus, $\hat{\lambda} = 3.08$ and $\hat{\zeta} = 0.64$.

The method of maximum likelihood.   A method of point estimation that is popular among statisticians is the *maximum likelihood* method. In contrast to the method of moments, the maximum likelihood method provides a procedure for deriving the point estimator of the parameter directly. Consider a random variable X with density function $f(x; \theta)$, in which θ is the parameter, such as the mean λ in the exponential distribution. On the basis of the sample values $x_1, \cdots, x_n$, one may inquire: "what is the most likely value of θ that produces the set of observations $x_1, \cdots, x_n$?" In other words, among the possible values of θ, what is the value that will maximize the likelihood of obtaining the set of observations? Such is the rationale underlying the *maximum likelihood* method of point estimation.

The likelihood of obtaining a particular sample value x_i can be assumed to be proportional to the value of the probability density function evaluated at x_i. Then, assuming random sampling, the likelihood of obtaining n independent observations $x_1, \cdots, x_n$ is

$$L(x_1, \cdots, x_n; \theta) = f(x_1; \theta)\, f(x_2; \theta) \cdots f(x_n; \theta) \tag{5.4}$$

which is the *likelihood function* of observing the set $x_1, \cdots, x_n$. The *maximum likelihood estimator* $\hat{\theta}$ is then the value of θ that maximizes the likelihood function $L(x_1, \cdots, x_n; \theta)$. This estimator may be obtained by differentiating $L(x_1, \cdots, x_n; \theta)$ with respect to θ and setting the derivative equal to zero, giving usually an absolute maximum (Hoel, 1962); that is, $\hat{\theta}$ is obtained as the solution to the following equation.

$$\frac{\partial L(x_1, \cdots, x_n; \theta)}{\partial \theta} = 0 \tag{5.5}$$

Because of the multiplicative nature of the likelihood function, it is frequently more convenient to maximize the logarithm of the likelihood function instead; that is,

$$\frac{\partial \log L(x_1, \cdots, x_n; \theta)}{\partial \theta} = 0 \tag{5.6}$$

The solution for $\hat{\theta}$ from Eq. 5.6 should be the same as that obtained with Eq. 5.5.

For density functions with two or more parameters, the likelihood function becomes

$$L(x_1, \cdots, x_n; \theta_1, \cdots, \theta_m) = \prod_{i=1}^{n} f(x_i; \theta_1, \cdots, \theta_m) \tag{5.7}$$

where $\theta_1, \cdots, \theta_m$ are the m parameters to be estimated. In this case, the maximum likelihood estimators would be obtained from the solution to the following set of simultaneous equations.

$$\frac{\partial L(x_1, \cdots, x_n; \theta_1, \cdots, \theta_m)}{\partial \theta_j} = 0; \qquad j = 1, \cdots, m \tag{5.8}$$

The maximum likelihood estimate (MLE) of a parameter possesses many of the desirable properties of an estimator mentioned earlier. In particular, for large sample size n, the maximum likelihood estimator is often considered the "best" estimate, in that it has the minimum variance (asymptotically) (Hoel, 1962).

EXAMPLE 5.3

The times between successive arrivals of vehicles in a traffic flow were observed as follows.

$$1.2, 3.0, 6.3, 10.1, 5.2, 2.4, 7.1 \text{ sec}$$

Suppose the interarrival time of vehicles follows an exponential distribution; that is,

$$f_T(t) = \frac{1}{\lambda} e^{-t/\lambda}$$

Determine the maximum likelihood estimate (MLE) for the mean interarrival time λ.

From Eq. 5.4, the likelihood function of the seven observed values is

$$L(t_1, \ldots, t_7; \lambda) = \prod_{i=1}^{7} \frac{1}{\lambda} \exp\left(-t_i/\lambda\right)$$

$$= (\lambda)^{-7} \exp\left(-\frac{1}{\lambda} \Sigma t_i\right)$$

where t_i is the ith observed interarrival time and $\Sigma = \sum_{i=1}^{7}$. Then, according to Eq. 5.5,

$$\frac{\partial L}{\partial \lambda} = -7\lambda^{-8} \exp\left(-\frac{1}{\lambda} \Sigma t_i\right) + \lambda^{-7} \exp\left(-\frac{1}{\lambda} \Sigma t_i\right) \frac{\Sigma t_i}{\lambda^2} = 0$$

or

$$\lambda^{-8}\left[-7 + \frac{\Sigma t_i}{\lambda}\right] \exp\left(-\frac{1}{\lambda}\Sigma t_i\right) = 0$$

From which we obtain

$$\hat{\lambda} = \frac{\sum_{i=1}^{7} t_i}{7} = 5.04 \text{ sec}$$

In general, therefore the MLE for λ from a sample of size n is

$$\hat{\lambda} = \frac{1}{n}\sum_{i=1}^{n} t_i$$

EXAMPLE 5.4

A triaxial specimen of saturated sand is subjected to cyclic vertical loads with a stress amplitude of ± 200 psf in a laboratory test. The number of load cycles applied until the sand specimen fails has been recorded for five independent specimens as follows.

25, 20, 28, 33, 26 cycles

Suppose the number of load cycles to failure for the sand is assumed to follow a log-normal distribution; estimate the parameters λ and ζ by the maximum likelihood method.

Let us first derive the general expressions of the maximum likelihood estimators $\hat{\lambda}$ and $\hat{\zeta}$ for a log-normal distribution. From Eq. 5.7, the likelihood function is given by

$$L(x_1, \ldots, x_n; \lambda, \zeta) = \prod_{i=1}^{n}\left\{\frac{1}{\sqrt{2\pi}\,\zeta x_i}\exp\left[-\frac{1}{2}\left(\frac{\ln x_i - \lambda}{\zeta}\right)^2\right]\right\}$$

$$= \left(\frac{1}{\sqrt{2\pi}\,\zeta}\right)^n \left(\prod_{i=1}^{n} x_i^{-1}\right)\exp\left[-\frac{1}{2\zeta^2}\sum_{i=1}^{n}(\ln x_i - \lambda)^2\right]$$

The presence of exponentials suggests that it is more convenient to work with the logarithmic form; thus

$$\ln L(x_i, \ldots, x_n; \lambda, \zeta) = -n\ln\sqrt{2\pi} - n\ln\zeta - \sum_{i=1}^{n}\ln x_i - \frac{1}{2\zeta^2}\sum_{i=1}^{n}(\ln x_i - \lambda)^2$$

To maximize the likelihood function, we have

$$\frac{\partial \ln L}{\partial \lambda} = \frac{1}{\hat{\zeta}^2}\Sigma(\ln x_i - \hat{\lambda}) = 0$$

$$\frac{\partial \ln L}{\partial \zeta} = -\frac{n}{\hat{\zeta}} + \frac{1}{\hat{\zeta}^3}\Sigma(\ln x_i - \hat{\lambda})^2 = 0$$

Since $\hat{\zeta} \neq 0$, the solution to these two simultaneous equations yields

$$\hat{\lambda} = \frac{\sum_{i=1}^{n}\ln x_i}{n}$$

and

$$\hat{\zeta}^2 = \frac{1}{n}\sum_{i=1}^{n}(\ln x_i - \hat{\lambda})^2$$

Substituting the values of the observed data, we obtain the MLE of the parameters as

$\hat{\lambda} = \frac{1}{5}(\ln 25 + \ln 20 + \ln 28 + \ln 33 + \ln 26) = 3.26$

$\hat{\zeta}^2 = \frac{1}{5}[(\ln 25 - 3.26)^2 + (\ln 20 - 3.26)^2 + (\ln 28 - 3.26)^2$
$\qquad\qquad\qquad\qquad + (\ln 33 - 3.26)^2 + (\ln 26 - 3.26)^2] = 0.027$

5.2.2. Interval estimation of the mean

How good is the estimator $\overline{X}$*?* So far, we have discussed the *point estimates* of the mean and variance; such estimates, however, do not convey information on the degree of accuracy of these estimates. For this reason, the interval over which a parameter may lie often is used to supplement the point estimate (a single number) of the same parameter. Such intervals are called the *confidence intervals*, and the method of estimation is known as *interval estimation*.

Since we are using the sample mean $\bar{x}$ to estimate the population mean μ, the accuracy of this estimate is naturally of concern. We examine this as follows.

First of all, for a random sample of size n, the values $x_1, x_2, \cdots, x_n$ can be conceived to be the respective sample values of a set of independent random variables $X_1, X_2, \cdots, X_n$. Moreover, in random sampling, the density functions of $X_1, \cdots, X_n$ are individually the same as that of the population X; that is,

$$f_{X_1}(x_1) = f_{X_2}(x_2) = \cdots = f_{X_n}(x_n) = f_X(x)$$

Then the sample mean is also a random variable

$$\overline{X} = \frac{1}{n} \sum_{i=1}^{n} X_i \qquad (5.9)$$

Its expected value is

$$E(\overline{X}) = E\left(\frac{1}{n} \sum_{i=1}^{n} X_i \right) = \frac{1}{n} \sum_{i=1}^{n} E(X_i)$$

Hence

$$E(\overline{X}) = \frac{1}{n} \cdot n\mu = \mu \qquad (5.10)$$

The expected value of the sample mean $\overline{X}$, therefore, is equal to the population mean; in view of this, $\overline{X}$ is said to be an "unbiased" estimator of the population mean μ.

Since $\overline{X}$ is a random variable, it also has a variance

$$\mathrm{Var}(\overline{X}) = \mathrm{Var}\left(\frac{1}{n} \sum_{i=1}^{n} X_i \right) = \frac{1}{n^2} \mathrm{Var}\left(\sum_{i=1}^{n} X_i \right)$$

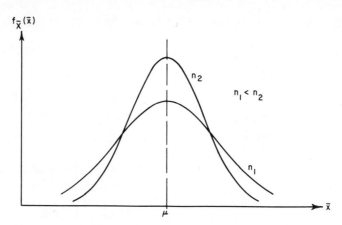

Figure 5.3 Distribution of sample mean

where X_1, X_2, $\cdots$, X_n are statistically independent (in random sampling); hence

$$\text{Var}(\bar{X}) = \frac{1}{n^2} \sum_{i=1}^{n} \text{Var}(X_i) = \frac{1}{n^2} (n\sigma^2) = \frac{\sigma^2}{n} \qquad (5.11)$$

Therefore, according to Eqs. 5.10 and 5.11, the sample mean $\bar{X}$ has a mean value μ and standard deviation $\sigma/\sqrt{n}$. These results apply so long as the X_i's are statistically independent and identically distributed as X; that is, if random sampling is assumed. In practice, of course, it is difficult to verify whether the assumptions of random sampling are satisfied; it is important, however, that every effort be made to ensure that samples taken from a population are sufficiently random to permit the use of the results derived above.

In Chapter 4 we saw that the sum of n independent normal variates is also a normal variate. Hence, if the underlying population X is Gaussian, the estimator $\bar{X}$ is also Gaussian. Moreover, if the sample size n is sufficiently large, the sample mean $\bar{X}$ will be approximately Gaussian (by virtue of the *central limit theorem*), even if the underlying population is not Gaussian.

Therefore, for large sample size n, we can generally assume that $\bar{X}$ has a normal distribution $N(\mu, \sigma/\sqrt{n})$. As the sample size n increases, the distribution of $\bar{X}$ becomes narrower as illustrated in Fig. 5.3, indicating that the quality of the estimate $\bar{x}$ improves with the sample size n. In other words, as n increases the sample mean $\bar{x}$ is more likely to be closer to the population mean μ. In the extreme case, as $n \to \infty$, $\bar{x} \to \mu$.

Confidence interval with known variance. Consider first the case in which there is prior knowledge of the variance or standard deviation of the population, and only the mean value is to be estimated. This condition is sometimes encountered; for example, in electronic distance measurement, the standard error is fairly constant for a given type of equipment; in such cases, therefore, σ may be assumed to be known from previous experience.

We have just seen that for large sample size, the sample mean $\bar{X}$ can be described with the normal distribution $N(\mu, \sigma/\sqrt{n})$. By a simple transformation (see Example 4.1), it can be shown that $(\bar{X} - \mu)/(\sigma/\sqrt{n})$ is a standard normal variate. Hence the probability that $(\bar{X} - \mu)/(\sigma/\sqrt{n})$ will be in a given interval, for example, between ± 1.96, is given by

$$P\left(-1.96 < \frac{\bar{X} - \mu}{\sigma/\sqrt{n}} \le 1.96\right) = \Phi(1.96) - \Phi(-1.96) = 0.95$$

For the concrete data in Example 5.1, if the standard deviation σ is known (for example, through years of experience and testing) to be equal to 0.65 ksi, the preceding statement becomes

$$P\left[-1.96 < \frac{\bar{X} - \mu}{0.65/\sqrt{25}} \le 1.96\right] = 0.95$$

or

$$P[-0.255 < \bar{X} - \mu \le 0.255] = 0.95$$

Physically, this statement implies that before obtaining the test results, it is expected that the sample mean $\bar{X}$ will lie within 0.255 ksi of the actual mean μ with 95% probability. After the test results are obtained, which give $\bar{x} = 5.6$ (from Example 5.1), the equation above yields

$$P[-0.255 < 5.6 - \mu \le 0.255] = 0.95$$

or

$$P[5.6 - 0.255 \le \mu < 5.6 + 0.255] = 0.95$$

Thus

$$P[5.345 \le \mu < 5.855] = 0.95$$

This appears to imply that "the mean value μ of the crushing strength of the concrete lies between 5.345 ksi and 5.855 ksi with probability 0.95."

Strictly speaking, however, this implication is not correct. In the first place, in the classical approach the population mean μ is a constant, not a random variable. Moreover, if another 25 concrete specimens were tested, the probability that μ will lie in the same interval may well be different.

Alternatively, if different sets of observed data were used to construct similar 95% probability intervals, we may say that on the average 95%

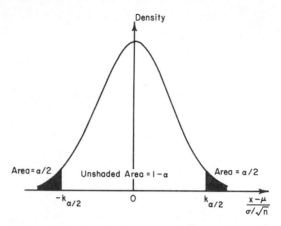

Figure 5.4   Density function of $(\overline{X} - \mu)/(\sigma/\sqrt{n})$

of these intervals will contain the population mean μ. Hence, in a general case, if we denote $(1 - \alpha)$ as the specified confidence level, and $\mp k_{\alpha/2}$ as the values of the standard normal variate with cumulative probability levels $\alpha/2$ and $(1 - \alpha/2)$, respectively, as shown in Fig. 5.4, we may write

$$P\left(-k_{\alpha/2} < \frac{\overline{X} - \mu}{\sigma/\sqrt{n}} \leq k_{\alpha/2}\right) = 1 - \alpha \qquad (5.12)$$

Upon rearrangement and substitution of the observed sample mean $\bar{x}$, Eq. 5.12 becomes

$$P\left(\bar{x} - k_{\alpha/2}\frac{\sigma}{\sqrt{n}} < \mu \leq \bar{x} + k_{\alpha/2}\frac{\sigma}{\sqrt{n}}\right) = 1 - \alpha$$

More properly, then, the interval estimated on the basis of a single sample of size n should be interpreted as follows: "There is a confidence of $(1 - \alpha)$ that the estimated interval contains the unknown μ." Thus such an interval is called the $(1 - \alpha)$ *confidence interval* for the mean μ, and is given by

$$\langle\mu\rangle_{1-\alpha} = \left(\bar{x} - k_{\alpha/2}\frac{\sigma}{\sqrt{n}} \; ; \quad \bar{x} + k_{\alpha/2}\frac{\sigma}{\sqrt{n}}\right) \qquad (5.13)$$

It should be emphasized that the confidence intervals so obtained would be exact for normal populations with known standard deviations. However, for nonnormal populations, confidence intervals of Eq. 5.13 are only approximate; the accuracy of the approximation, however, will increase with the sample size n.

The following steps summarize the general procedure for establishing the confidence interval of the mean μ when there is prior knowledge of the standard deviation σ.

1. Choose the confidence level $(1 - \alpha)$.
2. Determine the value $k_{\alpha/2}$ from a table of normal probability (for example, Table A.1 in Appendix A); specifically,

$$k_{\alpha/2} = \Phi^{-1}\left(1 - \frac{\alpha}{2}\right)$$

3. Apply Eq. 5.13 using the sample mean $\bar{x}$ estimated from the observed sample of size n, obtaining the $(1 - \alpha)$ confidence interval for the mean

$$\langle\mu\rangle_{1-\alpha} = \left(\bar{x} - \frac{\sigma}{\sqrt{n}} k_{\alpha/2}; \quad \bar{x} + \frac{\sigma}{\sqrt{n}} k_{\alpha/2}\right)$$

EXAMPLE 5.5

The daily dissolved oxygen (DO) concentration for a stream at a station has been recorded for 30 days. The daily level of DO concentration is known to vary with a standard deviation of $\sigma = 4.2$ mg/l. From the sample of 30 observations, the sample mean is calculated to be $\bar{x} = 2.52$ mg/l. Determine the 99% confidence interval for the mean daily DO concentration.

Following the steps outlined above, we obtain

$\qquad$ (a) $1 - \alpha = 0.99 \qquad$ or $\quad \alpha = 1 - 0.99 = 0.01$

$\qquad$ (b) from Table A.1, $k_{.005} = \Phi^{-1}(0.995) = 2.58$

$\qquad$ (c) $\dfrac{\sigma}{\sqrt{n}} k_{\alpha/2} = \dfrac{\sqrt{4.2}}{\sqrt{30}} 2.58 = 0.965$

The 99% confidence interval for μ, therefore, is $(2.52 - 0.965; 2.52 + 0.965)$ or $(1.56; 3.49)$ mg/l.

Similarly, we obtain the 95% confidence interval as follows:

$$k_{.025} = \Phi^{-1}(0.975) = 1.96$$

and,

$$\frac{\sigma}{\sqrt{n}} k_{.025} = 0.733$$

Hence, the 95% confidence interval is $(1.79; 3.25)$ mg/l.

Therefore, if the distribution of DO concentration is Gaussian, these results would be the exact 99% and 95% confidence intervals. If the underlying distribution is not Gaussian (or is unknown), the results obtained above would be approximate confidence intervals.

Comparing the 95% and 99% confidence intervals computed in Example 5.5, we observe that the 99% confidence interval is larger than that at the

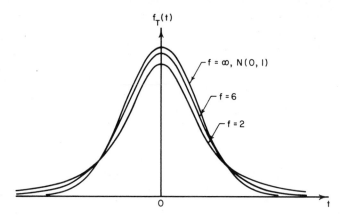

Figure 5.5 Student t-distribution

95% confidence level. This is reasonable because a larger interval is more likely to contain μ than a smaller one.

Furthermore, we observe that the confidence interval for the mean depends on the standard deviation σ and on the sample size n. From Eq. 5.13, it is clear that as σ decreases or as n increases, the confidence interval becomes narrower for the same confidence level $(1 - \alpha)$. This means that smaller population variance or larger sample size would increase the accuracy of the sample mean as the estimator of the population mean.

Confidence interval with unknown variance. In general, the value of σ is not known and must be estimated using Eq. 5.3. The foregoing procedure for determining the confidence interval for μ may still be used when the sample size n is large. That is, when n is large (for instance, > 20), the sample variance s^2 is a good estimator of the population variance σ^2 (see Section 5.2.4). Consequently, in such cases, using s for σ in Eq. 5.13, we obtain the confidence interval as if σ is known. It should be emphasized, however, that the confidence intervals so obtained will be very approximate if n is small (for example, < 10).

When there is no prior knowledge of the population variance, an exact confidence interval for μ can be determined if the underlying population is Gaussian. In this case, the probability distribution of $(\bar{X} - \mu)/(S/\sqrt{n})$ is required. This can be shown (for example, Freund, 1962) to have the t-distribution (or the Student's t-distribution) with $(n - 1)$ *degrees of freedom*, whose density function is

$$f_T(t) = \frac{\Gamma[(f + 1)/2]}{\sqrt{\pi f}\,\Gamma(f/2)}\left(1 + \frac{t^2}{f}\right)^{-(1/2)(f+1)} \qquad -\infty < t < \infty \quad (5.14)$$

where f is the *degree of freedom*. A family of t-distributions with various values of f is shown in Fig. 5.5. It may be observed that the t-distribution has a bell-shape density function similar to the normal curve and is symmetrical about the origin. For small values of f the density function of the t-distribution is flatter than the standard normal distribution; however, as f increases, it tends toward the standard normal distribution, as illustrated in Fig. 5.5.

On this basis, therefore, we can form the following probability statement for the random variable $(\bar{X} - \mu)/(S/\sqrt{n})$:

$$P\left(-t_{\alpha/2,n-1} < \frac{\bar{X} - \mu}{S/\sqrt{n}} \le t_{\alpha/2,n-1}\right) = 1 - \alpha \qquad (5.15)$$

where $t_{\alpha/2,n-1}$ has a similar interpretation as $k_{\alpha/2}$ of Eq. 5.12. In the present case, of course, $t_{\alpha/2,n-1}$ denotes the precentile value of the t-distribution with $(n - 1)$ degrees of freedom. In general, $t_{\alpha/2,f}$ is the value of the variate T at the cumulative probability $(1 - \alpha/2)$, as shown in Fig. 5.6. Values of $t_{\alpha/2,f}$ are tabulated for various probability levels $p\text{-}(1 - \alpha/2)$, with different degrees of freedom in Table A.2. Rearranging the terms in Eq. 5.15, the exact confidence interval for the mean (of Gaussian population), therefore, is

$$\langle\mu\rangle_{1-\alpha} = \left[\bar{x} - t_{\alpha/2,n-1}\frac{s}{\sqrt{n}} \; ; \; \bar{x} + t_{\alpha/2,n-1}\frac{s}{\sqrt{n}}\right] \qquad (5.16)$$

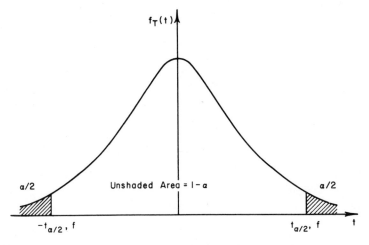

Figure 5.6 $(1 - \alpha)$ Confidence interval in t-distribution

where $\bar{x}$ and s are the sample mean and sample standard deviation, n is the sample size, and $(1 - \alpha)$ is the specified confidence level.

EXAMPLE 5.6

Suppose that the 30 observations of daily DO concentration presented in Example 5.5 give a sample mean $\bar{x} = 2.52$ mg/l and sample standard deviation $s = 4.2$ mg/l. Determine the confidence interval for the population mean μ.

The number of data points is 30; hence, $(\bar{X} - \mu)/(S/\sqrt{n})$ has a t-distribution with $f = n - 1 = 29$ degrees of freedom. If a 99% confidence level is desired, $\alpha = 1 - 0.99 = 0.01$; and $(\alpha/2) = 0.005$. From Table A.2, with $f = 29$, we obtain under the column for $(1 - \alpha/2) = 0.995$,

$$t_{\alpha/2,29} = t_{.005,29} = 2.756$$

Hence the 99% confidence interval for the mean daily DO concentration is

$$\langle \mu \rangle_{.99} = \left(2.52 - 2.756 \frac{\sqrt{4.2}}{\sqrt{30}} ; \; 2.52 + 2.756 \frac{\sqrt{4.2}}{\sqrt{30}} \right) = (1.49; 3.55) \text{ mg/l}$$

This is a larger interval than that obtained in Example 5.5 of $(1.56, 3.49)$ mg/l, which was obtained assuming that the standard deviation is known. This is to be expected, since not knowing the value of σ introduces additional uncertainty; hence, to maintain the same confidence level, a wider interval is required than if σ is known.

One-sided confidence limit for the mean. The confidence interval established above is called a *two-sided confidence interval*, because it includes the upper and lower limits that bound the value of the population mean μ. There are instances, in practice, in which only the lower limit or the upper limit is pertinent. In such cases, we would be interested in the *one-sided confidence limit* for the mean μ. For example, in the case of material strength, or capacity of a highway or of a flood channel, the lower limit of the mean μ will be of engineering interest.

For such purposes, the $(1 - \alpha)$ lower confidence limit, denoted $< \mu)_{1-\alpha}$, means that the population mean μ will be larger than this limit with a confidence level of $(1 - \alpha)$. Assuming prior knowledge of σ, such a limit is obtained by forming the following probability statement for the standard normal variate $(\bar{X} - \mu)/(\sigma/\sqrt{n})$:

$$P\left(\frac{\bar{X} - \mu}{\sigma/\sqrt{n}} \leq k_\alpha \right) = 1 - \alpha \qquad (5.17)$$

where $1 - \alpha$ is the specified confidence level, and $k_\alpha = \Phi^{-1}(1 - \alpha)$. Rearranging the terms in Eq. 5.17, we have

$$P\left(\mu \geq \bar{X} - k_\alpha \frac{\sigma}{\sqrt{n}} \right) = 1 - \alpha \qquad (5.17a)$$

Hence the $(1 - \alpha)$ *lower confidence limit* for the mean μ is

$$< \mu)_{1-\alpha} = \left(\bar{x} - k_\alpha \frac{\sigma}{\sqrt{n}} \right) \tag{5.18}$$

EXAMPLE 5.7

Test results for 100 randomly selected specimens of 1-cm-diameter A36 steel show that the sample mean and sample standard deviation of the yield strength are, respectively, $\bar{x} = 2200$ kgf (kilogram force) and $s = 220$ kgf. For specification purposes, the manufacturer is required to specify the 95% lower confidence limit of the mean yield strength. Because of the large sample size (n = 100), assume that σ is satisfactorily given by s, which is 220 kgf.
With
$$1 - \alpha = 0.95, \qquad \alpha = 0.05$$
and
$$k_{.05} = \Phi^{-1}(0.95) = 1.65$$
The lower confidence limit, therefore, is

$$\bar{x} - k_\alpha \frac{\sigma}{\sqrt{n}} = 2200 - 1.65 \frac{220}{\sqrt{100}}$$
$$= 2164 \text{ kgf}$$

In other words, it is 95% confident that the mean yield strength will be at least 2164 kgf.

Conversely, there are situations in which the upper confidence limits are required. For example, in determining the wind load on a structure, we would like to state with a high degree of confidence that the mean wind load will not exceed a certain limit. In such a case, the upper confidence limit on μ is desired. Following the same procedure as that of Eq. 5.17, it can be shown that the $(1 - \alpha)$ *upper confidence limit* is

$$(\mu >_{1-\alpha} = \left(\bar{x} + k_\alpha \frac{\sigma}{\sqrt{n}} \right) \tag{5.19}$$

If the sample size n is small, and the population standard deviation is not known, the t-distribution should be used to determine the corresponding upper and lower confidence limits. On this basis, the appropriate confidence limits are as follows:
$(1 - \alpha)$ lower confidence limit

$$< \mu)_{1-\alpha} = \bar{x} - t_{\alpha,n-1} \frac{s}{\sqrt{n}} \tag{5.20}$$

$(1 - \alpha)$ upper confidence limit

$$(\mu >_{1-\alpha} = \bar{x} + t_{\alpha,n-1} \frac{s}{\sqrt{n}} \tag{5.21}$$

It should be emphasized that the confidence intervals given in Eqs. 5.13 and 5.16, and the one-sided confidence limits of Eqs. 5.18 through 5.21, for the population mean μ are exact if the underlying population is Gaussian. However, for practical purposes, these results are applicable to non-Gaussian populations if the sample size is at least moderately large (for example, $n > 10$); for this reason, the preceding equations may be used to determine the (approximate) confidence intervals and limits of μ irrespective of the distribution of the underlying population.

EXAMPLE 5.8

Table E5.8 shows data for storms and associated runoffs on the Monocacy River at Jug Bridge, Maryland (data from Linsley and Franzini, 1964).

(a) Compute the sample mean and sample standard deviation for the precipitation and runoff, based on the data given in Table E5.8.

(b) Using the sample variance in place of the corresponding population variance, determine the 99.9% confidence interval for the mean precipitation. Also determine the corresponding 99.9% upper confidence limit.

Table E5.8. Precipitation and Runoff Data

Storm no.	Precipitation (in.)	Runoff (in.)
1	1.11	0.52
2	1.17	0.40
3	1.79	0.97
4	5.62	2.92
5	1.13	0.17
6	1.54	0.19
7	3.19	0.76
8	1.73	0.66
9	2.09	0.78
10	2.75	1.24
11	1.20	0.39
12	1.01	0.30
13	1.64	0.70
14	1.57	0.77
15	1.54	0.59
16	2.09	0.95
17	3.54	1.02
18	1.17	0.39
19	1.15	0.23
20	2.57	0.45
21	3.57	1.59
22	5.11	1.74
23	1.52	0.56
24	2.93	1.12
25	1.16	0.64

Solution

(a) Let x = precipitation (in inches); and y = runoff (in inches).

x_i	x_i^2	y_i	y_i^2
1.11	1.23	0.52	0.27
1.17	1.37	0.40	0.16
1.79	3.20	0.97	0.94
5.62	31.58	2.92	8.53
1.13	1.28	0.17	0.03
1.54	2.37	0.19	0.04
3.19	10.18	0.76	0.58
1.73	2.99	0.66	0.44
2.09	4.37	0.78	0.61
2.75	7.56	1.24	1.54
1.20	1.44	0.39	0.15
1.01	1.02	0.30	0.09
1.64	2.69	0.70	0.49
1.57	2.46	0.77	0.59
1.54	2.37	0.59	0.35
2.09	4.37	0.95	0.90
3.54	12.53	1.02	1.04
1.17	1.37	0.39	0.15
1.15	1.32	0.23	0.05
2.57	6.60	0.45	0.20
3.57	12.74	1.59	2.53
5.11	26.11	1.74	3.03
1.52	2.31	0.56	0.31
2.93	8.58	1.12	1.25
1.16	1.35	0.64	0.41
53.89	153.39	20.05	24.68

Thus

$$\bar{x} = \frac{53.89}{25} = 2.16 \text{ in.}$$

$$s_x^2 = \frac{1}{24}[153.39 - 25(2.16)^2] = 1.53$$

$$s_x = 1.24 \text{ in.}$$

$$\bar{y} = \frac{20.05}{25} = 0.80 \text{ in.}$$

$$s_y^2 = \frac{1}{24}[24.68 - 25(0.80)^2] = 0.36$$

$$s_y = 0.60 \text{ in.}$$

(b) With $\bar{x} = 2.16$, and assuming $\sigma_X \simeq s_x = 1.24$, the standard deviation of $\bar{X}$ is

$$\sigma_{\bar{x}} = \frac{\sigma_X}{\sqrt{n}} = \frac{1.24}{5} = 0.25$$

Although the precipitation is not Gaussian (see Example 6.3), $\bar{X}$ may be assumed to be approximately Gaussian since the sample size is relatively large ($n = 25$). Hence, according to Eq. 5.12, the 99.9% confidence interval for the mean precipitation is

$$\langle\mu_X\rangle_{.999} = \left(\bar{x} - k_{.0005}\frac{\sigma_X}{\sqrt{25}} \; ; \; \bar{x} + k_{.0005}\frac{\sigma_X}{\sqrt{25}}\right)$$
$$= [2.16 - 3.29(0.25); \; 2.16 + 3.29(0.25)]$$
$$= (1.34 \text{ in.}, \; 2.98 \text{ in.})$$

whereas the one-sided upper 99.9% confidence limit on the mean precipitation is

$$(\mu_X)_{.999} = \bar{x} + k_{.001}\frac{\sigma_X}{\sqrt{25}} = 2.16 + 3.09 \times 0.25 = 2.93 \text{ in.}$$

EXAMPLE 5.9

(a) In a traffic survey where speeds of vehicles are measured, it is desired to determine the mean vehicle speed to within ± 1 kph (kilometer per hour) with 99% confidence. From a preliminary study, the standard deviation of the vehicle speed is found to be 3.58 kph. Assume that all observations are independent; determine the number of observations required.

(b) If 150 observations were taken, what would be the confidence level associated with the interval of ± 1 kph of the mean speed? Assume that the standard deviation of vehicle speed is still 3.58 kph.

Solution

(a) Let n be the number of observations required. The confidence interval is given by $\bar{x} \pm k_{\alpha/2}(\sigma/\sqrt{n})$, where σ is the known standard deviation. For a 99% confidence level, $\alpha = 0.01$ and $k_{\alpha/2} = k_{.005} = 2.58$. Therefore, setting

$$k_{\alpha/2}\frac{\sigma}{\sqrt{n}} = 2.58 \times \frac{3.58}{\sqrt{n}} = 1$$

the number of observations required is

$$n = (2.58 \times 3.58)^2 = 85$$

(b) If 150 observations had been taken and the same confidence interval were desired, we would expect the confidence level to increase. In other words, the value of α would decrease. Setting

$$k_{\alpha/2} \times \frac{3.58}{\sqrt{150}} = 1$$

we obtain

$$k_{\alpha/2} = \frac{\sqrt{150}}{3.58} = 3.43$$

and

$$\frac{\alpha}{2} = 1 - \Phi(3.43) = 1 - 0.99969 = 0.00031$$

The confidence level, therefore, is

$$1 - \alpha = 1 - 0.00062 = 0.99938 \quad \text{or} \quad 99.938\%$$

5.2.3. Problems of measurement theory

One of the major applications of point and interval estimation is in the theory of measurements (Parratt, 1961; Barry, 1964). Problems involving measurements require estimation of a fixed (but unknown) quantity, which is therefore analogous to the estimation of the unknown population mean μ.

In measuring, for instance, a distance δ, several (for example n) measurements may be taken constituting a sample of size n. The object then is to estimate the actual distance δ from the sample measurements* $d_1, d_2, \cdots, d_n$. Point and interval estimations then may be used to estimate this true distance (not its mean value). In this regard, δ is analogous to μ; hence

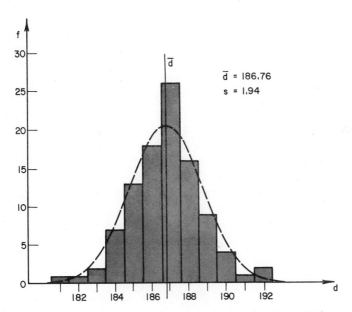

Figure 5.7   Histogram of measured distance (after Bachmann, 1973)

---

* Observed measurements will, in general, contain two types of measurement errors, namely, random errors and systematic errors (Parratt, 1961; Barry, 1964). It is assumed here that the sample measurements have been adjusted for systematic errors.

the methods developed for estimating the mean μ (which is a constant) can be used to estimate the distance δ (also a constant). In particular, the point estimate of δ is

$$\bar{d} = \frac{1}{n} \sum_{i=1}^{n} d_i \tag{5.22}$$

In other words, a series of measurements $d_1, d_2, \cdots, d_n$ are presumed to be the sample values of the independent random variables $D_1, D_2, \cdots, D_n$ representing the populations of possible measurements, so that the point estimator of δ is

$$\bar{D} = \frac{1}{n} \sum_{i=1}^{n} D_i \tag{5.23}$$

with expected value

$$E(\bar{D}) = \delta \tag{5.24}$$

and

$$\mathrm{Var}(\bar{D}) \simeq \frac{s^2}{n} \tag{5.25}$$

where

$$s^2 = \frac{1}{n-1} \sum_{i=1}^{n} (d_i - \bar{d})^2$$

In measurement theory, the standard deviation of $\bar{D}$, that is, $\sqrt{\mathrm{Var}\,\bar{D}} \simeq (s/\sqrt{n})$, is known as the *standard error*.

Implicit in Eqs. 5.22 through 5.25 are also the assumptions of random sampling; namely, in this case, that $D_1, D_2, \cdots, D_n$ are statistically independent and are identically distributed or $f_{D_1} = f_{D_2} = \cdots = f_{D_n}$. Moreover, these distributions are invariably assumed to be normal, as supported by observations (see, for example, Fig. 5.7).

It follows then that the variate $(\bar{D} - \delta)/(S/\sqrt{n})$ has a t-distribution with $(n-1)$ degrees of freedom; hence, the basis for the confidence interval for δ is

$$P\left(-t_{\alpha/2,n-1} < \frac{\bar{D} - \delta}{S/\sqrt{n}} \leq t_{\alpha/2,n-1}\right) = 1 - \alpha$$

and thus the $(1 - \alpha)$ confidence interval for δ is

$$\langle \delta \rangle_{1-\alpha} = \left(\bar{d} - t_{\alpha/2,n-1}\frac{s}{\sqrt{n}} \; ; \quad \bar{d} + t_{\alpha/2,n-1}\frac{s}{\sqrt{n}}\right) \tag{5.26}$$

When a function of one or more distances (or geometric dimensions) is involved, the value of the function is usually estimated on the basis of the mean measured distances. That is, if a function of several distances l_1, l_2, $\cdots$, l_k is

$$\zeta = Z(l_1, l_2, \cdots, l_k)$$

in which l_1, l_2, $\cdots$, l_k are estimated, respectively, by the mean measurements $\bar{l}_1, \bar{l}_2, \cdots, \bar{l}_k$, then the point estimator of ζ (using the approximation of Eq. 4.43) is

$$\bar{Z} \simeq Z(\bar{L}_1, \bar{L}_2, \cdots, \bar{L}_k) \tag{5.27}$$

where $\bar{L}_i$ is the estimator of l_i in accordance with Eq. 5.23. The estimator $\bar{Z}$, therefore, is also a random variable with

$$E(\bar{Z}) \simeq Z[E(\bar{L}_1), E(\bar{L}_2), \cdots, E(\bar{L}_k)] = \zeta \tag{5.28}$$

and, in view of the errors in $\bar{L}_i$, the standard error in $\bar{Z}$, assuming independent $\bar{L}_1, \bar{L}_2, \cdots, \bar{L}_k$, therefore, is obtained (applying Eq. 4.44) from

$$\mathrm{Var}(\bar{Z}) \simeq \sum_{i=1}^{k} \left(\frac{\partial \bar{Z}}{\partial \bar{L}_i}\right)^2 \sigma_{\bar{L}_i}^2 \tag{5.29}$$

which is known as the "propagation of errors" in measurement theory.

Thence, assuming $\bar{Z}$ to be Gaussian* with mean ζ and standard deviation $\sigma_{\bar{Z}} = \sqrt{\mathrm{Var}(\bar{Z})}$, we obtain the confidence interval for ζ as follows:

$$P\left(-k_{\alpha/2} < \frac{\bar{Z} - \zeta}{\sigma_{\bar{Z}}} \leq k_{\alpha/2}\right) = 1 - \alpha$$

Thus the $(1 - \alpha)$ confidence interval is

$$\langle \zeta \rangle_{1-\alpha} = (\bar{z} - k_{\alpha/2}\, \sigma_{\bar{Z}};\quad \bar{z} + k_{\alpha/2}\, \sigma_{\bar{Z}}) \tag{5.30}$$

where $\bar{z} = Z(\bar{l}_1, \bar{l}_2, \cdots, \bar{l}_k)$.

To clarify these, consider the following examples from surveying.

EXAMPLE 5.10

The straight-line distance between two geodetic stations A and B is measured with an electronic ranging instrument called a *tellerometer*. The following are ten

---

* This assumption would be consistent with the first-order approximation if Eq. 5.27 is linear; however, for nonlinear functions, this assumption would not be valid. The estimators $\bar{L}_1, \bar{L}_2, \ldots, \bar{L}_k$ are approximately Gaussian by virtue of the central limit theorem; hence $\bar{Z}$ will not be Gaussian unless Eq. 5.27 is linear.

independent measurements of the distance:

(1) 45479.4 m	(6) 45479.2 m
(2) 45479.6 m	(7) 45479.6 m
(3) 45479.3 m	(8) 45479.5 m
(4) 45479.5 m	(9) 45479.3 m
(5) 45479.8 m	(10) 45479.1 m

(a) Estimate the true distance δ.
(b) Compute the standard deviation of the measured distances.
(c) Determine the standard error of the estimated distance.
(d) Determine the 2-sided 90% confidence interval of the actual distance δ.

Solution

(a) Estimated distance,

$$\bar{d} = \frac{1}{10} \{45479.4 + 45479.6 + \cdots + 45479.1\}$$
$$= 45{,}479.43 \text{ m}$$

(b) Variance of the measured distances,

$$s^2 = \tfrac{1}{9}\{(45479.4 - 45479.43)^2 + (45479.6 - 45479.43)^2$$
$$+ \cdots + (45479.1 - 45479.43)^2\}$$
$$= \tfrac{1}{9}\{0.401\} = 0.0445 \text{ m}^2$$

Hence the standard deviation is $s = 0.21$ m.

(c) According to Eq. 5.25, the standard error of the estimated distance is

$$\sigma_{\bar{D}} = \frac{s}{\sqrt{n}} = \frac{0.21}{\sqrt{10}} = 0.0664 \text{ m}$$

(d) With $f = n - 1 = 9$, and $\alpha = 0.10$; $t_{.05,9} = 1.8331$ from Table A.2. Then using $\bar{d} = 45479.43$ m, and $s = 0.21$ m, we obtain the 90% confidence interval for δ,

$$\langle \delta \rangle_{.90} = \left[45479.43 - 1.833 \left(\frac{0.21}{\sqrt{10}} \right); 45479.43 + 1.833 \left(\frac{0.21}{\sqrt{10}} \right) \right]$$
$$= (45479.31; 45479.55) \text{ m}$$

EXAMPLE 5.11

The area of a rectangular tract of land is being considered. The sides of the rectangle are measured several times, with associated statistics summarized as follows (see also Fig. E5.11).

Length	No. of independent measurements	Mean measurement	Sample variance
D	9	60 m	0.81 m²
B	4	70 m	0.64 m²
C	4	30 m	0.32 m²

Determine the 95% confidence interval of the actual area of the tract.

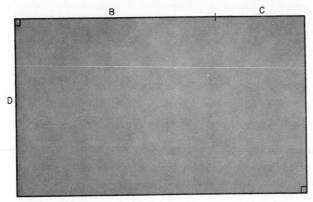

Figure E5.11

Solution

In this case, the area is

$$A = (B + C)D$$

According to Eq. 5.27, the area is estimated by

$$\bar{A} = (\bar{B} + \bar{C})\bar{D}$$

Substituting the mean measured distances, the estimated area therefore is

$$\bar{A} = (70 + 30)60 = 6{,}000 \text{ m}^2$$

The standard error of the estimator $\bar{A}$, according to Eq. 5.29, is

$$
\begin{aligned}
\sigma_{\bar{A}}^2 &= \bar{D}^2\sigma_{\bar{B}}^2 + \bar{D}^2\sigma_{\bar{C}}^2 + (\bar{B} + \bar{C})^2\sigma_{\bar{D}}^2 \\
&= (60)^2\left(\frac{s_B^{\,2}}{4}\right) + (60)^2\left(\frac{s_C^{\,2}}{4}\right) + (100)^2\left(\frac{s_D^{\,2}}{9}\right) \\
&= 3600(0.16) + 3600(0.08) + 10{,}000(0.09) \\
&= 1764 \text{ m}^4
\end{aligned}
$$

Thus

$$\sigma_{\bar{A}} = 42 \text{ m}^2$$

Finally, using Eq. 5.30, we obtain the 95% confidence interval for the area A as follows:

$$k_{.025} = 1.96$$

$$P\left(-1.96 < \frac{6000 - A}{42} \leq 1.96\right) = 0.95$$

Thus the required confidence interval is

$$
\begin{aligned}
\langle A \rangle_{.95} &= [6000 - 1.96(42); \ 6000 + 1.96(42)] \\
&= (5916.9, \ 6083.1) \text{ m}^2
\end{aligned}
$$

The problems illustrated in Examples 5.10 and 5.11 are quite common in surveying, photogrammetry, and geodetic engineering.

5.2.4. Interval estimation of the variance

How good is the estimator S^2? Using an approach similar to the establishment of the confidence interval for the population mean, the confidence interval for the population variance σ^2 may also be developed. For this purpose, we first observe that the sample variance (see Eq. 5.3)

$$S^2 = \frac{1}{n-1} \sum_{i=1}^{n} (X_i - \bar{X})^2 \tag{5.31}$$

is a random variable, with expected value

$$E(S^2) = \frac{1}{n-1} E\left[\sum_{i=1}^{n} (X_i - \bar{X})^2 \right]$$

$$= \frac{1}{n-1} E\left[\sum_{i=1}^{n} \{ (X_i - \mu) - (\bar{X} - \mu) \}^2 \right]$$

$$= \frac{1}{n-1} \left[\sum_{i=1}^{n} E(X_i - \mu)^2 - nE(\bar{X} - \mu)^2 \right]$$

but

$$E(X_i - \mu)^2 = \sigma^2$$

and it can be shown that

$$E(\bar{X} - \mu)^2 = \frac{\sigma^2}{n}$$

Hence

$$E(S^2) = \frac{1}{n-1} \left[\sum_{i=1}^{n} \sigma^2 - n \cdot \frac{\sigma^2}{n} \right] = \sigma^2 \tag{5.32}$$

For this reason, Eq. 5.31 is an unbiased estimator of σ^2, as we asserted earlier in Eq. 5.3.

The variance of S^2 is given (Hald, 1952) by

$$\text{Var}(S^2) = \frac{\sigma^4}{n} \left(\frac{\mu_4}{\sigma^4} - \frac{n-3}{n-1} \right) \tag{5.33}$$

where $\mu_4 = E(X - \mu)^4$ is the *fourth central moment* of the population random variable X. It may be observed that as n increases, the variance of S^2 decreases.

Confidence interval for σ^2. For large n, the sample variance of Eq.

5.31 may be assumed, on the basis of the central limit theorem, to have a normal distribution with the mean and variance of Eqs. 5.32 and 5.33. Then, at a confidence level of $(1 - \alpha)$,

$$P\left(-k_{\alpha/2} < \frac{S^2 - \sigma^2}{\sqrt{\mathrm{Var}(S^2)}} \leq k_{\alpha/2}\right) = 1 - \alpha \qquad (5.34)$$

where μ_4 in Eq. 5.33 can be evaluated using the fourth sample moment, or

$$\mu_4 \simeq \frac{1}{n} \sum_{i=1}^{n} (x_i - \bar{x})^4 \qquad (5.35)$$

Hence the $(1 - \alpha)$ confidence interval for the population variance (when n is large) may be obtained as

$$\langle\sigma^2\rangle_{1-\alpha} = [s^2 - k_{\alpha/2} \sqrt{\mathrm{Var}(S^2)}\,; \quad s^2 + k_{\alpha/2} \sqrt{\mathrm{Var}(S^2)}\,] \qquad (5.36)$$

If the population is Gaussian, the variance of S^2 is (Freund, 1962)

$$\mathrm{Var}(S^2) = \frac{2}{n-1} \sigma^4 \qquad (5.37)$$

Thence, the corresponding confidence interval becomes

$$\langle\sigma^2\rangle_{1-\alpha} = \left(\frac{s^2}{1 + k_{\alpha/2} \sqrt{2/(n-1)}}\,; \quad \frac{s^2}{1 - k_{\alpha/2} \sqrt{2/(n-1)}}\right) \qquad (5.38)$$

EXAMPLE 5.12

For the concrete strength data of Example 5.1, we had $s^2 = 0.44$ and $n = 25$. Assuming the strength of the concrete to be a normal variate, we obtain the 95% confidence interval for its variance σ^2, approximately on the basis of Eq. 5.38, as follows:

$$\langle\sigma^2\rangle_{.95} = \left(\frac{0.44}{1 + 1.96\sqrt{2/24}}\,; \quad \frac{0.44}{1 - 1.96\sqrt{2/24}}\right)$$
$$= (0.28; 1.01)$$

Exact confidence limits of δ^2 *for normal population.* The approximations given in Eqs. 5.36 and 5.38 can be quite poor when n is small. If the population is normal, however, exact confidence limits can be obtained; the basis for such exact estimates is as follows.

Rewriting Eq. 5.31, we have

$$(n - 1)S^2 = \sum_{i=1}^{n} [(X_i - \mu) - (\bar{X} - \mu)]^2$$

$$= \sum_{i=1}^{n} (X_i - \mu)^2 - n(\bar{X} - \mu)^2$$

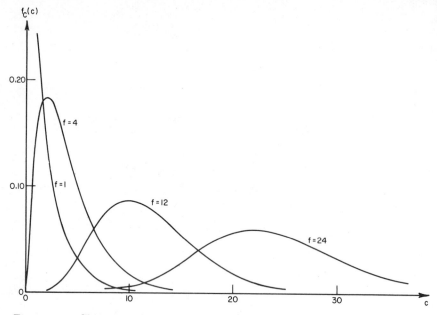

Figure 5.8   Chi-square-distribution with different f

Dividing through by σ^2, we obtain

$$\frac{(n-1)S^2}{\sigma^2} = \sum_{i=1}^{n} \left(\frac{X_i - \mu}{\sigma}\right)^2 - \left(\frac{\bar{X} - \mu}{\sigma/\sqrt{n}}\right)^2 \qquad (5.39)$$

where X_i are presumed to be normal, and thus $\bar{X}$ is also normal. Then the first term on the right side of Eq. 5.39 is the sum of squares of n independent standard normal variates; it can be shown, by generalizing the result of Example 4.6, that this has a chi-square distribution with n degrees of freedom (to be denoted χ_n^2). Similarly, the second term on the right side of Eq. 5.39 is also the square of a standard normal variate and therefore has a chi-square distribution with one degree of freedom. Moreover, it can be shown (Hoel, 1962) that the sum of two chi-square variates with p and q degrees of freedom is also a chi-square variate with $(p + q)$ degrees of freedom. On these bases, therefore, $(n - 1)S^2/\sigma^2$ of Eq. 5.39 has a χ_{n-1}^2 distribution; that is, a chi-square distribution with $(n - 1)$ degrees of freedom.

In general, the density function of the χ_f^2 distribution with f degrees of freedom is given by

$$f_C(c) = \frac{1}{2^{f/2}\Gamma(f/2)} c^{(f/2 - 1)} e^{-c/2} \qquad c \geq 0 \qquad (5.40)$$

Such a distribution is shown in Fig. 5.8 for different degrees-of-freedom f. As would be expected, by virtue of the central limit theorem, the χ_f^2 distribution approaches the normal distribution as $f \to \infty$; this may be observed also in Fig. 5.8. Because $f = n - 1$, this gives us a basis (albeit crude) for determining the sample size n necessary to ensure reasonable approximations of Eqs. 5.36 and 5.38. Visually, from Fig. 5.8, it appears that $n \geq 25$ may be sufficient sample size to permit the applications of Eqs. 5.36 and 5.38.

If the population is Gaussian, the upper confidence limit for the variance σ^2, according to the foregoing chi-square distribution, is given by

$$P\left[\frac{(n-1)S^2}{\sigma^2} \geq c_{\alpha,n-1}\right] = 1 - \alpha \qquad (5.41)$$

where $c_{\alpha,n-1}$ denotes the value of the χ_{n-1}^2 variate at the cumulative probability of α; that is,

$$P(C \leq c_{\alpha,n-1}) = \alpha$$

as illustrated graphically in Fig. 5.9. Values of $c_{\alpha,n-1}$ are tabulated in Table A.3 of Appendix A for specified values of α and $n - 1 = f$.

Then the exact $(1 - \alpha)$ upper confidence limit for σ^2 of a normal population is

$$(\sigma^2 >_{1-\alpha} = \frac{(n-1)s^2}{c_{\alpha,n-1}} \qquad (5.42)$$

Although two-sided confidence intervals for σ^2 may similarly be developed, the one-sided (upper) confidence limit of Eq. 5.42 is more useful in the case of variances.

EXAMPLE 5.13

For the DO data in Example 5.5, we had $s^2 = 4.2$ and $n = 30$. If a 95% upper confidence limit on σ^2 is desired, then from Table A.3 (with $\alpha = 0.05$) we obtain

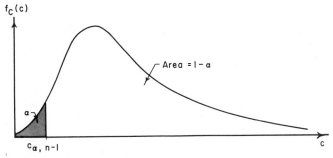

Figure 5.9 Chi-square distribution with $(n - 1)$ degrees of freedom

$c_{\alpha, n-1} = c_{.05, 29} = 17.708$. Hence the (upper) confidence limit for the variance of DC is

$$\langle \sigma^2 \rangle_{.95} = \frac{29 \times 4.2}{17.708} = 6.89 \ (mg/l)^2$$

5.2.5. Estimation of proportion

In many engineering problems requiring probabilistic formulations, the necessary probability measures must be estimated on the basis of experimental observations; for example, the probability of hurricane-intensity wind occurring in a year, the proportion of vehicular traffic making left turns at an intersection, or the proportion of embankment material meeting specified compaction standards.

In such cases, the required probability may be estimated as the proportion of occurrences (of an event) in a Bernoulli sequence (see Section 3.2.3). Suppose that we have a sequence of n independent trials $X_1, X_2, \cdots, X_n$, where every X_i is a two-valued random variable; specifically, $X_i = 1$ or 0 denotes the occurrence or nonoccurrence of an event in the ith trial. Then the sequence $X_1, X_2, \cdots, X_n$ constitutes a *random sample* of size n.

The probability p of occurrence of an event in a trial is the parameter in the binomial distribution. The *maximum likelihood estimator* of this probability can be shown to be

$$\hat{P} = \frac{1}{n} \sum_{i=1}^{n} X_i \tag{5.43}$$

In other words, the estimate of p is the proportion of occurrences among the sequence of n trials.

Confidence interval for p can also be developed as follows. Observe first that

$$E(\hat{P}) = E\left(\frac{1}{n} \sum_{i=1}^{n} X_i\right) = \frac{1}{n} \sum_{i=1}^{n} E(X_i)$$

but

$$E(X_i) = 1(p) + 0(1 - p) = p$$

Hence

$$E(\hat{P}) = \frac{1}{n}(np) = p \tag{5.44}$$

And

$$\text{Var}(\hat{P}) = \frac{1}{n^2} \sum_{i=1}^{n} \text{Var}(X_i)$$

$$= \frac{1}{n^2} \sum_{i=1}^{n} [E(X_i^2) - E^2(X_i)]$$

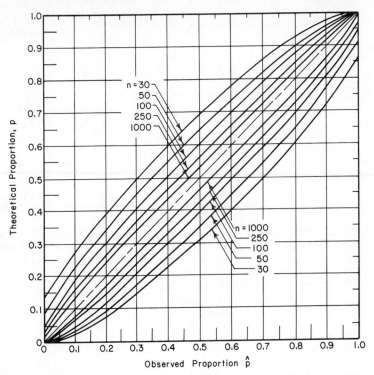

Figure 5.10   95% confidence interval of proportion (after Clopper and Pearson, 1934)

where $E(X_i^2) = p$. Thence,

$$\text{Var } (\hat{P}) = \frac{1}{n^2} n(p - p^2) = \frac{p(1 - p)}{n} \tag{5.45}$$

Therefore the estimator $\hat{P}$ is centered around p with a variance that decreases with the sample size n.

For large n, $\hat{P}$ will be approximately Gaussian by virtue of the central limit theorem; furthermore, the variance of Eq. 5.45 may be approximated by

$$\text{Var } (\hat{P}) \simeq \frac{\hat{p}(1 - \hat{p})}{n} \tag{5.45a}$$

where $\hat{p}$ is the observed proportion obtained from the sample data.

Then the confidence interval for p is obtained from

$$P\left(-k_{\alpha/2} < \frac{\hat{p} - p}{\sqrt{\hat{p}(1 - \hat{p})/n}} \leq k_{\alpha/2}\right) = 1 - \alpha \qquad (5.46)$$

giving the $(1 - \alpha)$ confidence interval as

$$\langle p \rangle_{1 - \alpha} = \left(\hat{p} - k_{\alpha/2}\sqrt{\frac{\hat{p}(1 - \hat{p})}{n}} \; ; \; \hat{p} + k_{\alpha/2}\sqrt{\frac{\hat{p}(1 - \hat{p})}{n}}\right) \qquad (5.47)$$

Figure 5.10 is a graph showing the 95% confidence interval for p as a function of the observed proportion $\hat{p}$ for different sample size n.

EXAMPLE 5.14

In inspecting the quality of soil compaction in a highway project, 10 out of 50 specimens inspected do not pass the CBR requirement. It is desired to estimate the actual proportion p of embankment that will be well compacted (that is, meet CBR requirement) and also establish a 95% confidence interval on p.

The point estimate for p is given by Eq. 5.43 as

$$\hat{p} = \frac{40}{50} = 0.8$$

The corresponding 95% confidence interval is, according to Eq. 5.47,

$$\langle p \rangle_{0.95} = \left\{0.8 - 1.96\sqrt{\frac{0.8(1 - 0.8)}{50}} \; ; \; 0.8 + 1.96\sqrt{\frac{0.8(1 - 0.8)}{50}}\right\}$$

$$= \{0.69; 0.91\}$$

5.3. CONCLUDING REMARKS

In modeling real-world situations, the form of the probability distribution of a random variable may be deduced theoretically on the basis of physical considerations or inferred empirically on the basis of observational data. However, the parameters of the distribution or the main descriptors (mean and variance) of the random variable must necessarily be related empirically to the real world; therefore, estimation based on factual data is required. Classical methods of parameter estimation are presented in this chapter; in Chapter 6 empirical and inferential methods for determining probability distributions are described.

Classical methods of estimation are of two types—*point and interval* estimations. The common methods of point estimation are the *method of maximum likelihood* and the *method of moments*; the former derives the estimator directly; the latter evaluates a parameter by first estimating the moments (usually the mean and variance) of the variate through the corresponding sample moments. Interval estimation includes a determination of the interval that contains the parameter value with a prescribed level of confidence.

It should be recognized that when population parameters are estimated on the basis of finite samples, errors of estimation are unavoidable. Within the classical methods of estimation, the significance of such errors are not reflected in the (point) estimates of the parameters; they can only be expressed in terms of appropriate confidence intervals. Explicit consideration of these errors is embodied in the Bayesian approach to estimation, which is the subject of Chapter 8.

PROBLEMS

5.1 In the measurement of daily dissolved oxygen (DO) concentrations in a stream, let p denote the probability that the DO concentration will fall below the required standard on a single day. DO concentration is measured daily until unsatisfactory stream quality is encountered, and the number of days in this sequence of measurement is recorded. Suppose 10 sequences have been observed and the length of each sequence is

$$2, 5, 6, 4, 6, 6, 8, 5, 10, 1 \quad \text{(days)}$$

(a) Determine the maximum likelihood estimator for p, and estimate p on the basis of the observed data.

(b) Estimate p by the method of moments. (*Hint.* Use the relations in Table 5.1).

5.2 For the concrete crushing strength data tabulated in Table E5.1 in Example 5.1, determine the point estimates for μ and σ by the method of maximum likelihood. Assume that concrete strength follows a Gaussian distribution.

5.3 The distribution of wave height has been suggested to follow a Rayleigh density function,

$$f_H(h) = \frac{h}{\alpha^2} e^{-\frac{1}{2}(h/\alpha)^2} \quad h \geq 0$$

with parameter α. Suppose the following measurements on wave heights were recorded: 1.5, 2.8, 2.5, 3.2, 1.9, 4.1, 3.6, 2.6, 2.9, 2.3 m.
Estimate the parameter α by the method of maximum likelihood.

5.4 Data on rainfall intensities (in inches) collected between 1918 and 1946 for the watershed of Esopus Creek, N.Y., are tabulated below as follows:

1918—43.30	1925—43.93	1932—50.37	1939—42.96
1919—53.02	1926—46.77	1933—54.91	1940—55.77
1920—63.52	1927—59.12	1934—51.28	1941—41.31
1921—45.93	1928—54.49	1935—39.91	1942—58.83
1922—48.26	1929—47.38	1936—53.29	1943—48.21
1923—50.51	1930—40.78	1937—67.59	1944—44.67
1924—49.57	1931—45.05	1938—58.71	1945—67.72
			1946—43.11

(a) Determine the point estimates for the mean μ and variance σ^2.

(b) Determine the 95% confidence interval for the mean μ. Assume the annual rainfall intensity is Gaussian, and $\sigma \simeq s$.

5.5 Consider the annual maximum wind velocity (V) data given in Problem 3.25.
  (a) Calculate the sample mean and sample variance of V.
  (b) Determine the 99% confidence interval for the mean velocity. Assume that the true standard deviation of V, σ_V, is satisfactorily given by the sample standard deviation s_v.
  (c) Assume that V has a log-normal distribution; determine the point estimates for the corresponding parameters λ_V and ζ_V.

5.6 A structure is designed to rest on 100 piles. Nine test piles were driven at random locations into the supporting soil stratum and loaded until failure occurred. Results are tabulated as follows.

Test pile	Pile capacity (tons)
1	82
2	75
3	95
4	90
5	88
6	92
7	78
8	85
9	80

  (a) Estimate the mean and standard deviation of the individual pile capacity to be used at the site.
  (b) Establish the 98% confidence interval for the mean pile capacity, assuming known $\sigma = s$.
  (c) Determine the 98% confidence interval for the mean pile capacity on the basis of unknown variance.

5.7 The daily dissolved oxygen concentration (DO) for a location A downstream from an industrial plant has been recorded for 10 consecutive days.

Day	DO (mg/l)
1	1.8
2	2.0
3	2.1
4	1.7
5	1.2
6	2.3
7	2.5
8	2.9
9	1.6
10	2.2

  (a) Assume that the daily DO concentration has a normal distribution $N(\mu, \sigma)$; estimate the values of μ and σ.

(b) Determine the 95% confidence interval for the true mean μ.

(c) Determine the 95% lower confidence limit of μ.

5.8 A river has the following record on the levels of floods that occurred each year between 1960 through 1970.

Year	Flood level (m) (above mean flow)
1960	3.7
1961	2.3
1962	5.1, 3.5
1963	5.2
1964	4.7, 6.1, 5.2
1965	3.4, 7.2, 1.5
1966	1.5, 3.6
1967	5.2, 1.4
1968	1.3, 4.5
1969	3.4
1970	4.4, 2.4

(a) Draw the histogram of flood levels at 1-m interval.

(b) Draw the histogram of the *annual maximum flood levels* at 1-m interval.

(c) Based on the histogram, what is the return period for a 7-m flood?

(d) Compute sample mean and sample variance of the annual maximum flood.

(e) Establish the 99% 2-sided confidence interval for the mean annual maximum flood.

(f) Assume that the annual maximum flood level has a log-normal distribution with the mean and variance computed in part (d); on this basis, determine the return period for a 7-m flood of this river.

5.9 From a set of data on the daily BOD level at a certain station for 30 days, the following have been computed:

$$\bar{x} = 3.5 \ (\text{mg/l})$$
$$s^2 = 0.184 \ (\text{mg/l})^2$$

Assume that the daily BOD level is a Gaussian variable.

(a) Estimate the mean and standard deviation of the BOD level.

(b) Determine the 99.5% confidence interval for the mean BOD.

(c) If the engineer is not satisfied with the width of the confidence interval established in part (b), and would like to reduce this interval by 10%, keeping the 99.5% confidence level, how many *additional* daily measurements have to be gathered? *Ans. 7.*

5.10 Suppose that a sample of 9 steel reinforcing bars were tested for yield strength, and the sample mean was found to be 20 kips.

(a) What is the 90% confidence interval for the population mean, if the standard deviation is assumed to be equal to 3 kips?

(b) How many additional bars must be tested to increase the confidence of the same interval to 95%? *Ans. 4.*

(c) If the standard deviation is not known, but the 9 measurements yielded

$$\sum_{i=1}^{9} (x_i - 20)^2 = 84.5$$

then what would be the 90% confidence interval for the mean yield strength? Assume that the yield strength is a normal variate.

5.11 A 20-year data series for the annual maximum wind velocity V for a city in Illinois yielded the following quantities:

$$\bar{v} = 76.5 \text{ mph}$$

$$\sum_{i=1}^{20} (v_i - \bar{v})^2 = 2640 \text{ (mph)}^2$$

(a) Determine the sample standard deviation s_v.
(b) Determine the 95% upper confidence limit for μ_V; that is,

$$P(\mu_V < \text{Limit}) = 0.95$$

assume $\sigma_V = s_v$ from part (a).
(c) Assume that the annual maximum wind velocity is a log-normal variate with $\mu_V = 76.5$ mph and $\sigma_V = s_v$ from part (a). Estimate the distribution parameters λ_V and ζ_V.

5.12 The height H of a radio tower is being determined by measuring the horizontal distance L from the center of its base to the instrument and the vertical angle β as shown in Fig. P5.12.

(a) The distance L is measured 3 times, and the readings are: 124.3, 124.2, 124.4 ft.
Determine the estimated distance, and its standard error. *Ans. 124.3 ft; 0.0577 ft.*
(b) The angle β is measured 5 times and the readings are: $40° 24.6'$, $40° 25.0'$, $40° 25.5'$, $40° 24.7'$, $40° 25.2'$.
Determine the estimated angle, and its standard error. *Ans. 40°25'; 0.164'.*

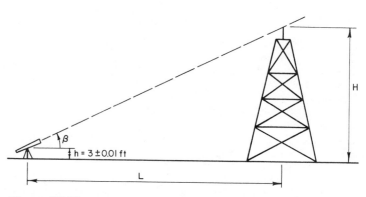

Figure P5.12

(c) Estimate the height of the tower $\bar{H}$. Assume the instrument is 3 ft high with a standard deviation of 0.01 ft. *Ans. 108.85 ft.*

(d) Compute the standard error of the estimated height of the tower, $\sigma_{\bar{H}}$. *Ans. 0.051 ft.*

(e) Determine the 98% confidence interval of the actual height of the tower H. Assume that $\bar{H}$ is normally distributed about the actual height H. *Ans. (108.73 ft; 108.97 ft).*

5.13 To determine the area of a rectangular tract of land shown in Fig. P5.13, the sides b and c were measured 5 times each. Following are the 5 independent measurements made on b and c:

Side b (m)	Side c (m)
500.5	299.8
499.5	300.3
500.0	300.2
500.2	299.7
499.8	300.0

The area of the tract is computed as

$$\bar{A} = \bar{b} \cdot \bar{c}$$

where $\bar{b}$ and $\bar{c}$ are the sample means of the respective measurements. Estimate the 95% confidence interval for the actual area A.

5.14 The following five repeated independent observations (measurements) were made on each of the outer and inner radii of a circular ring shown in Fig. P5.14.

outer radius r_1: 2.5, 2.4, 2.6, 2.6, 2.4 cm

inner radius r_2: 1.6, 1.5, 1.6, 1.4, 1.4 cm

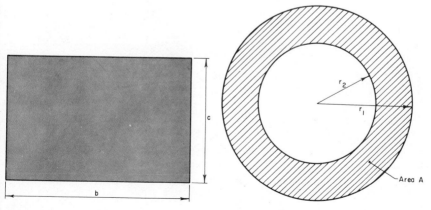

Figure P5.13   *Figure P5.14*

(a) Determine the best estimates of the outer and inner radii, and corresponding standard errors.

(b) The shaded area between the two concentric circles is computed based on the mean values of the measured outer and inner radii; namely $\bar{A} = \pi(\bar{r}_1{}^2 - \bar{r}_2{}^2)$. What is the computed area? *Ans. 12.57* cm².

(c) Determine the standard deviation (standard error) of the computed area. *Ans. 0.819* cm².

(d) If it is desired to determine the sample mean of r_1 within ± 0.07 cm with 99% confidence, how many additional independent measurements must be made on r_1? Assume that all measurements are independent and taken with the same care and skill. *Ans. 12.*

5.15 The distance between A and C is measured in 2 stages: namely, AB and BC as shown in Fig. P5.15. Measurements on AB and BC are recorded as follows:

$$AB: 100.5, \ 99.6, \ 100.1, \ 100.3, \ 99.5 \text{ ft}$$

$$BC: 50.2, \ 49.8, \ 50.0 \text{ ft}$$

(a) Compute the sample mean and sample variance of the measured distances for AB.

(b) Compute the standard error of the estimated distance of AB, that is, $s_{\overline{AB}}$.

(c) Establish the 98% confidence interval for the actual distance AB.

(d) If the distance AC is given by the sum of the estimated distances $\overline{AB}$ and $\overline{BC}$, that is,

$$AC = \overline{AB} + \overline{BC}$$

what is the standard error of the estimated total distance between A and C?

(e) Establish the 98% confidence interval on the actual length AC.

Figure P5.15

6. Empirical Determination of Distribution Models

6.1. INTRODUCTION

The probabilistic characteristics of a random phenomenon is sometimes difficult to discern or define, such that the appropriate probability model needed to describe these characteristics is not readily amenable to theoretical deduction or formulation. In particular, the functional form of the required probability distribution may not be easy to derive or ascertain. Under certain circumstances, the basis or properties of the physical process may suggest the form of the required distribution. For example, if a process is composed of the sum of many individual effects, the Gaussian distribution may be appropriate on the basis of the central limit theorem; whereas, if the extremal conditions of a physical process are of interest, an extreme-value distribution may be a suitable model.

Nevertheless, there are occasions when the required probability distribution has to be determined empirically (that is, based entirely on available observational data). For example, if the frequency diagram for a set of data can be constructed, the required distribution model may be determined by visually comparing a density function with the frequency diagram (see for example, Figs. 1.5 through 1.7). Alternatively, the data may be plotted on *probability papers* prepared for specific distributions (see Section 6.2 below). If the data points plot approximately on a straight line on one of these papers, the distribution corresponding to this paper may be an appropriate distribution model.

Furthermore, an assumed probability distribution (perhaps determined empirically as described above, or developed theoretically on the basis of prior assumptions) may be verified, or disproved, in the light of available data using certain statistical tests, known as *goodness-of-fit* tests for distribution. Moreover, when two or more distributions appear to be plausible probability distribution models, such tests can be used to delineate the relative degree of validity of the different distributions. Two such tests are commonly used for these purposes—the *chi-square* (χ^2) and the *Kolmogorov-Smirnov* (K-S) tests.

In practice, the choice of the probability distribution may also be dictated by mathematical tractability or convenience. For example, because of the mathematical simplifications possible with the normal distribution, and the wide availability of information (probability tables) associated with this distribution, the normal (or log-normal) distribution is frequently used to model nondeterministic problems—at times, even when there is no clear basis for such a model. Probabilistic information derived on the basis of such prescribed distributions could be useful, especially when the information is needed only for relative purposes.

6.2. PROBABILITY PAPER

Graph papers for plotting observed experimental data and their corresponding cumulative frequencies (or probabilities) are called *probability papers*. Probability papers are constructed such that a given probability paper is associated with a specific probability distribution; that is, different probability papers correspond to different probability distributions.

Preferably, a probability paper should be constructed using a transformed probability scale in such a manner as to obtain a linear graph between the cumulative probabilities of the underlying distribution and the corresponding values of the variate. For example, in the case of the uniform distribution, the cumulative distribution function is linearly related to the values of the variate; thus the probability paper for this distribution would be constructed using arithmetic scales for the values of the variate and the associated cumulative probabilities (between 0 and 1.0). For other distributions, however, special scales are required for the cumulative probabilities in order to achieve the desired linear relationship.

The linearity, or lack of linearity, of a set of sample data plotted on a particular probability paper, therefore, can be used as a basis for determining whether the distribution of the underlying population is the same as that of the probability paper. On this basis, then, probability papers may be used to establish or explore the possible distribution(s) of the underlying population. In Sections 6.2.1 and 6.2.2 we illustrate the construction and application of two commonly used probability papers—the normal and the log-normal probability papers, and in Section 6.2.3 we describe the construction of probability papers for a general distribution.

Experimental data may be plotted on any probability paper; the *plotting position* of each data point is determined as follows.

If there are N observations x_1, x_2, $\cdots$, x_N, the mth value among the N observations (arranged in increasing order) is plotted at the cumulative probability $m/(N + 1)$.

This plotting position applies to all probability papers; its basis is discussed in Gumbel (1954). Although there are other plotting positions, such as $(m - \frac{1}{2})/N$, which was advocated by Hazen (1930) and has been

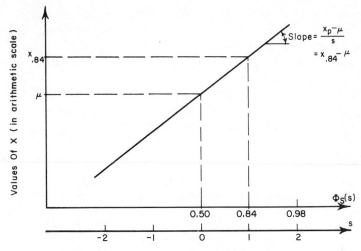

Figure 6.1 Construction of normal probability paper

used widely, this plotting position has certain theoretical weakness; in particular, when there are N observations, the plotting position $(m - \frac{1}{2})/N$ would yield a return period of $2N$ for the largest observation instead of N (Gumbel, 1954). Still other plotting positions have been suggested (for example, Kimball, 1946); however, none seems to have the theoretical attributes and the computational simplicity of $m/(N + 1)$.

6.2.1. The normal probability paper

The *normal (or Gaussian) probability paper* is constructed on the basis of the standard normal distribution function as follows. One axis (in arithmetic scale) represents the values of the variate X, as illustrated in Fig. 6.1. On the other axis are two parallel scales; one in arithmetic scale represents values of the standard normal variate s, whereas the other shows the cumulative probabilities $\Phi_S(s)$ corresponding to the indicated values of s as shown in Fig. 6.1. A normal variate X with distribution $N(\mu, \sigma)$ would then be represented on this paper by a straight line passing through $\Phi_S(s) = 0.50$ and $X = \mu$, with a slope $(x_p - \mu)/s = \sigma$; where x_p is the value of the variate at probability p. In particular, at $p = 0.84$, $s = 1$; hence, the slope is $(x_{.84} - \mu)$.

Such normal probability papers are available commercially. The scale for the standard variate s, however, is usually omitted in such commercial papers.

Any set of data may be plotted on the normal probability paper; however, if the resulting graph of data points shows a lack of linearity, this would suggest that the underlying population is not Gaussian. Conversely, if the data points plotted on this paper show a linear or approximately linear trend, the straight line through these data points represents a specific normal distribution applicable to the data set (at least within the range of observations). The mean value and standard deviation of the underlying population may also be determined graphically from this straight line—the value of X on this line corresponding to $\Phi_S(s) = 0.50$ is the estimate of the mean value μ_X, whereas the slope of the straight line is the estimate of the standard deviation σ_X; thus $\sigma_X \simeq x_{.84} - \mu_X$ (see Fig. 6.1).

EXAMPLE 6.1

The data for fracture toughness of steel plate, given in Table E6.1, are plotted on the normal probability paper in Fig. E6.1.

Table E6.1. Fracture Toughness of Base Plate of 18% Nickel Maraging Steel (Data From Kies et al., 1965)

m	K_{Ic} (ksi $\sqrt{\text{in.}}$)	$\dfrac{m}{N+1}$
1	69.5	0.0370
2	71.9	0.0741
3	72.6	0.1111
4	73.1	0.1418
5	73.3	0.1852
6	73.5	0.2222
7	74.1	0.2592
8	74.2	0.2963
9	75.3	0.3333
10	75.5	0.3704
11	75.7	0.4074
12	75.8	0.4444
13	76.1	0.4815
14	76.2	0.5185
15	76.2	0.5556
16	76.9	0.5926
17	77.0	0.6296
18	77.9	0.6667
19	78.1	0.7037
20	79.6	0.7407
21	79.7	0.7778
22	79.9	0.8148
23	80.1	0.8518
24	82.2	0.8889
25	83.7	0.9259
26	93.7	0.9630

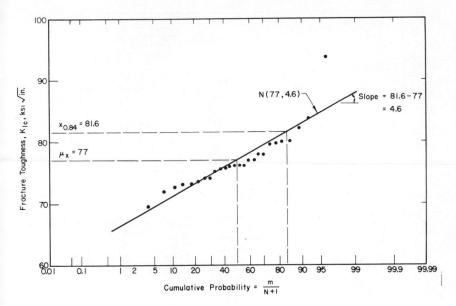

Figure E6.1   Fracture toughness plotted on normal probability paper

In Fig. E6.1, values of the fracture toughness K_{Ic} are plotted against the plotting positions $m/(N + 1)$, with $N = 26$.

The straight line shown in Fig. E6.1 is drawn (by eye) through the data points, from which we find $\mu_{K_{Ic}} = 77$ ksi $\sqrt{\text{in}}$. Also, we observe that the value of K_{Ic} at the 84% probability level is 81.6; thus $\sigma_{K_{Ic}} = 81.6 - 77 = 4.6$ ksi $\sqrt{\text{in}}$.

6.2.2.   The log-normal probability paper

The logarithmic normal probability paper can be obtained from the normal probability paper by simply changing the arithmetic scale for values of the variate X (on the normal probability paper) to a logarithmic scale. The resulting paper would be as shown in Fig. E6.2. In this case, the standard normal variate becomes

$$S = \frac{\ln (X/x_m)}{\zeta},$$

where x_m is the median of X.

If a random phenomenon can be modeled approximately with a log-normal distribution, then experimental data obtained therefrom should be approximately linear when the mth value among N observations and their plotting positions $m/(N + 1)$ are plotted on the log-normal probability paper. If the plotted data points yield a straight line, this line represents the particular log-normal distribution for the underlying population.

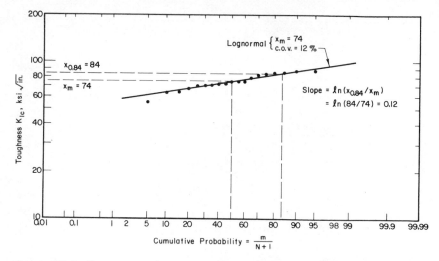

Figure E6.2 Fracture toughness of welds plotted on log-normal probability paper

Table E6.2. Fracture Toughness of MIG Welds (Data from Kies et al., 1965)

m	K_{Ic} (ksi $\sqrt{\text{in.}}$)	$\dfrac{m}{N+1}$
1	54.4	0.05
2	62.6	0.10
3	63.2	0.15
4	67.0	0.20
5	70.2	0.25
6	70.5	0.30
7	70.6	0.35
8	71.4	0.40
9	71.8	0.45
10	74.1	0.50
11	74.1	0.55
12	74.3	0.60
13	78.8	0.65
14	81.8	0.70
15	83.0	0.75
16	84.4	0.80
17	85.3	0.85
18	86.9	0.90
19	87.3	0.95

Accordingly, the median x_m is simply the value of the variate on this line corresponding to $\Phi_S(s) = 0.50$; whereas the parameter ζ is given by the slope of the line, that is,

$$\zeta = \frac{1}{s} \ln \left(\frac{x_p}{x_m}\right) = \ln \left(\frac{x_{.84}}{x_m}\right)$$

EXAMPLE 6.2

Data for the fracture toughness of MIG welds are tabulated in Table E6.2.

The plotting positions $m/(N + 1)$ are shown in column 3 of Table E6.2; these are plotted against the fracture toughness K_{Ic} on log-normal probability paper in Fig. E6.2.

On the basis of the linear graph of the plotted data shown in Fig. E6.2, we may say that the fracture toughness of such welds has a log-normal distribution. Specifically, the straight line drawn (by eye) through the data points represents a log-normal distribution with a median of 74 ksi $\sqrt{\text{in.}}$ and a COV of 12%.

Table E6.3.   Precipitation and Runoff Data for Example 6.3

m	Precipitation X (in.)	Runoff Y (in.)	$\dfrac{m}{N + 1}$
1	1.01	0.17	0.038
2	1.11	0.19	0.077
3	1.13	0.23	0.115
4	1.15	0.33	0.154
5	1.16	0.39	0.192
6	1.17	0.39	0.231
7	1.17	0.40	0.269
8	1.20	0.45	0.308
9	1.52	0.52	0.346
10	1.54	0.56	0.385
11	1.54	0.59	0.423
12	1.57	0.64	0.462
13	1.64	0.66	0.500
14	1.73	0.70	0.538
15	1.79	0.76	0.577
16	2.09	0.77	0.615
17	2.09	0.78	0.654
18	2.57	0.95	0.692
19	2.75	0.97	0.731
20	2.93	1.02	0.769
21	3.19	1.12	0.808
22	3.54	1.24	0.846
23	3.57	1.59	0.885
24	5.11	1.74	0.923
25	5.62	2.92	0.962

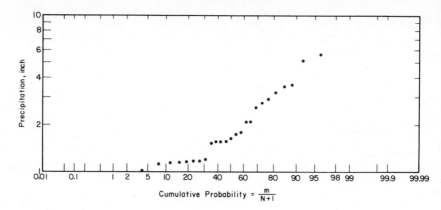

Figure E6.3a   Precipitation plotted on log-normal probability paper

EXAMPLE 6.3

Plot on log-normal probability papers the precipitation and runoff data of the Monocacy River described in Example 5.8.

Rearranging the data given in Example 5.8 in increasing order, we obtain Table E6.3.

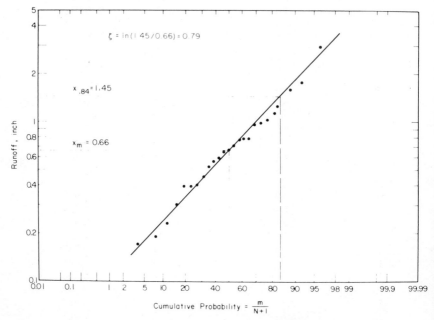

Figure E6.3b   Runoff plotted on log-normal probability paper

The precipitation values are plotted against the respective plotting positions $m/(N + 1)$ on log-normal probability paper in Fig. E6.3a. Similarly, the runoff data are plotted on log-normal probability paper in Fig. E6.3b. From these two figures, the following may be inferred.

(a) The absence of linearity in the graph of Fig. E6.3a means that the distribution of precipitation in the Monocacy River basin is not log-normal.

(b) On the basis of Fig. E6.3b, the runoff of the Monocacy River may be described with a log-normal distribution, with median $x_m = 0.66$ in. and parameter $\zeta = \ln(1.45/0.66) = 0.79$.

6.2.3. Construction of general probability paper

As indicated earlier, probability papers are constructed in such a way that the values of the variate and the associated cumulative probabilities yield a straight line on a two-dimensional graph; conversely, therefore, a straight line on a specific probability paper represents a particular distribution (consistent with that of the probability paper) with given values of the parameters. For this purpose, a probability paper should be constructed so that it is independent of the values of the parameters of the distribution. This is accomplished by defining a *standard variate* (if one exists) appropriate for the given distribution.

In the last two sections, we illustrated the construction and application of the normal and log-normal probability papers; similar papers may be constructed for other distributions. We illustrate this with the following examples.

EXAMPLE 6.4

The density function of the (shifted) exponential distribution, Eq. 3.43, is

$$f_X(x) = \lambda e^{-\lambda(x-a)}; \quad x \geq a$$
$$= 0; \quad x < a$$

where λ is the parameter, and a is the minimum value of X. In this case, the standard variate is $S = \lambda(X - a)$. The density function of S then, according to Eq. 4.6, is

$$f_S(s) = f_X\left(\frac{s}{\lambda} + a\right)\left|\frac{1}{\lambda}\right| = e^{-s}; \quad s \geq 0$$
$$= 0; \quad s < 0$$

with corresponding CDF
$$F_S(s) = 1 - e^{-s}; \quad s \geq 0$$

On this basis, therefore, we construct the exponential probability paper as follows.

On one axis, scale values (in convenient arithmetic scale) of the standard variate s; on the same (or a parallel) axis, mark the corresponding cumulative probabilities $F_S(s) = 1 - e^{-s}$. The other (perpendicular) axis will represent values of the variate X (in arithmetic scale). For illustration, specific values of s and $F_S(s)$ have been calculated as summarized in Table E6.4a.

Drawing grid lines for given $F_S(s)$ at the indicated values of s shown in Table

Table E6.4.a   Specific Values of s and $F_S(s)$

s	$F_S(s)$	s	$F_S(s)$
0.11	0.10	2.53	0.92
0.22	0.20	2.66	0.93
0.36	0.30	2.81	0.94
0.51	0.40	3.00	0.95
0.69	0.50	3.10	0.955
0.80	0.55	3.22	0.96
0.92	0.60	3.35	0.965
1.05	0.65	3.51	0.97
1.20	0.70	3.69	0.975
1.39	0.75	3.91	0.98
1.61	0.80	4.20	0.985
1.90	0.85	4.61	0.99
2.30	0.90		

E6.4a, we obtain the resulting paper, as shown in Fig. E6.4a. A straight line (with positive slope) on this paper represents a particular exponential distribution, in which its intercept on the x-axis is the value of a, and its slope is $1/\lambda$.

Sample values from an exponential population should plot, using plotting position $m/(N + 1)$, approximately on a straight line on this probability paper. To illustrate this, consider the hypothetical set of data in Table E6.4b for a random variable X. The mth observed value and corresponding plotting position $m(N + 1)$ are shown plotted in Fig. E6.4b. From the straight line drawn (by eye) through the data points in Fig. E6.4b, we obtain estimates of the parameters $a \simeq 150$ and $1/\lambda \simeq 2000/2.69.$ = 743.

In the present case, however, the purposes of a probability paper can be accomplished also with a semilogarithmic paper. Observe that for the exponential distribution,

$$1 - F_X(x) = e^{-\lambda x}$$

thus,

$$\ln [1 - F_X(x)] = -\lambda x$$

Therefore, by scaling $1 - F_X(x)$ on one axis (in logarithmic scale) and x on the other axis (in arithmetic scale), the graph between $1 - F_X(x)$ and x is a straight line on this paper with a slope of λ. On this paper, however, sample data should be plotted at the plotting positions $[1 - m/(N + 1)]$.

EXAMPLE 6.5 *(The Gumbel probability paper)*

One of the extreme-value distributions (see Vol. II) is the Type I asymptotic distribution of extremes, known also as the *Gumbel distribution*. Its CDF for the largest value is given by the double exponential function.

$$F_X(x) = \exp [-e^{-\alpha(x-u)}] \qquad -\infty < x < \infty$$

in which u is the *characteristic largest value*, and $1/\alpha$ is a measure of dispersion.

Table E6.4.b Sample Values of X

x	m	Plotting position $\dfrac{m}{N+1}$
200	1	0.024
201	2	0.049
203	3	0.073
212	5	0.122
248	7	0.171
389	16	0.390
1331	35	0.854
1031	33	0.805
208	4	0.098
226	6	0.146
289	10	0.244
543	20	0.488
360	15	0.366
1635	37	0.902
559	21	0.512
909	28	0.683
408	17	0.415
2497	39	0.951
774	24	0.585
946	29	0.707
2781	40	0.976
308	12	0.293
274	9	0.220
531	19	0.463
460	18	0.439
791	26	0.634
952	30	0.732
1844	38	0.927
952	31	0.756
1427	36	0.878
306	11	0.268
787	25	0.610
254	8	0.195
772	23	0.561
842	27	0.659
981	32	0.780
1122	34	0.829
611	22	0.537
332	13	0.317
343	14	0.341

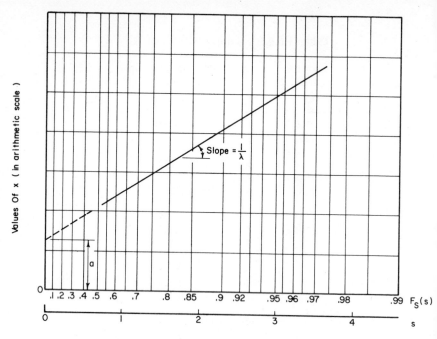

Figure E6.4a   Construction of the exponential probability paper

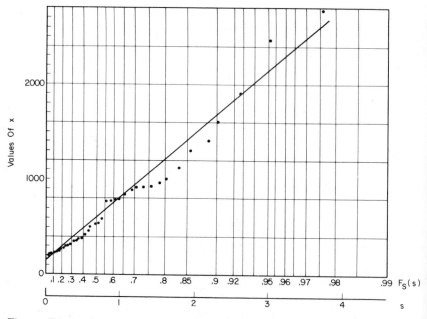

Figure E6.4b   Sample values of X plotted on exponential probability paper

272

Table E6.5. Specific Values of s and $F_S(s)$

s	$F_S(s)$	s	$F_S(s)$	s	$F_S(s)$
−1.53	0.01	0.37	0.50	2.48	0.92
−1.10	0.05	0.51	0.55	2.62	0.93
−0.83	0.10	0.67	0.60	2.78	0.94
−0.64	0.15	0.84	0.65	2.97	0.95
−0.48	0.20	1.03	0.70	3.08	0.955
−0.33	0.25	1.25	0.75	3.20	0.96
−0.19	0.30	1.50	0.80	3.33	0.965
−0.05	0.35	1.82	0.85	3.49	0.97
0.09	0.40	2.25	0.90	3.68	0.975
0.23	0.45	2.36	0.91	3.90	0.98

In this case, the standard variate can be defined as

$$S = \alpha(X - u)$$

Then

$$F_S(s) = \exp\left(-e^{-s}\right)$$

Using the specific values of s and corresponding probabilities $F_S(s)$ calculated as summarized in Table E6.5, we construct the Gumbel probability paper as follows.

Scale the values of s on one axis and the associated probabilities $F_S(s)$ on the same (or a parallel) axis as shown in Fig. E6.5. The other axis in Fig. E6.5 represents

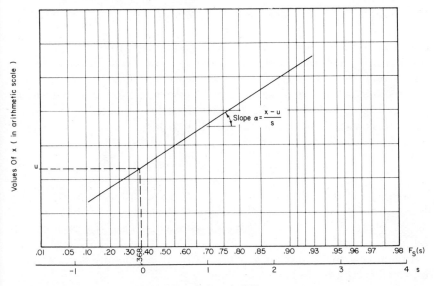

Figure E6.5 Construction of Gumbel probability paper

values of X, in arithmetic scale. The result is the Type I extremal probability paper, also known as the *Gumbel probability paper*.

Again, a straight line on this paper represents a particular Type I extremal distribution—the value of X on this line at $F_S(s) = e^{-1} = 0.368$ (or $s = 0$) is the characteristic largest value u, whereas the slope of the straight line is α, as illustrated in Fig. E6.5.

6.3.   TESTING VALIDITY OF ASSUMED DISTRIBUTION

When a theoretical distribution has been assumed, perhaps determined on the basis of the general shape of the histogram or on the basis of the data plotted on a given probability paper, the validity of the assumed distribution may be verified or disproved statistically by *goodness-of-fit* tests. Two such tests for distribution are available—the *chi-square* and the *Kolmogorov-Smirnov* methods; one or the other of these is generally used to test the validity of an assumed distribution model.

6.3.1.   Chi-square test for distribution

Consider a sample of n observed values of a random variable. The chi-square goodness-of-fit test compares the observed frequencies $n_1, n_2, \cdots, n_k$ of k values (or in k intervals) of the variate with the corresponding frequencies $e_1, e_2, \cdots, e_k$ from an assumed theoretical distribution. The basis for appraising the goodness of this comparison is the distribution of the quantity

$$\sum_{i=1}^{k} \frac{(n_i - e_i)^2}{e_i}$$

which approaches the chi-square (χ_f^2) distribution with $(f = k - 1)$ degrees of freedom as $n \to \infty$ (Hoel, 1962). However, if the parameters of the theoretical model are unknown and must be estimated from the data, the above statement remains valid if the degree of freedom is reduced by one for every unknown parameter that must be estimated.

On this basis, if an assumed distribution yields

$$\sum_{i=1}^{k} \frac{(n_i - e_i)^2}{e_i} < c_{1-\alpha, f} \qquad (6.1)$$

where $c_{1-\alpha, f}$ is the value of the appropriate χ_f^2 distribution at the cumulative probability $(1 - \alpha)$, the assumed theoretical distribution is an acceptable model, at the *significance level* α. Otherwise, the assumed disbution is not substantiated by the data at the α significance level.

In applying the χ^2 test for goodness of fit, it is generally necessary (for satisfactory results) to have $k \geq 5$ and $e_i \geq 5$.

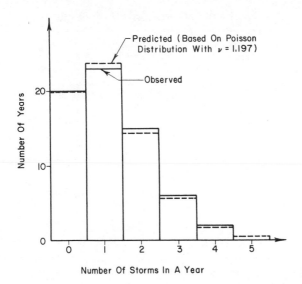

Figure E6.6   Histogram and Poisson model for storm occurrences

EXAMPLE 6.6

Suppose that severe rainstorms have been recorded at a given station over a period of 66 years. During this period, there were 20 years without severe rainstorms; and 23, 15, 6, and 2 years, respectively, with 1, 2, 3, and 4 rainstorms annually. The histogram for the annual number of rainstorms recorded at the station is shown in Fig. E6.6. Because the occurrence of severe rainstorms is random, and judging from the shape of the histogram, a Poisson distribution seems an appropriate model for the annual number of rainstorms at the given location (station). In particular, on the basis of the data, we estimate the average occurrence rate of rainstorms annually as

$$\hat{\nu} = \frac{1}{66}(23 + 15 \times 2 + 6 \times 3 + 2 \times 4)$$

$$= \frac{79}{66} = 1.20 \text{ rainstorms/year}$$

We now apply the chi-square test to determine whether the Poisson distribution is a suitable model, at the 5% significance level. In this case, since four storms in a year was observed only twice, this data is combined with the data for three storms a year; thus, $k = 4$. Since the parameter ν is estimated by $\hat{\nu}$, the quantity $\sum_{i=1}^{4} (n_i - e)^2/e_i$ has a χ^2 distribution with $f = k - 2 = 2$ degrees of freedom. Based on the computations summarized in Table E6.6, $\Sigma (n_i - e_i)^2/e_i = 0.068$, which is less than $c_{.95,2} = 5.99$. Hence the Poisson distribution is a valid model for the annual number of rainstorms, at the 5% significance level.

Table E6.6. χ^2 Test for Storm Occurrence

No. of storms at station per year	Observed frequency n_i	Theoretical frequency e_i	$(n_i - e_i)^2$	$\dfrac{(n_i - e_i)^2}{e_i}$
0	20	19.94	0.0036	0.0002
1	23	23.87	0.7569	0.0317
2	15	14.29	0.5041	0.0353
≥3	8	7.90	0.0100	0.0013
	66	66.00		0.0685

EXAMPLE 6.7

Consider the frequency diagram for the crushing strength of concrete cubes shown in Fig. E6.7. Visually, on the basis of the frequency diagram and the theoretical distributions shown in Fig. E6.7, both the normal and the log-normal density functions appear to be suitable models for the concrete strength.

In this case, the χ^2-test will be used to determine the relative goodness of fit between the two candidate distributions.

For the purpose of this example, 8 intervals of the strength are considered as shown in Table E6.7.

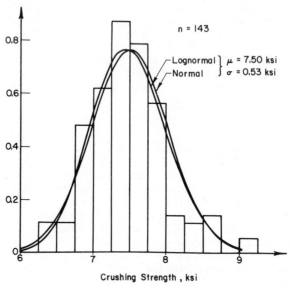

Figure E6.7   Frequency diagram of crushing strengths of concrete cubes (data from Cusens and Wettern, 1959)

Table E6.7. Chi-Square Test for Relative Goodness-of-fit

Interval (ksi)	Observed frequency n_i	Theoretical frequency e_i		$\dfrac{(n_i - e_i)^2}{e_i}$	
		Normal	Log normal	Normal	Log normal
<6.75	9	11.1	9.9	0.40	0.09
6.74–7.00	17	13.2	14.0	1.09	0.92
7.00–7.25	22	21.1	22.1	0.04	0.00
7.25–7.50	31	26.1	26.9	0.92	0.62
7.50–7.75	28	26.1	25.6	0.14	0.23
7.75–8.00	20	21.0	19.8	0.05	0.00
8.00–8.50	9	20.2	19.4	6.22	5.57
<8.50	7	4.2	5.3	1.87	0.54
		143.0	143.0	10.73	7.97

The observed and theoretical frequencies within the indicated intervals are summarized in Table E6.7.

In both cases (that is, the normal and the log-normal distributions), the respective parameters were estimated by the sample mean and sample variance; hence the net number of degrees of freedom for either distribution is $f = 8 - 3 = 5$. At the significance level $\alpha = 5\%$, we obtain from Table A.3, $c_{.95,5} = 11.07$. Comparing this with the values of $\Sigma\,(n_i - e_i)^2/e_i$ calculated in Table E6.7, we observe that although both distributions appear to be valid models for the concrete strength (on the basis of the frequency diagram of Fig. E6.7), the log-normal model is superior to the normal model according to this test, because $7.97 < 10.73$.

It should be emphasized that because there is arbitrariness in the choice of the significance level α, the χ^2 goodness-of-fit test (as well as the Kolmogorov-Smirnov method described subsequently) may not provide absolute information on the validity of a specific distribution. For example, it is conceivable that a distribution that is acceptable at one significance level may be unacceptable at another significance level. In spite of this, however, such tests remain useful, especially for determining the relative goodness of fit of two or more theoretical distributions, as illustrated in Example 6.7.

6.3.2. Kolmogorov-Smirnov test for distribution

Another widely used goodness-of-fit test is the *Kolmogorov-Smirnov* (K-S) test. The basic procedure involves the comparison between the experimental cumulative frequency and an assumed theoretical distribution function. If the discrepancy is large with respect to what is normally expected from a given sample size, the theoretical model is rejected.

For a sample of size n, rearrange the set of observed data in increasing

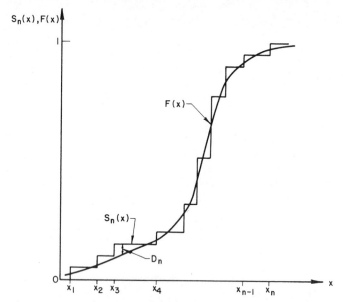

Figure 6.2   Empirical cumulative frequency vs. theoretical distribution function

order. From these ordered sample data we develop a stepwise cumulative frequency function as follows:

$$
S_n(x) = \begin{cases} 0 & x < x_1 \\[2mm] \dfrac{k}{n} & x_k \le x < x_{k+1} \\[2mm] 1 & x \ge x_n \end{cases} \tag{6.2}
$$

where x_1, x_2, $\cdots$, x_n are the values of the ordered sample data, and n is the sample size. Figure 6.2 shows a plot of $S_n(x)$ and also the proposed theoretical distribution function $F(x)$. In the Kolmogorov-Smirnov test, the maximum difference between $S_n(x)$ and $F(x)$ over the entire range of X is the measure of discrepancy between the theoretical model and the observed data. Let this maximum difference be denoted by

$$
D_n = \max_x \mid F(x) - S_n(x) \mid \tag{6.3}
$$

Theoretically, D_n is a random variable whose distribution depends on n. For a specified significance level α, the K-S test compares the observed maximum difference of Eq. 6.3 with the critical value D_n^α, which is defined by

$$P(D_n \leq D_n^\alpha) = 1 - \alpha$$

Critical values D_n^α at various significance levels α are tabulated in Table A.4 for various values of n. If the observed D_n is less than the critical value D_n^α, the proposed distribution is acceptable at the specified significance level α; otherwise, the assumed distribution would be rejected.

The advantage of the Kolmogorov-Smirnov (K-S) test over the chi-square (χ^2) test is that it is not necessary to divide the data into intervals; hence the problems associated with the chi-square approximation for small e_i and/or small number of intervals k would not appear with the K-S test.

EXAMPLE 6.8

The data for fracture toughness of steel plate in Example 6.1 have been plotted on normal probability paper as shown in Fig. E6.1. The data appear to fall approximately on a straight line corresponding to a normal model N (77, 4.6). Perform a Kolmogorov-Smirnov test to evaluate the appropriateness of this model relative to the given data, at the 5% significance level.

The sample cumulative frequencies are plotted according to Eq. 6.2 in Fig. E6.8.

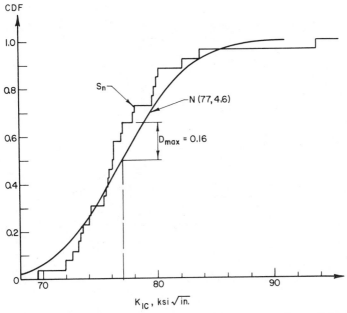

Figure E6.8 Cumulative distribution of fracture toughness data

The theoretical distribution function for the proposed normal model N (77, 4.6) is also shown in the same figure. The maximum discrepancy between the two functions is $D_n = 0.16$ and occurs at $K_{Ic} = 77$ ksi $\sqrt{\text{in}}$. In this case, there are 26 observed data points; hence the critical value of D_n^{α} at the 5% significance level is $D_n^{.05} = 0.265$ (obtained from Table A.4). Since the maximum discrepancy of 0.16 is less than $D_{26}^{.05} = 0.265$, the normal model N (77, 4.6) is verified at the 5% significance level.

EXAMPLE 6.9

Data for stream temperature at mile 41.83 of the Little Deschutes River in Oregon measured at 3-hr intervals over a 3-day period (August 1–3, 1969), are shown plotted in Fig. E6.9 in accordance with Eq. 6.2. The distribution function of the proposed theoretical model is also shown in the same figure.

In this case the maximum difference between $S_n(x)$ and $F(x)$ is observed to be $D_n = 0.174$ at the temperature of 70.9°F.

With $n = 23$, the critical value D_n^{α} at the 5% significance level, obtained from Table A.4, is $D_{23}^{.05} = 0.273$. Since $D_n < D_n^{\alpha}$, the proposed theoretical distribution is suitable for modeling the stream temperature of this river at the significance level of $\alpha = 5\%$.

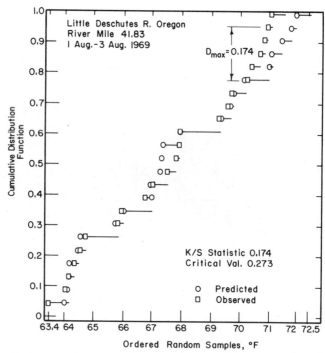

Figure E6.9 Kolmogorov-Smirnov test for proposed stream temperature prediction model (after Morse, 1972)

6.4. CONCLUDING REMARKS

Whereas Chapter 5 was concerned with the statistical estimation of parameters of a distribution, this chapter is concerned with the determination of the probability distribution for a random variable, and with questions related to the validity of an assumed distribution, based on finite samples of the population. Unless developed theoretically from physical considerations, the required distribution model may be determined empirically. One way of doing this is through the use of probability papers constructed for specific distributions. The linearity, or lack of linearity, of sample data plotted on such papers would suggest the appropriateness of a given distribution for modeling the population.

The validity of an assumed distribution may also be appraised by goodness-of-fit tests, including specifically the *chi-square* (χ^2) and the *Kolmogorov-Smirnov* (K-S) tests. Such tests, however, depend on the prescribed level of significance, the choice of which is largely a subjective matter. Nevertheless, these tests are useful for determining (in the light of sample data) the relative appropriateness of several potentially possible distribution models.

PROBLEMS

6.1 Plot the data in Example 6.1 on log-normal probability paper. Estimate the median and COV from the straight line drawn through the data points.

6.2 The ultimate strains (ε_u, in %) of 15 No. 5 steel reinforcing bars were measured. The results are as follows (data from Allen, 1972):

19.4	17.9	16.1
16.0	17.8	16.8
16.6	18.8	17.0
17.3	20.1	18.1
18.4	19.1	18.6

Plot these data on both the normal and the log-normal probability papers, and discuss the results.

6.3 The shear strengths (in kips per square feet, ksf) of 13 undisturbed samples of clay from the Chicago subway project are tabulated as follows (data from Peck, 1940):

0.35	0.42	0.49	0.70	0.96
0.40	0.43	0.58	0.75	
0.41	0.48	0.68	0.87	

Plot the data on log-normal probability paper. Estimate the parameters of the log-normal distribution to describe the shear strength of Chicago clay.

6.4 For the wind velocity data in Problem 3.25 (of Chapter 3), plot the data on normal probability paper. Determine the normal distribution for describing the wind velocity.

6.5 A random variable with a triangular distribution between a and $a + r$, as shown in Fig. P6.5, is given by the density function

$$f_X(x) = \frac{2(x-a)}{r^2} ; \qquad a \le x \le a + r$$

$$= 0; \qquad\qquad \text{elsewhere.}$$

(a) Determine the appropriate standard variate S for this distribution.
(b) Construct the corresponding probability paper. What do the values of X at $F_S(0)$ and $F_S(1.0)$ on this paper mean?
(c) Suppose the following sample values were observed for X.

36	32	34	71
18	69	45	66
56	71	53	58
64	50	55	53
72	28	62	48
			75

Plot the above set of data on the triangular probability paper. From this plot estimate the minimum and maximum values of X.

6.6 The density function of the Rayleigh distribution is given by

$$f_X(x) = \frac{x}{\alpha^2} e^{-\frac{1}{2}(x/\alpha)^2}; \qquad x \ge 0$$

$$= 0; \qquad\qquad x < 0$$

in which the parameter α is the modal (or most probable) value of X.

(a) Construct the probability paper for this distribution. What does the slope of a straight line on this paper represent?
(b) The following is a set of data for stress range induced by vehicle loads on highway bridge members.

Strain Range, in micro in./in. (Data courtesy of W. H. Walker)

48.4	52.7	42.4
47.1	44.5	146.2
49.5	84.8	115.2
116.0	52.6	43.0
84.1	53.6	103.6
99.3	33.5	64.7
108.1	43.8	69.8
47.3	56.3	44.0
93.7	34.5	36.2
36.3	62.8	50.6
122.5	180.5	167.0

Plot this set of data on the Rayleigh probability paper constructed in part (a).

(c) What inference can you draw regarding the Rayleigh distribution as a model for live-load stress range in highway bridges, in light of the data plotted above? Determine the most probable stress range (if possible) from the results of part (b).

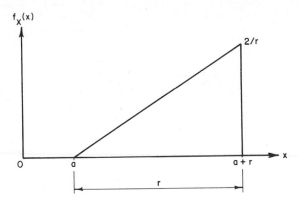

Figure P6.5

6.7 Time-to-failure (or malfunction) of a certain type of diesel engine has been recorded as follows (in hours).

0.13	121.58	2959.47	102.34
0.78	672.87	124.09	393.37
3.55	62.09	85.28	184.09
14.29	656.04	380.00	1646.01
54.85	735.89	298.58	412.03
216.40	895.80	678.13	813.00
1296.93	1057.57	861.93	239.10
952.65	470.97	1885.22	2633.98
8.82	151.44	862.93	658.38
29.75	163.95	1407.52	855.95

   (a) Construct the exponential probability paper (see Example 6.4) and plot on it the data given above.

   (b) On the basis of the results of part (a), estimate the minimum and mean time-to-failure of such engines.

   (c) Perform a chi-square test to determine the validity of the exponential distribution at the 1 % significance level.

6.8 The following are observations of the number of vehicles per minute arriving at an intersection from a one-way street:

$$0, 3, 1, 2, 0, 1, 1, 1, 2, 0, 1, 4, 3, 1, 1, 0, 0, 1, 0, 2$$

Perform a chi-square test to determine if the arrivals can be modeled by a Poisson process, at the 1 % significance level.

6.9 Cars coming toward an intersection are required to stop at the stop sign before they find a gap large enough to cross or to make a turn. This acceptance gap G, measured in seconds, varies from driver to driver, since some drivers may be more alert or more risk-taking than others. The following are measurements taken for several similar intersections.

Acceptance gap size (sec.)	Observed number of drivers
0.5–1.5	0
1.5–2.5	6
2.5–3.5	34
3.5–4.5	132
4.5–5.5	179
5.5–6.5	218
6.5–7.5	183
7.5–8.5	146
8.5–9.5	69
9.5–10.5	30
10.5–11.5	3
11.5–12.5	0

(a) Plot a histogram for the acceptance gap size.

(b) Assume that the distribution of G is normal; estimate its mean and variance. You may assume that all observations in each interval have the gap length equal to the average gap length for that interval. For example, for the interval 1.5–2.5, it may be assumed that there are 6 observations with gap length of 2.0 sec.

(c) Perform a chi-square goodness-of-fit test at the 1% significance level.

6.10 An extensive series of ultimate load tests on reinforced concrete columns was carried out at the University of Illinois (Hognestad, 1951). The ratio ϕ of the actual ultimate load to that computed by the appropriate ACI 318-63 formula, without consideration of the understrength factor in the ACI code, is tabulated below (for part of the 84 square tied columns tested).

Table of ratio ϕ	
0.79976	0.99410
0.82395	0.93811
0.99938	0.81649
0.78017	0.87551
0.91342	0.95705
0.90304	0.92863
0.86011	0.93054
1.01836	1.03065
0.90133	

(a) Plot the data on normal probability paper, and estimate (if possible) the mean and standard deviation from this plot.

(b) Perform a chi-square test at the 5% significance level on the fitted normal distribution.

(c) Repeat part (b) using the Kolmogorov-Smirnov test.

6.11 Data on the rate of oxygenation K in streams at 20°C have been observed at the Cincinnati Pool, Ohio River, and summarized in the following table (data from Kothandaraman, 1968).

K (per day)	Observed frequency
0.000 to 0.049	1
0.050 to 0.099	11
0.100 to 0.149	20
0.150 to 0.199	23
0.200 to 0.249	15
0.250 to 0.299	11
0.300 to 0.349	2

A normal distribution with a mean oxygenation rate of 0.173 per day and a standard deviation of 0.066 per day (both values are estimated from observed data) is proposed to model the oxygenation rate at the Cincinnati Pool, Ohio River.

Perform a chi-square test on the goodness of fit of the proposed distribution at the 5% significance level.

6.12 On the basis of the data given in Problem 6.2 and using the Kolmogorov-Smirnov method, determine which of the two distributions (normal and lognormal) considered in Problem 6.2 is a better model for the distribution of the ultimate strains of steel reinforcing bars.

7. Regression and Correlation Analyses

When dealing with two or more variables, the functional relation between the variables is often of interest. However, if one or both variables (in a two-variable case) are random, there will be no unique relationship between the values of the two variables—given a value of one variable (the controlled variable), there is a range of possible values of the other—and thus a probabilistic description is required. If the probabilistic relationship between the variables is described in terms of the mean and variance of one random variable as a function of the value of the other variable, we have what is known as *regression analysis*. When the analysis is limited to linear mean-value functions, it is called *linear regression*. In general, however, regression may be nonlinear. In some cases, nonlinear regression problems may be converted to linear ones by appropriate transformation of the original variables.

In the following, we present the concepts of regression analysis (including nonlinear regression and multiple linear regression), and their applications to engineering problems.

7.1. BASIC FORMULATION OF LINEAR REGRESSION

7.1.1. Regression with constant variance

When pairwise data for two variables, say X and Y, are plotted on a two-dimensional graph, such as shown in Fig. 7.1, the possible values of one variable, for example, Y, may depend on the value of the other variable X. For this reason, it would be inappropriate to analyze the data, say for Y (for example, in determining the mean and variance of Y), without due consideration of X. In the case of Fig. 7.1, we observe that there is a general tendency for the values of Y to increase with increasing values of X (X may be deterministic or random). Hence the mean value of Y will also increase with increasing values of X; the actual values of Y, of course, may not always increase with increasing values of X. In general, the mean value of Y will depend on the value of X. Suppose that this

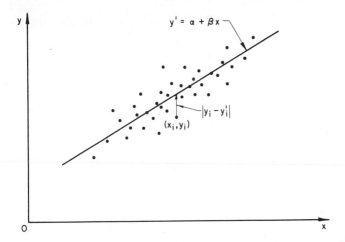

Figure 7.1 Linear regression analysis of data for two variables

relationship is linear; that is,

$$E(Y \mid X = x) = \alpha + \beta x \qquad (7.1)$$

where α and β are constants, and the variance of Y may be independent or a function of x. This is known as the *linear regression of Y on X*. Consider first the case with $\mathrm{Var}(Y \mid x) = $ constant.

Conceivably, there could be many straight lines, depending on the values of α and β, that might qualify as the mean-value function of Y in the light of the data. The "best" line may be the one that passes through the data points with the least error. To obtain this, we see from Fig. 7.1 that the difference between each observed value y_i and the straight line $y_i' = \alpha + \beta x_i$ is $\mid y_i - y_i' \mid$. Therefore the line with the least total error can be obtained by minimizing the sum of the squared errors—that is, by minimizing

$$\Delta^2 = \sum_{i=1}^{n} (y_i - y_i')^2 = \sum_{i=1}^{n} (y_i - \alpha - \beta x_i)^2$$

to obtain α and β, where n is the number of data points. This is known as the *method of least squares*, which leads to the following:

$$\frac{\partial \Delta^2}{\partial \alpha} = \sum_{i=1}^{n} 2(y_i - \alpha - \beta x_i)(-1) = 0$$

$$\frac{\partial \Delta^2}{\partial \beta} = \sum_{i=1}^{n} 2(y_i - \alpha - \beta x_i)(-x_i) = 0$$

From which the *least-squares estimates* of α and β are as follows:

$$\hat{\alpha} = \frac{1}{n} \Sigma y_i - \frac{\hat{\beta}}{n} \Sigma x_i = \bar{y} - \hat{\beta}\bar{x} \tag{7.2}$$

and

$$\hat{\beta} = \frac{\Sigma x_i y_i - n\bar{x}\bar{y}}{\Sigma x_i^2 - n\bar{x}^2} = \frac{\Sigma (x_i - \bar{x})(y_i - \bar{y})}{\Sigma (x_i - \bar{x})^2} \tag{7.3}$$

where $\Sigma = \sum_{i=1}^{n}$.

Therefore the least-squares regression line is

$$E(Y \mid x) = \hat{\alpha} + \hat{\beta}x \tag{7.4}$$

It should be emphasized that, strictly speaking, this regression line is valid only over the range of values of x for which data had been observed.

Equations 7.1 and 7.4 are referred to as the *regression of Y on X*. If X and Y are both random variables, we may also obtain the least-squares *regression of X on Y* using the same procedure; in this latter case, we would obtain the regression equation for $E(X \mid y)$. In general, this is a different linear equation from that of $E(Y \mid x)$; the two regression lines, however, always intersect at $(\bar{x}, \bar{y})$. For example, Meadows et al. (1972) discussed the per capita energy consumption Y versus the per capita GNP output X of different countries. If we are interested in predicting the energy consumption for a given GNP output of a country, a regression analysis of Y on X would be appropriate. Alternatively, the GNP output of a country may be estimated on the basis of the energy consumption; in this case, the regression of X on Y is required (see Problem 7.5).

Since the general trend is accounted for through the regression line of Eq. 7.4, the variance about this line is the measure of dispersion of interest, which is the *conditional variance* $\text{Var}(Y \mid x)$. For the case where the conditional variance $\text{Var}(Y \mid x)$ is assumed to be constant within the range of x of interest, an unbiased estimate of this variance is

$$s_{Y|x}^2 = \frac{1}{n-2} \sum_{i=1}^{n} (y_i - y_i')^2$$

$$= \frac{1}{n-2} \left[\sum_{i=1}^{n} (y_i - \bar{y})^2 - \hat{\beta}^2 \sum_{i=1}^{n} (x_i - \bar{x})^2 \right] \tag{7.5}$$

Observe that this is

$$s_{Y|x}^2 = \frac{\Delta^2}{n-2} \tag{7.5a}$$

Thus the corresponding *conditional standard deviation* is $s_{Y|x}$.

The coefficients $\hat{\alpha}$ and $\hat{\beta}$, and $s^2_{Y|x}$, are estimates of the respective true values of α, β, and $\text{Var}(Y \mid x)$. Confidence intervals may also be established on the basis of available data. For this purpose, if we assume that Y has a normal distribution about the regression line $E(Y \mid x)$ for all values of x, then $\hat{\alpha}$ and $\hat{\beta}$ individually follow the t-distribution (Hald, 1952). In such a case, the regression values

$$E(Y \mid x) = \hat{\alpha} + \hat{\beta}x$$

will also be t-distributed. On these bases, the required confidence intervals can be determined. It is worth noting here that these intervals for α, β, $E(Y \mid x)$, and Var $(Y \mid x)$ will decrease with increasing n.

The physical effect of the linear regression of Y on X can be measured by the reduction of the original variance of Y, $s_Y{}^2$, resulting from taking into account the general trend with X. This reduction is represented by,

$$r^2 = 1 - \frac{s^2_{Y|x}}{s_Y{}^2} \tag{7.6}$$

where

$$s_Y{}^2 = \frac{1}{n-1} \sum_{i=1}^{n} (y_i - \bar{y})^2$$

is the sample variance of Y. It will be shown later (in Section 7.5) that (for large n) r is approximately equal to a point estimate of the correlation coefficient.

Regression of normal variates. The assumptions of linear model and constancy of variance underlying linear regression are, in fact, inherent properties of populations that are jointly normal. We recall from Example 3.25 that if X and Y are jointly normally distributed, the conditional mean and variance of Y given $X = x$ are as follows:

$$E(Y \mid x) = \mu_Y + \rho \frac{\sigma_Y}{\sigma_X} (x - \mu_X)$$

and,

$$\text{Var}(Y \mid x) = \sigma_Y{}^2(1 - \rho^2)$$

where ρ is the correlation coefficient. These results mean that if two variates are jointly normal the regression of Y on X is linear with constant conditional variance (that is, independent of x); specifically, in this case, the linear equation is of the form of Eq. 7.1 with

$$\beta = \rho \frac{\sigma_Y}{\sigma_X}$$

and,

$$\alpha = \mu_Y - \beta\mu_X$$

Therefore, if the underlying populations are jointly normal, linear regression should, properly, be used.

The expressions for α and β given above may be compared with the least-squares estimates $\hat{\alpha}$ and $\hat{\beta}$ of Eqs. 7.2 and 7.3 (and its subsequent extension in Eq. 7.23). Also, the above conditional variance may be compared with the corresponding estimate $s_{Y|x}^2$ given subsequently in Eq. 7.24.

EXAMPLE 7.1

Tabulated in the first three columns of Table E7.1 are shear strengths, in kips per square foot (ksf), obtained from 10 specimens taken at various depths of a clay stratum. Determine the mean and variance of the shear strength as a linear function of depth. Assume that the variance is constant with depth.

Table E7.1 summarizes the computations in the regression analysis.

On the basis of the calculations in Table E7.1, the least-squares mean shear strength (in ksf) as a function of depth x is given by

$$E(Y \mid x) = 0.015 + 0.0517x$$

whereas the variance of the shear strength at a given depth is estimated to be 0.0368 (ksf)^2, giving $s_{Y|x} = 0.192$ ksf. If the linear trend with depth is not taken into account, the unconditional variance of the shear strength would be 0.197 (ksf)^2, and $s_Y = 0.44$ ksf. Hence the conditional standard deviation $s_{Y|x}$ is considerably smaller than s_Y.

The regression equation obtained above may be used to predict the shear strength

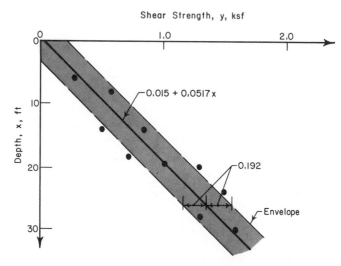

Figure E7.1 Regression line for shear strength with depth

Table E7.1. Computational Tableau for Example 7.1

Specimen no.	Depth (ft) x_i	Strength (ksf) y_i	$x_i y_i$	x_i^2	y_i^2	$y_i' = \alpha + \beta x_i$	$y_i - y_i'$	$(y_i - y_i')^2$
						To determine $\hat{\alpha}$ and $\hat{\beta}$ →	To determine $s_{Y\mid x}$ →	
1	6	0.28	1.68	36	0.078	0.325	−0.045	0.0020
2	8	0.58	4.64	64	0.336	0.429	0.151	0.0228
3	14	0.50	7.00	196	0.250	0.739	−0.239	0.0571
4	14	0.83	11.63	196	0.689	0.739	0.091	0.0083
5	18	0.71	12.78	324	0.504	0.946	−0.236	0.0557
6	20	1.01	20.20	400	1.020	1.049	−0.039	0.0015
7	20	1.29	25.80	400	1.662	1.049	0.241	0.0580
8	24	1.50	36.00	576	2.250	1.257	0.243	0.0590
9	28	1.29	36.10	784	1.662	1.463	−0.173	0.0299
10	30	1.58	47.40	900	2.495	1.566	0.014	0.0002
Σ	182	9.57	203.23	3876	10.946			$\Delta^2 = 0.2945$

$$\bar{x} = 18.2, \quad \bar{y} = 0.957$$

$$\hat{\beta} = \frac{203.23 - (10)(18.2)(0.957)}{3876 - (10)(18.2)^2} = 0.0517$$

$$\hat{\alpha} = 0.957 - (0.0517)(18.2) = 0.015$$

$$s_Y^2 = \tfrac{1}{9}[10.946 - (10)(0.957)^2] = 0.197$$

$$s_{Y\mid x}^2 = \frac{0.2945}{10 - 2} = 0.0368$$

$$s_{Y\mid x} = \sqrt{0.0368} = 0.192$$

$$r^2 = 1 - \frac{0.0386}{0.197} = 0.813$$

from 6 ft to 30 ft deep. It may not apply to depths beyond 30 ft, unless the linear trend can be justified beyond this depth on physical ground (for example, the same soil type).

Graphically, the regression line obtained above is shown in Fig. E7.1; also shown is the envelope with $\pm s_{Y|x}$ from the regression line. This represents a band width of one (conditional) standard deviation from either side of the regression line.

Table E7.2.   Computational Tableau for Example 7.2

	Precipitation x_i (in.)	Runoff y_i (in.)	$x_i y_i$	x_i^2	$y_i' = \hat{\alpha} + \hat{\beta} x_i$	$y_i - y_i'$	$(y_i - y_i')^2$
1	1.11	0.52	0.58	1.23	0.343	0.177	0.0313
2	1.17	0.40	0.47	1.37	0.369	0.031	0.0009
3	1.79	0.97	1.74	3.20	0.637	0.333	0.1110
4	5.62	2.92	16.40	31.60	2.280	0.640	0.4000
5	1.13	0.17	0.19	1.28	0.351	−0.181	0.0328
6	1.54	0.19	0.29	2.37	0.530	−0.340	0.1158
7	3.19	0.76	2.43	10.15	1.245	−0.485	0.2360
8	1.73	0.66	1.14	2.99	0.612	0.048	0.0023
9	2.09	0.78	1.63	4.37	0.770	0.010	0.0001
10	2.75	1.24	3.41	7.55	1.059	0.181	0.0328
11	1.20	0.39	0.47	1.44	0.381	0.009	0.0001
12	1.01	0.30	0.30	1.02	0.299	0.001	0.0000
13	1.64	0.70	1.15	2.69	0.574	0.126	0.0158
14	1.57	0.77	1.21	2.46	0.544	0.226	0.0511
15	1.54	0.59	0.91	2.37	0.530	0.060	0.0036
16	2.09	0.95	1.99	4.36	0.770	0.180	0.0326
17	3.54	1.02	3.62	12.55	1.400	−0.380	0.1442
18	1.17	0.39	0.46	1.37	0.368	0.022	0.0004
19	1.15	0.23	0.26	1.32	0.360	−0.130	0.0169
20	2.57	0.45	1.16	6.60	0.980	−0.530	0.2810
21	3.57	1.59	5.66	12.74	1.415	0.175	0.0306
22	5.11	1.74	8.90	26.18	2.084	−0.344	0.1185
23	1.52	0.56	0.85	2.31	0.521	0.039	0.0015
24	2.93	1.12	3.28	8.58	1.135	−0.015	0.0002
25	1.16	0.64	0.74	1.34	0.365	0.275	0.0755
Σ	53.89	20.05	59.24	153.44			$\Delta^2 = 1.7350$

$$\bar{x} = \frac{53.89}{25} = 2.16$$

$$\bar{y} = \frac{20.05}{25} = 0.80$$

$$\hat{\beta} = \frac{59.24 - 25(2.16)(0.80)}{153.44 - 25(2.16)^2} = 0.435$$

$$\hat{\alpha} = 0.80 - (0.435)(2.16) = -0.14$$

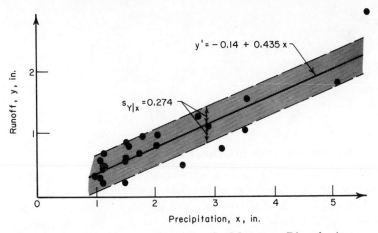

Figure E7.2   Runoff vs. precipitation for Monocacy River basin

EXAMPLE 7.2

The precipitation and runoff data for the 25 storms on the Monocacy River, were given earlier in Example 5.6.

(a) Plot the observed data for runoff vs. precipitation.

(b) Determine and draw the regression line of runoff on precipitation (that is, the mean runoff for given value of precipitation).

(c) Estimate the variance of runoff for a given precipitation. Assume that the variance of the runoff is constant with precipitation.

(d) Assume that the runoff corresponding to a given precipitation is a normal variate; what is the probability that the runoff will exceed 2 in. during a storm with 4-in. precipitation?

Solution

(a) The plot of data is shown in Fig. E7.2.

(b) For the regression analysis, see Table E7.2. From the table, we obtain

$$E(Y \mid x) = -0.14 + 0.435x$$

It should be emphasized that this regression equation is applicable only within the range of the data; in particular, it should not be used for precipitation less than 1 in.

(c) For given precipitation, the variance of runoff is

$$s_{Y|x}^2 = \frac{\Delta^2}{n-2} = \frac{1.735}{25-2} = 0.075 \text{ in.}^2$$

$$s_{Y|x} = \sqrt{0.075} = 0.274 \text{ in.}$$

(d) When the precipitation is 4 in., the mean runoff

$$E(Y \mid X = 4) = -0.14 + 0.435(4)$$
$$= 1.6 \text{ in.}$$

Therefore the normal distribution for the runoff Y in this storm is $N(1.6, 0.274)$ in.

Hence

$$P(Y > 2 \mid X = 4) = 1 - \Phi\left(\frac{2 - 1.6}{0.274}\right)$$
$$= 1 - \Phi(1.46)$$
$$= 1 - 0.9279$$
$$= 0.0721$$

7.1.2.   Regression with nonconstant variance

Conceivably, the conditional variance about the regression line may be a function of the independent (controlled) variable; this would be the case when the scattergram of the data shows a significant variation in the degree of scatter with values of the controlled variable. In such cases, the regression analysis presented in Section 7.1.1 can be modified to take account of the variation in the conditional variance. This variation may be expressed as

$$\mathrm{Var}(Y|x) = \sigma^2 g^2(x)$$

where $g(x)$ is a predetermined function, and σ is an unknown constant. Again, for linear regression,

$$E(Y \mid x) = \alpha + \beta x$$

In determining the regression equation, it would seem reasonable to assume that data points in regions of small variance should have more "weight" than those in regions of large variance. On this premise, we therefore assign weights inversely proportional to the variance; or

$$w_i' = \frac{1}{\mathrm{Var}(Y \mid x_i)} = \frac{1}{\sigma^2 g^2(x_i)}$$

Then the squared error is

$$\Delta^2 = \sum_{i=1}^{n} w_i'(y_i - \alpha - \beta x_i)^2$$

from which the least-squares estimates of α and β become

$$\hat{\alpha} = \frac{\Sigma w_i y_i - \hat{\beta}\Sigma w_i x_i}{\Sigma w_i} \tag{7.7}$$

and

$$\hat{\beta} = \frac{\Sigma w_i(\Sigma w_i y_i x_i) - (\Sigma w_i y_i)(\Sigma w_i x_i)}{\Sigma w_i(\Sigma w_i x_i^2) - (\Sigma w_i x_i)^2} \tag{7.8}$$

where

$$w_i = \sigma^2 w_i' = \frac{1}{g^2(x_i)}$$

An unbiased estimate of the unknown σ^2 is

$$s^2 = \frac{\Sigma w_i(y_i - \hat{\alpha} - \hat{\beta} x_i)^2}{n - 2}$$

Hence an estimate of the conditional variance is

$$s^2_{Y|x} = s^2 g^2(x) \tag{7.9}$$

and

$$s_{Y|x} = s g(x) \tag{7.9a}$$

EXAMPLE 7.3

The maximum settlements and maximum differential settlements of 18 storage tanks in Libya have been observed; the data are plotted as shown in Fig. E7.3.

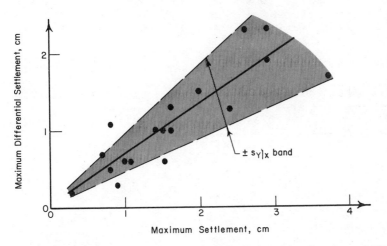

Figure E7.3 Settlement of tanks on Libyan sand (data after Lambe and Whitman, 1969)

From Fig. E7.3, the scatter of the differential settlement appears to increase with the maximum settlement. Physically, this would be expected, since the maximum differential settlement would ordinarily not exceed the maximum settlement. For these reasons, the conditional standard deviation of the differential settlement Y may be assumed to increase linearly with the maximum settlement X, or

$$\text{Var}\,(Y\,|\,x) = \sigma^2 x^2$$

Table E7.3. Computational Tableau for Example 7.3

Tank no. i	Maximum settlement (cm) x_i	Maximum differential settlement (cm) y_i	w_i	$w_i x_i$	$w_i y_i$	$w_i x_i y_i$	$w_i x_i^2$	$w_i(y_i - \hat{\alpha} - \hat{\beta} x_i)^2$
1	0.3	0.2	11.11	3.33	2.22	0.67	1.0	0.0178
2	0.7	0.7	2.04	1.43	1.43	1.00	1.0	0.0816
3	0.8	0.5	1.56	1.25	0.78	0.62	1.0	0.0066
4	0.8	1.1	1.56	1.25	1.72	1.37	1.0	0.4465
5	0.9	0.3	1.23	1.11	0.37	0.33	1.0	0.1339
6	1.0	0.6	1.00	1.00	0.60	0.60	1.0	0.0090
7	1.1	0.6	0.83	0.91	0.50	0.55	1.0	0.0212
8	1.4	1.0	0.51	0.71	0.51	0.71	1.0	0.0010
9	1.5	1.0	0.44	0.67	0.44	0.66	1.0	0.0002
10	1.6	1.0	0.39	0.63	0.39	0.62	1.0	0.0028
11	1.6	1.3	0.39	0.63	0.51	0.81	1.0	0.0180
12	2.0	1.5	0.25	0.50	0.38	0.75	1.0	0.0060
13	2.4	1.3	0.17	0.42	0.22	0.53	1.0	0.0158
14	2.6	2.3	0.15	0.38	0.35	0.90	1.0	0.0479
15	2.9	1.9	0.12	0.34	0.23	0.66	1.0	0.0001
16	2.9	2.3	0.12	0.34	0.28	0.80	1.0	0.0164
17	3.7	1.7	0.07	0.27	0.12	0.44	1.0	0.0394
18	1.5	0.6	0.44	0.67	0.26	0.40	1.0	0.0776
Σ			22.38	15.84	11.31	12.42	18.0	0.9418

$$\hat{\beta} = \frac{(22.38)(12.42) - (11.31)(15.84)}{(22.38)(18) - (15.84)^2} = 0.65$$

$$\hat{\alpha} = \frac{11.31 - (0.65)(15.84)}{22.38} = 0.045$$

$$s^2 = \frac{0.9418}{16} = 0.0589$$

$$s_{Y|x} = 0.243x$$

Thus

$$w_i = \frac{1}{x_i^2}$$

The required computations are summarized in Table E7.3, from which we obtain the regression equation for estimating the expected maximum differential settlement Y (in centimeters) on the basis of information for the maximum settlement X of the tanks (in cm) as

$$E(Y \mid x) = 0.045 + 0.65x$$

The corresponding standard deviation is

$$s_{Y|x} = 0.243x$$

7.2. MULTIPLE LINEAR REGRESSION

The value of an engineering variable may depend on several factors. In such cases, the mean and variance of the dependent variable will be a function of the values of several variables. When the mean-value function is assumed to be linear, the resulting analysis is known as *multiple linear regression*.

Linear regression analysis for more than two variables is simply a generalization of the regression analysis discussed in Section 7.1. Suppose that the dependent variable of interest is Y, and that it is a linear function of m variables $X_1, X_2, \ldots, X_m$. The assumptions underlying multiple regression analysis are as follows.

1. The mean value of Y is a linear function of $x_1, x_2, \ldots, x_m$; that is,

$$E(Y \mid x_1, \ldots, x_m) = \beta_o + \beta_1 x_1 + \cdots \beta_m x_m \qquad (7.10)$$

   where $\beta_o, \beta_1, \ldots, \beta_m$ are constants, to be determined from observed data.
2. The conditional variance of Y given $x_1, \cdots, x_m$ is constant; that is,

$$\mathrm{Var}(Y \mid x_1, \ldots, x_m) = \sigma^2$$

   or proportional to a given function of $x_1, \ldots, x_m$; that is,

$$\mathrm{Var}(Y \mid x_1, \ldots, x_m) = \sigma^2 g^2(x_1, \ldots, x_m)$$

The regression analysis then determines estimates for $\beta_0, \beta_1, \ldots, \beta_m$ and σ^2 based on a set of observed data $(x_{1i}, x_{2i}, \ldots, x_{mi}, y_i)$, $i = 1, \ldots, n$. Equation 7.10 can be written also as

$$E(Y \mid x_1, \ldots, x_m) = \alpha + \beta_1(x_1 - \bar{x}_1) + \cdots + \beta_m(x_m - \bar{x}_m) \qquad (7.11)$$

in which the $\bar{x}_i$'s are the sample means of X_i and α is simply a readjusted constant.

Again we restrict our derivation for the case in which the conditional variance $\mathrm{Var}(Y \mid x_1, \ldots, x_m)$ is constant. The sum of squared errors for a set of n data points, then, is

$$\Delta^2 = \sum_{i=1}^{n} (y_i - y_i')^2$$

$$= \sum_{i=1}^{n} [y_i - \alpha - \beta_1(x_{1i} - \bar{x}_1) - \cdots - \beta_m(x_{mi} - \bar{x}_m)]^2 \qquad (7.12)$$

By the least-squares criterion, we minimize Δ^2 to obtain the following set of equations for determining the estimates of α and β_j, $j = 1, 2, \ldots, m$:

$$\frac{\partial \Delta^2}{\partial \alpha} = 2 \sum [y_i - \hat{\alpha} - \hat{\beta}_1(x_{1i} - \bar{x}_1) - \cdots - \hat{\beta}_m(x_{mi} - \bar{x}_m)] = 0$$

$$\frac{\partial \Delta^2}{\partial \beta_1} = 2 \sum \{[y_i - \hat{\alpha} - \hat{\beta}_1(x_{1i} - \bar{x}_1) - \cdots$$

$$- \hat{\beta}_m(x_{mi} - \bar{x}_m)](x_{1i} - \bar{x}_1)\} = 0$$

$$\vdots \qquad \qquad \vdots \qquad \qquad (7.13)$$

$$\frac{\partial \Delta^2}{\partial \beta_m} = 2 \sum \{[y_i - \hat{\alpha} - \hat{\beta}_1(x_{1i} - \bar{x}_1) - \cdots$$

$$- \hat{\beta}_m(x_{mi} - \bar{x}_m)](x_{mi} - \bar{x}_m)\} = 0$$

where $\Sigma = \sum_{i=1}^{n}$. From the first of these equations, we have

$$\Sigma y_i - n\hat{\alpha} - \hat{\beta}_1 \Sigma(x_{1i} - \bar{x}_1) - \cdots - \hat{\beta}_m \Sigma(x_{mi} - \bar{x}_m) = 0$$

but $\Sigma(x_{1i} - \bar{x}_1) = \cdots = \Sigma(x_{mi} - \bar{x}_m) = 0$; thus

$$\hat{\alpha} = \frac{\Sigma y_i}{n} = \bar{y} \qquad (7.14)$$

Substituting this value of $\hat{\alpha}$ into the remaining equations in Eq. 7.13, we obtain

$$\hat{\beta}_1 \Sigma(x_{1i} - \bar{x}_1)^2 + \hat{\beta}_2 \Sigma(x_{1i} - \bar{x}_1)(x_{2i} - \bar{x}_2) + \cdots$$

$$+ \hat{\beta}_m \Sigma(x_{1i} - \bar{x}_1)(x_{mi} - \bar{x}_m) = \Sigma(x_{1i} - \bar{x}_1)(y_i - \bar{y})$$

$$\vdots \qquad \qquad \vdots \qquad \qquad (7.15)$$

$$\hat{\beta}_1 \Sigma(x_{mi} - \bar{x}_m)(x_{1i} - \bar{x}_1) + \hat{\beta}_2 \Sigma(x_{mi} - \bar{x}_m)(x_{2i} - \bar{x}_2) + \cdots$$

$$+ \hat{\beta}_m \Sigma(x_{mi} - \bar{x}_m)^2 = \Sigma(x_{mi} - \bar{x}_m)(y_i - \bar{y})$$

It can be observed that this represents a set of m linear simultaneous equations involving the m unknowns $\hat{\beta}_1, \ldots, \hat{\beta}_m$. The solution of Eq. 7.15 yields the required coefficients $\hat{\beta}_1, \ldots, \hat{\beta}_m$, from which we obtain the least-squares regression equation

$$E(Y \mid x_1, \ldots x_n) = \hat{\alpha} + \hat{\beta}_1 (x_1 - \bar{x}_1) + \cdots + \hat{\beta}_m(x_m - \bar{x}_m)$$

$$= \hat{\beta}_0 + \hat{\beta}_1 x_1 + \cdots + \hat{\beta}_m x_m \qquad (7.16)$$

where

$$\hat{\beta}_0 = \hat{\alpha} - \hat{\beta}_1 \bar{x}_1 - \cdots - \hat{\beta}_m \bar{x}_m$$

The variance of Y about this mean-value function, namely, $\mathrm{Var}(Y \mid x_1, \ldots, x_m)$, is a measure of the conditional dispersion about the regression equation. An unbiased estimate of this conditional variance is

$$s^2_{Y|x_1,\ldots,x_m} = \frac{\Delta^2}{n-m-1}$$

$$= \frac{\Sigma[y_i - \hat{\alpha} - \hat{\beta}_1(x_{1i} - \bar{x}_1) - \cdots - \hat{\beta}_m(x_{mi} - \bar{x}_m)]^2}{n-m-1} \qquad (7.17)$$

The corresponding conditional standard deviation, therefore, is

$$s_{Y|x_1,\ldots,x_m} = \frac{\Delta}{\sqrt{n-m-1}} \qquad (7.17a)$$

Obviously, Eq. 7.17 is valid only if the sample size n is larger than $m + 1$. Again, the assumption of normal distribution for Y may be invoked for the purpose of establishing confidence intervals on the regression coefficients and for computing probabilities associated with the random variable Y.

EXAMPLE 7.4

An important factor in the prediction of frost depth for highway pavement design is the mean annual temperature for the site under consideration. The mean annual temperature records at 10 different weather stations in West Virginia are summarized in Table E7.4a.

Since a pavement may be constructed in various locations over the state, where temperature records may not be available, it is desired to predict the mean annual temperature of a locality on the basis of its elevation and latitude, using the information in Table E7.4a. The following equation is assumed:

$$E(Y \mid x_1, x_2) = \beta_0 + \beta_1 x_1 + \beta_2 x_2$$

where

Y = mean annual temperature, in °F

x_1 = elevation, in feet, above sea level

x_2 = north latitude, in degrees

Determine estimates for β_0, β_1, and β_2 and evaluate the conditional variance Var $(Y \mid x_1, x_2)$, which is assumed to be constant. Table E7.4b summarizes the computations required in the multiple linear regression analysis. From these results, the mean annual temperature for a locality in West Virginia, with elevation x_1 and latitude x_2, is given by

$$E(Y \mid x_1, x_2) = 121.3 - 0.0034x_1 - 1.65x_2$$

whereas the standard deviation $s_{Y|x_1,x_2}$ of the mean annual temperature at any locality is estimated to be $\sqrt{0.547} = 0.74$°F. It may be observed from Table E7.4b that by taking the elevation and latitude of a locality into account in estimating its mean annual temperature, the variance of Y is reduced by 94.5% of the unconditional variance.

At Gary, West Virginia, which is located at an elevation of 1426 ft and latitude of

Table E7.4a.   Mean Annual Temperature in West Virginia—Data From Moulton and Schaub (1969)

Weather stations	Elevation (ft)	North latitude (deg)	Mean annual temperature (°F)
Bayard	2375	39.27	47.5
Buckhannon	1459	39.00	52.3
Charleston	604	38.35	56.8
Flat Top	3242	37.58	48.4
Kearneysville	550	39.38	54.2
Madison	675	38.05	55.1
New Martinsville	635	39.65	54.4
Pickens	2727	38.66	48.8
Rainelle	2424	37.97	50.5
Wheeling	659	40.10	52.7

$37.37°N$, the expected mean annual temperature would be

$$E(Y \mid 1426, 37.37) = 121.3 - 0.0034 \times 1426 - 1.65 \times 37.37$$
$$= 54.80°F$$

and a conditional standard deviation of $0.74°F$. If the mean annual temperature is assumed to be Gaussian, the 10-percentile value of the mean annual temperature $y_{.1}$ is determined as follows:

$$P(Y < y_{.1}) = \Phi\left(\frac{y_{.1} - 54.8}{0.74}\right) = 0.1$$

or

$$y_{.1} = 54.8 + 0.74\Phi^{-1}(0.1)$$
$$= 53.9°$$

7.3.   NONLINEAR REGRESSION

Relationships between engineering variables are not always linear, or may not always be adequately described by linear models. Experimental data for such variables may show a nonlinear trend between the observed values of the variables. For example, Fig. 7.2 shows a plot of the average all-day parking cost in a central business district versus the urban population for various cities in the United States. Similarly, data for the average dissolved oxygen (DO) in a pool measured at various temperatures are shown in Fig. 7.3. Although a linear relation may be used to describe the general trend between each pair of variables, predictions based on such linear relationships may overestimate (in certain ranges of the variables) or underestimate (in other ranges of the variables) the expected result. For example, the linear regression line in Fig. 7.3 will underestimate the average DO at temperatures between 23° and 24°C, but may overestimate the

Table E7.4b Computational Tableau for Example 7.4

Station no. i	Elevation x_{1i}	Latitude x_{2i}	Mean temperature y_i	$(x_{1i}-\bar{x}_1)^2 \times 10^3$	$(x_{2i}-\bar{x}_2)^2$	$(x_{1i}-\bar{x}_1) \times (x_{2i}-\bar{x}_2)$	$(x_{1i}-\bar{x}_1) \times (y_i-\bar{y})$	$(x_{2i}-\bar{x}_2) \times (y_i-\bar{y})$	$y_i' = \hat{\beta}_0 + \hat{\beta}_1 x_1 + \hat{\beta}_2 x_2$	$(y_i-y_i')^2$
1	2375	39.27	47.5	706	0.221	395	−383	−2.14	48.5	1.00
2	1459	39.00	52.3	6	0.040	−15	−18	0.05	52.0	0.09
3	604	38.35	56.8	867	0.203	419	−4404	−2.13	56.0	0.64
4	3242	37.58	48.4	2914	1.488	−2083	−6265	4.48	48.3	0.01
5	550	39.38	54.2	970	0.336	−571	−2098	1.23	54.5	0.09
6	675	38.05	55.1	740	0.563	645	−2606	−2.28	56.2	1.21
7	635	39.65	54.4	810	0.723	−765	−2097	1.98	53.7	0.49
8	2727	38.66	48.8	1421	0.020	−167	−3898	0.46	48.3	0.25
9	2424	37.97	50.5	790	0.689	−738	−1396	1.30	50.4	0.01
10	659	40.10	52.7	767	1.690	−1139	−552	0.81	52.9	0.04
Σ	15350	388.01	520.7	9991	5.973	−4019	−27171	3.76		3.83

$$\bar{x}_1 = \frac{15350}{10} = 1535$$

$$\bar{x}_2 = \frac{388.01}{10} = 38.80$$

$$\hat{\beta}_0 = \bar{y} = \frac{520.7}{10} = 52.07$$

$$s_Y^2 = \frac{1}{9}\Sigma(y_i-\bar{y})^2 = \frac{89.48}{9} = 9.94$$

$$\left.\begin{array}{l} 9991000\hat{\beta}_1 - 4019\hat{\beta}_2 = -27171 \\[4pt] -4019\hat{\beta}_1 + 5.973\hat{\beta}_2 = 3.76 \end{array}\right\}$$

$$\hat{\beta}_1 = \frac{\det\begin{bmatrix} -27171 & -4019 \\ 3.76 & 5.973 \end{bmatrix}}{\det\begin{bmatrix} 9991000 & -4019 \\ -4019 & 5.973 \end{bmatrix}} = -0.0034$$

$$\hat{\beta}_2 = \frac{\det\begin{bmatrix} 9991000 & -27171 \\ -4019 & 3.76 \end{bmatrix}}{43520000} = -1.65$$

$$\hat{\beta}_0 = 52.07 + 0.0034 \times 1535 + 1.65 \times 38.80 = 121.3$$

$$s_{Y|x_1,x_2}^2 = \frac{3.83}{10-2-1} = \frac{3.83}{7} = 0.547$$

$$s_{Y|x_1,x_2} = \sqrt{0.547} = 0.74$$

$$r^2 = 1 - \frac{0.547}{9.94} = 0.945$$

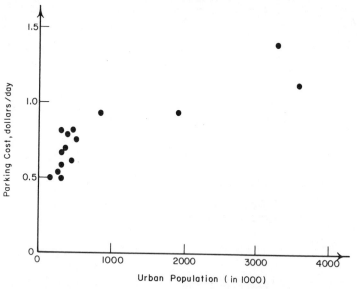

Figure 7.2 Average all-day off-street parking rates in central business district (data from Wynn, 1969)

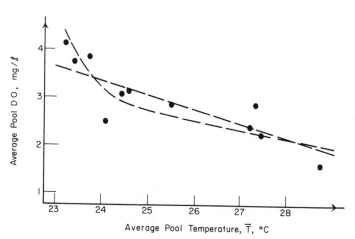

Figure 7.3 Average pool DO and temperature (data from Butts, Schnepper, and Evans, 1970)

average DO between 24° and 26°C. In such cases, a nonlinear relationship between the variables would be more appropriate. The determination of such nonlinear relationships on the basis of observational data involves *nonlinear regression analysis.*

Nonlinear regression is usually based on an assumed nonlinear (mean-value) function with certain undetermined coefficients that would be evaluated from the experimental data. The simplest type of nonlinear function for the regression of Y on X is

$$E(Y \mid x) = \alpha + \beta g(x) \qquad (7.18)$$

where $g(x)$ is a predetermined nonlinear function of x. For example, $g(x)$ may be $x + x^2$, e^x, $\ln x$, or any other function of x. Finally, nonlinear regression analysis is usually based on the assumption of a constant variance Var $(Y \mid x)$, or a variance that is a function of $g(x)$.

By defining a new variable $x' = g(x)$, Eq. 7.18 becomes

$$E(Y \mid x') = \alpha + \beta x' \qquad (7.19)$$

which is of the same mathematical form as the linear regression equation of Eq. 7.1. If the observed data pair (x_i, y_i) is also transformed to $[g(x_i), y_i]$ or (x_i', y_i), the original problem of nonlinear regression between x and y is thus converted to a linear regression between the variables x' and y. The corresponding regression coefficients α and β and Var$(Y \mid x)$ can then be estimated from Eqs. 7.2, 7.3, and 7.5, respectively.

EXAMPLE 7.5

The average all-day parking cost in the central business district of United States cities may be expressed in terms of the logarithm of the urban population; that is, modeled with the following nonlinear regression equation:

$$E(Y \mid x) = \alpha + \beta \ln x$$

with a constant Var $(Y \mid x)$, where

$$Y = \text{average all-day parking cost (in dollars)}$$
$$x = \text{urban population (in thousands)}$$

Determine the estimates for α, β, and Var $(Y \mid x)$ on the basis of the observed data referred to in Fig. 7.2 and given in Table E7.5.

The required computations for the regression analysis are summarized in Table E7.5; from these results we obtain the mean-value function

$$E(Y \mid x) = -0.773 + 0.244 \ln x$$

and

$$s_{Y \mid x}^2 = 0.013$$

or

$$s_{Y \mid x} = 0.11$$

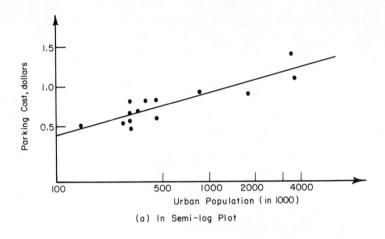

(a) In Semi-log Plot

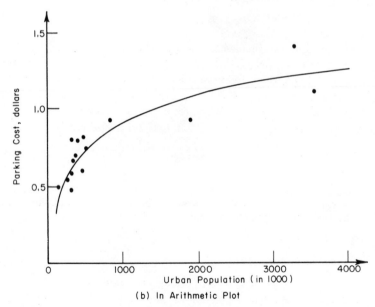

(b) In Arithmetic Plot

Figure E7.5 Average all-day parking cost vs. urban population. (*a*) In semilog plot. (*b*) In arithmetic plot. (Data from Wynn, 1969)

Table E7.5. Computational Tableau for Example 7.5

City	x_i	y_i	$x_i' = \ln x_i$	$x_i'y_i$	$x_i'^2$	y_i^2	$y_i' = \alpha + \beta x_i'$	$(y_i - y_i')^2$
			To determine $\hat\alpha$ and $\hat\beta$				To determine $s_{Y\mid x}$	
1	190	0.50	5.25	2.62	27.5	0.25	0.51	0.000
2	310	0.48	5.74	2.75	32.9	0.23	0.63	0.023
3	270	0.53	5.60	2.97	31.3	0.28	0.60	0.005
4	320	0.58	5.77	3.35	33.3	0.34	0.63	0.003
5	460	0.60	6.13	3.68	37.6	0.36	0.72	0.014
6	340	0.67	5.83	3.91	34.0	0.45	0.65	0.000
7	380	0.69	5.94	4.10	35.3	0.48	0.68	0.000
8	520	0.75	6.25	4.69	39.1	0.56	0.75	0.000
9	310	0.80	5.74	4.59	32.9	0.64	0.63	0.029
10	400	0.80	5.99	4.79	35.9	0.64	0.69	0.012
11	470	0.81	6.15	4.98	37.9	0.66	0.73	0.006
12	840	0.92	6.73	6.19	45.3	0.85	0.87	0.003
13	1910	0.92	7.56	6.95	57.1	0.85	1.07	0.023
14	3290	1.40	8.10	11.34	65.6	1.96	1.21	0.036
15	3600	1.12	8.19	9.17	67.1	1.25	1.23	0.012
Σ		11.57	94.97	76.08	612.8	9.80		$\Delta^2 = 0.166$

$$\bar{x}' = \frac{94.97}{15} = 6.33$$

$$\bar{y} = \frac{11.57}{15} = 0.771$$

$$\hat\beta = \frac{76.08 - 15 \times 6.33 \times 0.771}{612.8 - 15 \times (6.33)^2} = 0.244$$

$$\hat\alpha = 0.771 - 0.244 \times 6.33 = -0.773$$

$$s_Y{}^2 = \tfrac{1}{14}[9.80 - 15(0.771)^2] = 0.06$$

$$s_{Y\mid x}^2 = \frac{0.166}{15 - 2} = 0.013$$

$$s_{Y\mid x} = \sqrt{0.013} = 0.11$$

$$r^2 = 1 - \frac{0.013}{0.06} = 0.78$$

Figures E7.5*a* and E7.5*b* show this regression curve in semi-logarithmic and arithmetic scales, respectively.

EXAMPLE 7.6

An exponential model may be used for predicting the average DO concentration from the average pool temperature $\bar{T}$; that is,

$$\overline{DO} = \alpha e^{-\beta \bar{T}}$$

Estimate the coefficients α and β based on the data portrayed in Fig. 7.3.

Taking the logarithm on both sides of the equation given above, we have

$$\ln \overline{DO} = \ln \alpha - \beta \bar{T}$$

It may be observed that the right side of this equation is a linear function of T. Introducing the variables,

$$Y = \ln \overline{DO}$$
$$x = -\bar{T}$$

the nonlinear problem, therefore, is reduced to that of a linear regression. That is,

$$E(\ln \overline{DO} \mid \bar{T}) = \ln \alpha - \beta \bar{T}$$

or

$$E(Y \mid x) = \ln \alpha + \beta x$$

In this case, therefore, the original data are first converted from $(\overline{DO}_i, \bar{T}_i)$ to $(\ln \overline{DO}_i, -\bar{T}_i)$ and are then used in the linear regression analysis. On this basis, the regression

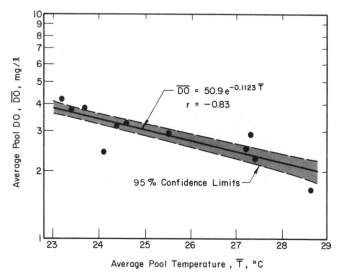

Figure E7.6 Weighted average pool DO and temperature relationship (after Butts, Schnepper, and Evans, 1970)

coefficients are estimated to be

$$\ln \hat{\alpha} = 3.93 \quad \text{and} \quad \hat{\beta} = 0.1123$$

Hence the exponential model is obtained as

$$\ln \overline{DO} = 3.93 - 0.1123\bar{T}$$

or

$$\overline{DO} = e^{3.93 - 0.1123T} = 50.9e^{-0.1123T}$$

This regression equation and the associated 95% confidence interval are shown in Fig. E7.6.

The form of nonlinear functions assumed in Eq. 7.18 can be generalized as follows:

$$E(Y \mid x) = \alpha + \beta_1 g_1(x) + \beta_2 g_2(x) + \cdots + \beta_m g_m(x) \qquad (7.20)$$

where $g_j(x)$, $j = 1, 2, \ldots, m$ are predetermined functions of the independent variable x. An example of Eq. 7.20 is the following general polynomial relation:

$$E(Y \mid x) = \alpha + \beta_1 x + \beta_2 x^2 + \cdots + \beta_m x^m \qquad (7.21)$$

We now observe that by the conversion $z_j = g_j(x)$, Eq. 7.20 becomes

$$E(Y \mid x) = \alpha + \beta_1 z_1 + \cdots + \beta_m z_m$$

Hence, by considering each of the functions $g_j(x_i)$ evaluated from the original data set x_i, the nonlinear problem of Eq. 7.21 is reduced to that of a multiple linear regression, presented earlier in Section 7.2.

7.4. APPLICATIONS OF REGRESSION ANALYSIS IN ENGINEERING

Regression analyses have been used widely in practically all branches of engineering for obtaining empirical relations between two (or more) variables. Sometimes the necessary relationship between two engineering variables cannot be derived on the basis of theoretical considerations; in these cases the required relationship may be determined empirically on the basis of experimental observations. For example, by plotting the logarithm of the observed fatigue life N of a material versus the logarithm of the applied stress range S, a linear trend is observed as shown in Fig. 7.4. This trend can be represented by

$$\log N = a - b \log S$$

Linear regression of $\log N$ on $\log S$ would then yield the constants a and b. This equation regression also suggests an S-N relation of the form

$$N S^b = a$$

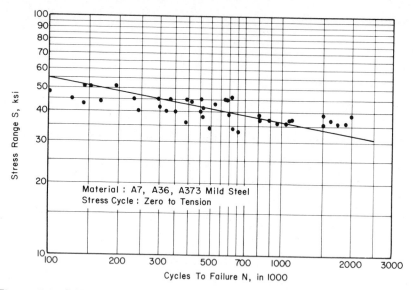

Figure 7.4 *S-N* relation for fatigue of material (data courtesy of W. H. Munse)

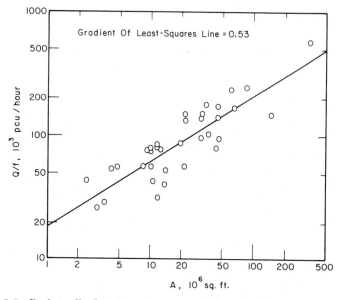

Figure 7.5 Peak traffic flow into city center with area *A* (after Miller, 1970)

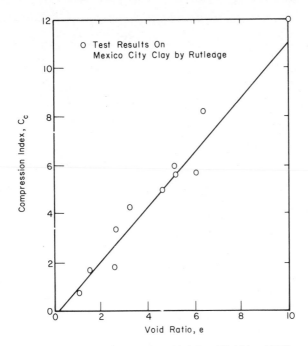

Figure 7.6   Compression index vs. void ratio (after Nishida, 1956)

In other situations the mathematical form of a required relationship may be derived or postulated from physical considerations; regression analysis may then be used to determine the values of the parameters, or to assess the validity of the theoretical equation, on the basis of observational data.

For example, Smeed (1968) postulated that the peak flow of traffic (in *passenger car unit*, pcu) into the center of a city is

$$Q = \alpha f A^{1/2}$$

where f is the fraction of the city center that is occupied by roadways; A is the area of the city center in square feet; and α is a constant depending on the speed of the traffic and the efficiency of the road system. Basically, this equation is based on the hypothesis that the volume of traffic that can enter the central area is proportional to its circumference. Data from 35 cities, including 20 from Britain, are shown plotted on log-log paper in Fig. 7.5. Least-squares linear regression of $\log(Q/f)$ on $\log A$ yields a slope of 0.53 for the regression line; Smeed's equation would be equivalent to a slope of 0.5. Also, from the regression line of Fig. 7.5, the constant α can be determined as the value of Q/f at $A = 1$.

Some engineering variables can be measured more readily and economically than others; for example, initial void ratios of clay samples can be

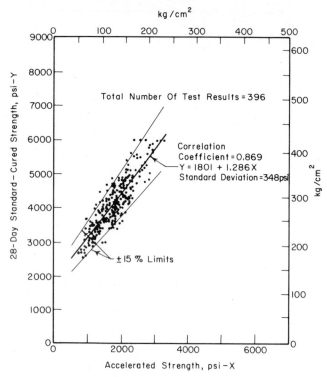

Figure 7.7   Relationship of accelerated to 28-day strength; combined field data from nine jobs across Canada (after Malhotra and Zoldners, 1969)

inexpensively measured in the laboratory whereas the direct determination of the compression index may require considerable labor and time. Consequently, if an empirical relation is established between the void ratio and the compression index of soils, such as the relation shown in Fig. 7.6, we can simply measure the void ratio and predict the compression index by using the regression equation. Another example is the determination of concrete strength. Normally the compressive strength of concrete specimens is tested after 28 days of curing. At the present rate of construction, 28 days would be a relatively long period; methods of early determination of concrete strength have been suggested, such as using an accelerated strength (based on an accelerated curing process). Figure 7.7 shows the results obtained by Malhotra and Zoldners (1969), which indicate a linear calibration between the two strengths. In traffic engineering, Heathington and Tuft (1971) have also succeeded in linearly calibrating the short-interval traffic volume with long-interval traffic volume in six Texas cities; some of these results are shown in Fig. 7.8.

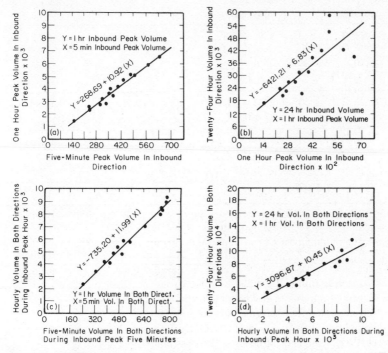

Figure 7.8 Relations between traffic volumes for different directions, in 6 Texas counties (after Heathington and Tutt, 1971)

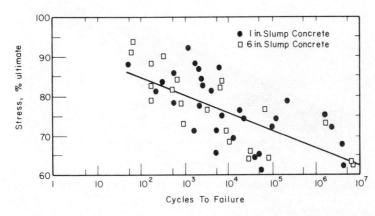

Figure 7.9 *S-N* diagram for concrete beams (after Murdock and Kesler, 1958)

Table 7.1. Multiple Regression for Estimating Trip Generation

Independent variables	Regression equations	$S_{Y|x_1, \ldots, x_n}$	r
X_1, X_2, X_3, X_4	$\begin{aligned}Y &= 4.33 + 3.89\,X_1 \\ &\quad - 0.005\,X_2 \\ &\quad - 0.128\,X_3 \\ &\quad - 0.012\,X_4\end{aligned}$	0.87	0.837
X_1, X_2	$\begin{aligned}Y &= 3.80 + 3.79\,X_1 \\ &\quad - 0.003\,X_2\end{aligned}$	0.87	0.835
X_2, X_4	$\begin{aligned}Y &= 5.49 \\ &\quad - 0.0089\,X_2 \\ &\quad + 0.227\,X_4\end{aligned}$	1.02	0.764
X_1	$Y = 2.88 + 4.60\,X_1$	0.89	0.827
X_2	$\begin{aligned}Y &= 7.22 \\ &\quad - 0.013\,X_2\end{aligned}$	1.10	0.718
X_4	$Y = 3.07 + 0.44\,X_4$	1.20	0.655
X_3	$Y = 3.55 + 0.74\,X_3$	1.30	0.575

Y = Expected number of resident's trips per dwelling unit
X_1 = Automobile ownership (no. per dwelling unit)
X_2 = Population density (no. per net residential acre)
X_3 = Distance from central business district (miles)
X_4 = Family income (thousand dollars)

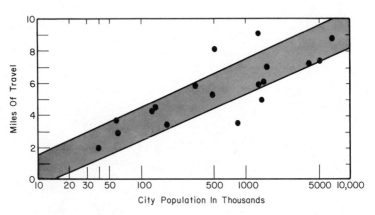

Figure 7.10 Work trip distance by city size (after Voorhees, 1966)

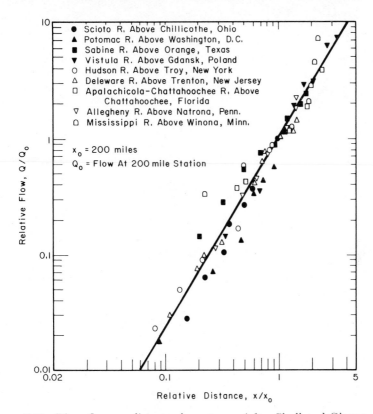

Figure 7.11 River flow vs. distance downstream (after Shull and Gloyna, 1969)

Multiple linear regression also finds many applications; for example, Martin et al. (1963) used multiple linear regression to obtain the expected number of trip generations Y per dwelling unit in a community as a function of automobile ownership X_1, population density X_2, distance from the central business district X_3, and family income X_4:

$$Y = 4.33 + 3.89\,X_1 - 0.005\,X_2 - 0.128\,X_3$$
$$- 0.012\,X_4$$

Other multiple linear regression analyses were also performed for Y based on fewer independent variables. The results of these analyses are summarized in Table 7.1. It can be observed from the values of r that the proportion of the variance reduced by taking account of the linear trend generally increases with the number of variables included in the regression analysis.

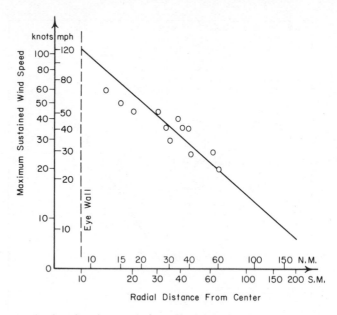

Figure 7.12 Surface hurricane wind profile (after Goldman and Ushijima, 1974)

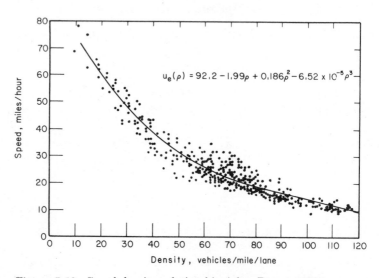

$$u_e(\rho) = 92.2 - 1.99\rho + 0.186\rho^2 - 6.52 \times 10^{-5}\rho^3$$

Figure 7.13 Speed-density relationship (after Payne, 1973)

Nonlinear regression is also widely used in engineering. Aside from those described earlier in Examples 7.5 and 7.6, Fig. 7.9 shows an application involving the logarithmic transformation, in which the average stress per cycle of repeated loading is plotted against the logarithm of the number of cycles to failure of concrete beams. This is another example of the *S-N* diagram for the average fatigue life. In this case, because of large variability in concrete strengths, a wide scatter is observed. Figure 7.10 shows that the average distance of travel to work may be linearly related to the logarithm of the city population; as the population in a city increases, the city spreads out to the suburbs and the average distance of travel to work also increases. An example of a double logarithmic transformation is shown in Fig. 7.11, where the logarithm of river flow increases linearly with the logarithm of distance downstream; similarly, Fig. 7.12 shows that the maximum sustained wind speed and the radial distance from the center of a hurricane also follow approximately a log-log relationship.

Polynomial functions are also often used in nonlinear regression. A third-degree polynomial curve is shown in Fig. 7.13 to describe the mean vehicle speed as a function of the traffic density.

7.5. CORRELATION ANALYSIS

7.5.1. Estimation of correlation coefficient

The study of the degree of linear interrelation between random variables is called *correlation analysis*. Recall that in regression analysis, we are interested in predicting the value of a variable (or estimating associated probability) for given values of the other variables. However, the accuracy of a linear prediction will depend on the correlation between the variables.

Mathematically, the correlation between two random variables X and Y is measured by the *correlation coefficient* defined in Eq. 3.73 as

$$\rho = \frac{\text{Cov}(X, Y)}{\sigma_X \sigma_Y} = \frac{E[(X - \mu_X)(Y - \mu_Y)]}{\sigma_X \sigma_Y}$$

Based on a set of observed values of X and Y, the correlation coefficient may be estimated by

$$\hat{\rho} = \frac{1}{n-1} \frac{\sum_{i=1}^{n}(x_i - \bar{x})(y_i - \bar{y})}{s_x s_y} = \frac{1}{n-1} \frac{\sum_{i=1}^{n} x_i y_i - n\bar{x}\bar{y}}{s_x s_y} \quad (7.22)$$

where $\bar{x}$, $\bar{y}$, s_x, and s_y are, respectively, the sample means and sample standard deviations of X and Y. The value of $\hat{\rho}$ also ranges from -1 to $+1$ and is a measure of the strength of *linear* relationship between the two variables X and Y. If the estimate $\hat{\rho}$ is close to $+1$ or -1, there is strong linear relationship between X and Y, and linear regression analysis may be

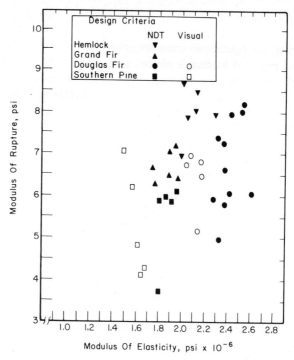

Figure 7.14 Comparison of MOE and MOR values of 40-ft laminated beams (from Galligan and Snodgrass, 1970)

carried out to obtain the regression equations. On the other hand, if $\hat{\rho} \simeq 0$, this would indicate a lack of linear relationship between the variables; such a case is illustrated in Fig. 7.14 between the modulus of rupture and the modulus of elasticity of laminated wood beams.

From Eqs. 7.3 and 7.22 it can be shown that

$$\hat{\rho} = \frac{\Sigma (x_i - \bar{x})(y_i - \bar{y})}{\Sigma (x_i - \bar{x})^2} \cdot \frac{s_x}{s_y}$$

$$= \hat{\beta} \frac{s_x}{s_y} \tag{7.23}$$

This is a useful relationship between the estimate of ρ and the regression coefficient $\hat{\beta}$. Furthermore, by substituting Eq. 7.23 into Eq. 7.5, we obtain

$$\hat{\mathrm{Var}}(Y \mid x) = \frac{1}{n-2} \left[\Sigma (y_i - \bar{y})^2 - \hat{\rho}^2 \frac{s_y^2}{s_x^2} \Sigma (x_i - \bar{x})^2 \right]$$

$$= \frac{n-1}{n-2} s_y^2 (1 - \hat{\rho}^2) \tag{7.24}$$

from which we also have

$$\hat{\rho}^2 = 1 - \frac{n-2}{n-1} \frac{s_{Y|x}^2}{s_y^2} \tag{7.25}$$

which is equal to r^2 of Eq. 7.6 for large n. On this basis, therefore, we can say that the larger the value of $|\hat{\rho}|$ the greater will be the reduction in the variance when the trend between the variables is taken into account, and hence the more accurate will be the prediction based on the regression equation.

EXAMPLE 7.7

In Table E7.7 are shown the data on blow counts N_i and corresponding unconfined compressive strength of very stiff clay q_i. These data are also shown in Fig. E7.7.

On the basis of these data, estimate the correlation coefficient $\hat{\rho}$ between the blow count and the unconfined strength of stiff clay.

The required calculations are indicated and summarized in Table E7.7. From these results, we estimate the correlation coefficient using Eq. 7.22 to be

$$\hat{\rho} = \frac{\frac{1}{9}[492.77 - 10(18.7)(2.12)]}{\sqrt{95.65}\sqrt{1.24}} = 0.98$$

This indicates that there is very high correlation between blow counts and the un-

Table E7.7. Computations for Example 7.7

Blow counts N_i	Compressive strength (tsf) q_i	N_i^2	q_i^2	$N_i q_i$
4	0.33	16	0.11	1.32
8	0.90	64	0.81	7.20
11	1.41	126	1.99	15.51
16	1.99	256	3.96	31.84
17	1.70	289	2.89	28.90
19	2.25	361	5.06	42.75
21	2.60	441	6.76	54.60
25	2.71	625	7.34	67.75
32	3.33	1024	11.09	106.56
34	4.01	1156	16.08	136.34
187	21.23	4358	56.09	492.77

$$\bar{N} = \frac{187}{10} = 18.7; \quad \bar{q} = \frac{21.23}{10} = 2.12$$

$$s_N^2 = \tfrac{1}{9}[4358 - 10(18.7)^2] = 95.65; \quad s_q^2 = \tfrac{1}{9}[56.09 - 10(2.12)^2] = 1.24$$

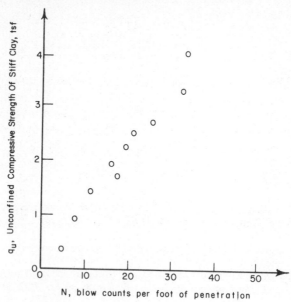

Figure E7.7   Unconfined compressive strength vs. blow counts for stiff clay

confined compressive strength of stiff clay; on this basis, therefore, the blow count may be used to estimate the unconfined strength of stiff clay.

EXAMPLE 7.8

For the data recorded on the Monocacy River (described earlier in Examples 5.8 and 7.2), estimate the correlation coefficient between runoff and precipitation.

Based on the computations tabulated earlier for Example 5.8, we obtain the sample variance of precipitation $s_x^2 = 1.53$; and the sample variance of runoff $s_y^2 = 0.36$. From the calculations in Example 7.2, we also have

$$\sum_{i=1}^{25} x_i y_i - 25\bar{x}\bar{y} = 59.24 - 25(2.16)(0.80) = 16.04$$

Hence

$$\hat{\rho} = \frac{(1/24)(16.04)}{\sqrt{1.53}\sqrt{0.36}} = 0.90$$

The correlation coefficient is required when calculating the joint probabilities of two or more random variables that are jointly normal (see Example 3.25). However, for non-normal variates the quantitative role of the correlation coefficient in the computation of joint probabilities is seldom defined. Nevertheless, the correlation coefficient is a measure of linear interdependency between two random variables irrespective of their distributions.

Multiple correlation. When more than two random variables are involved, as in the case of multiple linear regression of Eq. 7.16, any pair of variables may be mutually correlated, for example, between X_i and X_j, or between Y and X_i; the corresponding correlation coefficients are

$$\rho_{X_i,X_j} = \frac{E[(X_i - \mu_{X_i})(X_j - \mu_{X_j})]}{\sigma_{X_i}\sigma_{X_j}} \qquad (7.25)$$

and can be estimated as

$$\hat{\rho}_{X_i,X_j} = \frac{1}{n-1} \frac{(\sum_{k=1}^{n} x_{ik}x_{jk} - n\bar{x}_i\bar{x}_j)}{s_{x_i}s_{x_j}} \qquad (7.26)$$

7.5. CONCLUDING REMARKS

The statistical method for determining the mean and variance of one random variable as a function of the values of other variables is known as *regression analysis*. On the basis of the least-squares criterion, regression analysis provides a systematic approach for the empirical determination of the underlying relationships among the random variables. Furthermore, the associated correlation analysis determines the degree of linear interrelationship between the variables (in terms of the correlation coefficient); a high correlation means the existence of a strong linear relationship between the variables, whereas a low correlation would mean the lack of linear relationship (however, there could be a nonlinear relationship).

Regression and correlation analyses have applications in many areas of engineering, and are especially significant in situations where the necessary relationships must be developed empirically.

PROBLEMS

7.1 Assume hypothetically that the concentration of dissolved solids and the turbidity of a stream are measured simultaneously for five separate days, selected at random throughout a year. The data are as follows.

Day	Dissolved solids (mg/l)	Turbidity (JTU)
1	400	5
2	550	30
3	700	32
4	800	58
5	500	20

Because turbidity is easier to measure, a regression equation may be used to predict the concentration of dissolved solids on the basis of known turbidity. Assume that the variance of dissolved solid concentration is constant with turbidity.

(a) What are the values of the intercept and slope parameters (α and β) of the regression line? *Ans. 364.1; 7.79.*

(b) Estimate the standard deviation of dissolved solid concentration about the regression line. *Ans. 58.8.*

7.2 Suppose that data on the consumption of water per capita per day have been collected for four towns in the Midwest and tabulated as follows (see also Fig. P7.2).

Town	x Population (in 10^4)	y Per capita water consumption (in 100 gal/day)
1	1.0	1.0
2	4.0	1.3
3	6.0	1.3
4	9.0	1.4

(a) If the effect of population size of a town on the per capita consumption is neglected, determine the sample variance $s_y{}^2$.

(b) From the observed data, there seems to be a general trend that the per capita water consumption increases with the population of the town. Suppose it is assumed that

$$E(Y \mid x) = \alpha + \beta x$$

and Var $(Y \mid x)$ is constant for all x.

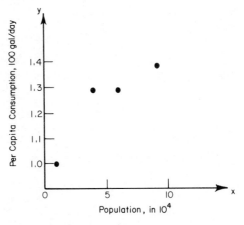

Figure P7.2

    (i) Determine the least-squares estimates for α and β.

    (ii) Estimate $s^2_{Y|x}$.

(c) An engineer is interested in studying the consumption of water in Urbana (a town with 50,000 population). Assume a normal distribution for Y; determine the probability that the demand for water in Urbana will exceed 7,000,000 gal/day.

7.3 Dissolved oxygen (DO) concentration (in parts per million, ppm) in a stream is found to decrease with the time of travel downstream (Thayer and Krutchkoff, 1966). Assume a linear relationship between the mean DO and the time of travel t. Determine the least-squares regression equation and estimate the standard deviation about the regression line from the following set of observations.

DO (ppm)	Time of travel t (days)
0.28	0.5
0.29	1.0
0.29	1.6
0.18	1.8
0.17	2.6
0.18	3.2
0.10	3.8
0.12	4.7

7.4 From a survey of the effect of fare increase on the loss in ridership for transit systems throughout the United States, the following data were obtained.

X Fare increase (%)	Y Loss in ridership (%)
5	1.5
35	12.0
20	7.5
15	6.3
4	1.2
6	1.7
18	7.2
23	8.0
38	11.1
8	3.6
12	3.7
17	6.6
17	4.4
13	4.5
7	2.8
23	8.0

(a) Plot the percent loss in ridership versus the percent fare increase.
(b) Perform a linear regression analysis to determine the expected percent loss in ridership as a function of the percent fare increase.
(c) Estimate the conditional standard deviation $s_{Y|x}$. *Ans. 0.82.*
(d) Evaluate the correlation coefficient between X and Y. *Ans. 0.97.*

7.5 Data for per capita energy consumption and per capita GNP output for eight different countries are tabulated below (data extracted from Meadows et al., 1972).

X	Y
600	1,000
2,700	700
2,900	1,400
4,200	2,000
3,100	2,500
5,400	2,700
8,600	2,500
10,300	4,000

Note that

X = GNP in U.S. dollar equivalent per person per year
Y = energy consumption in kilograms of coal equivalent per person per year

(a) Plot the data given above in a two-dimensional graph.
(b) Determine the correlation between GNP and energy consumption.
(c) Determine the regression for predicting energy consumption on the basis of per capita GNP output. Draw the regression line on the graph of part (a).
(d) Evaluate the conditional standard deviation $s_{Y|x}$, and sketch the $\pm s_{Y|x}$ band about the regression line of part (c).
(e) Similarly, determine the regression equation for predicting GNP on the basis of energy consumption, and display this graphically with the corresponding $\pm s_{X|y}$ band.

7.6 A tensile load test was performed on an aluminum specimen. The applied tensile force and the corresponding elongation of the specimen at various stages of the test are recorded as follows.

Tensile force (kips) x	Elongation $(10^{-3}$ in.) y
1	9
2	20
3	28
4	41
5	52
6	63

(a) Assume that the force-elongation relation of aluminum over this range of loads is linear. Determine the least-squares estimate for the Young's modulus of this aluminum specimen. The cross-sectional area of the specimen is 0.1 sq in., and the length of the specimen is 10 in. Young's modulus is given by the slope of the stress-strain curve.

(b) In addition to the assumption of a linear relationship between force and elongation, suppose zero elongation should correspond to zero tensile force; that is, the regression line is assumed to be

$$E(Y \mid x) = \beta x$$

What would be the best estimate of Young's modulus in this case?

7.7 The population in a community for the years 1962 to 1972 is tabulated as follows.

Year	Population
1962	24,010
1963	24,540
1964	24,750
1965	25,100
1966	25,340
1967	25,820
1968	26,100
1969	26,200
1970	26,500
1971	26,800
1972	27,450

It is suggested that the population in a given year will depend on the population of the previous year, as predicted from the following model:

$$x_t = a + bx_{t-1} + \varepsilon$$

where x_t and x_{t-1} are the population in the tth and $(t - 1)$th year, respectively, and ε is a normal random variable with zero mean and standard deviation σ.

(a) Based on the given population data, determine the estimates for a, b, and σ.

(b) Based on the population model and the estimates from part (a), estimate the population in 1973. What is the probability that the population in 1973 will exceed 28,000?

7.8 The peak-hour traffic volume and the 24-hour daily traffic volume on a toll bridge have been recorded for 14 days. The observed data are tabulated as follows.

X Peak-hour traffic volume (in 10^3)	Y 24-hr traffic volume (in 10^4)
1.4	1.6
2.2	2.3
2.4	2.0
2.7	2.2
2.9	2.6
3.1	2.6
3.6	2.1
4.1	3.0
3.4	3.0
4.3	3.8
5.1	5.1
5.9	4.2
6.4	3.8
4.6	4.2

Assume that the conditional standard deviation $s_{Y|x}$ varies quadratically with x from the origin.

(a) Determine the regression line $E(Y \mid x) = \hat{\alpha} + \hat{\beta}x$.

(b) Estimate the prediction error about the regression line, that is, $s_{Y|x}$.

(c) If the peak-hour traffic volume on a certain day is measured to be 3500 vehicles, what is the probability that more than 30,000 vehicles will be crossing the toll bridge that day?

7.9 The average durations (in days) of frost condition each year at 20 stations in West Virginia were compiled as follows (from Moulton and Schaub, 1969).

Weather station	Elevation (ft)	North latitude (deg)	Average duration of frost (days)
Bayard	2375	39.27	73.0
Brandywine	1586	38.63	29.0
Buckhannon	1459	39.00	28.0
Cairo	680	39.17	25.0
Charleston	604	38.35	11.5
Fairmont	1298	39.47	32.5
Flat Top	3242	37.58	64.0
Gary	1426	37.37	13.0
Kearneysville	550	39.38	23.0
Lewisburg	2250	37.80	37.0
Madison	675	38.05	26.0
Marlington	2135	38.23	73.0
New Martinsville	635	39.65	24.7

Continuation

Weather station	Elevation (ft)	North latitude (deg)	Average duration of frost (days)
Parsons	1,649	39.10	41.0
Pickens	2727	38.66	56.0
Piedmont	1053	39.48	34.0
Rainelle	2424	37.97	37.0
Spencer	789	38.80	16.0
Wheeling	659	40.10	41.0
Williamson	673	37.67	12.0

Perform a multiple linear regression analysis to predict the average duration of frost at a locality in terms of its elevation and latitude.

7.10 The difference between the photogrammetrically triangulated elevation—before adjustment—and the terrestrially determined elevation is an example of measurement error in photogrammetry. This error in elevation E has been observed and theoretically shown to be a nonlinear function of the distance X along the centerline of a triangulated strip as follows:

$$E = a + bx + cx^2$$

Estimate the least-squares values of a, b, and c on the basis of the following measurements. *Ans.* -0.023; 0.235; -0.347.

Distance along centerline of triangulated strip X (km)	Error in elevation E (m)
0	0
0.5	0
1.2	−0.3
1.7	−0.6
2.4	−1.4
2.7	−2.0
3.4	−3.1
3.7	−4.0

7.11 The average distance Y required for stopping a vehicle is a function of the speed of travel of the vehicle. The following set of data were observed for 10 cars at varying speeds.

Car	Speed (mph)	Stopping distance (ft)
1	25	46
2	5	6
3	60	110
4	30	46
5	10	13
6	45	75
7	15	16
8	40	70
9	45	90
10	20	20

(a) Plot the stopping distance vs. speed.
(b) Assume that the mean stopping distance varies linearly with the speed, that is,

$$E(Y \mid x) = \alpha + \beta x$$

Estimate α and β; and $s_{Y|x}$, which may be assumed to be constant.
(c) A nonlinear function is suggested to model the stopping distance-speed relationship as follows:

$$E(Y \mid x) = a + bx + cx^2$$

Estimate a, b, and c; and $s_{Y|x}$, which is assumed to be constant with x.
(d) Plot the two regression curves obtained from parts (b) and (c). Compare the relative accuracy of prediction between these two models.

7.12 The mean rate of oxygenation from the atmospheric reaeration process for a stream depends on the mean velocity of stream flow and average depth of the stream bed. The following are data recorded in 12 experiments (Kothandaraman, 1968).

Mean oxygenation rate X (ppm per day)	Mean velocity V (ft/sec)	Mean depth H (ft)
2.272	3.07	3.27
1.440	3.69	5.09
0.981	2.10	4.42
0.496	2.68	6.14
0.743	2.78	5.66
1.129	2.64	7.17
0.281	2.92	11.41
3.361	2.47	2.12
2.794	3.44	2.93
1.568	4.65	4.54
0.455	2.94	9.50
0.389	2.51	6.29

Suppose that the following relationship is used to estimate the mean oxygenation rate

$$E(X \mid V, H) = \alpha V^{\beta_1} H^{\beta_2}$$

Estimate α, β_1, and β_2 on the basis of the observed data.

7.13 The compressive and flexural strengths of nonbloated burned clay aggregate concrete are measured for 30 specimens after 7 days of curing. The data are (from Martin et al., 1972) as follows.

	7-day compressive strength, X (psi)	7-day flexural strength, Y (psi)
1	1400	257
2	1932	327
3	2200	317
4	2935	300
5	2665	340
6	2800	340
7	3065	343
8	3200	374
9	2200	377
10	2530	386
11	3000	383
12	2735	393
13	2000	407
14	3000	407
15	3235	407
16	2630	434
17	3030	427
18	3065	440
19	2735	450
20	3835	440
21	3065	456
22	3465	460
23	3600	456
24	3260	476
25	3500	480
26	3365	490
27	3335	497
28	3170	526
29	3600	546
30	4460	700

(a) Plot the compressive strength vs. flexural strength on a two-dimensional graph.
(b) Determine the correlation coefficient between the two strengths.

7.14 The settlement of a footing depends on that of the adjacent footing since they are subjected to similar load and soil conditions. Therefore some correlation

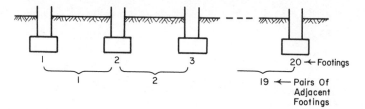

Figure P7.14

exists between the settlement behavior of two adjacent footings. The following is a set of data on the settlement of a series of footings on sand.

Footing	Settlement (in.)	Footing	Settlement (in.)
1	0.59	11	0.93
2	0.60	12	0.78
3	0.54	13	0.78
4	0.70	14	0.77
5	0.75	15	0.79
6	0.80	16	0.79
7	0.79	17	0.78
8	0.95	18	0.77
9	1.00	19	0.63
10	0.92	20	0.73

From a row of 20 footings, 19 pairs of adjacent footings can be obtained as shown in Fig. P7.14. The degree of dependence between the settlements of adjacent footings is described by the correlation coefficient.

(a) Estimate this correlation based on the 19 pairs of data. *Ans. 0.766.*

(b) Estimate the coefficient of variation of the settlement of a footing. *Ans. 0.157.*

8. *The Bayesian Approach*

8.1. INTRODUCTION

In Chapter 5 we presented the methods of point and interval estimation of distribution parameters, based on the *classical statistical* approach. Such an approach assumes that the parameters are constants (but unknown) and that sample statistics are used as estimators of these parameters. Because the estimators are invariably imperfect, errors of estimation are unavoidable; in the classical approach, confidence intervals are used to express the degree of these errors.

As implied earlier, accurate estimates of parameters require large amounts of data. When the observed data are limited, as is often the case in engineering, the statistical estimates have to be supplemented (or may even be superseded) by judgmental information. With the classical statistical approach there is no provision for combining judgmental information with observational data in the estimation of the parameters.

For illustration, consider a case in which a traffic engineer wishes to determine the effectiveness of the road improvement at an intersection. Based on his experience with similar sites and traffic conditions, and on a traffic-accident model, he estimated that the average occurrences of accidents at the improved intersection would be about twice a year. However, during the first week after the improved intersection is opened to traffic, an accident occurs at the intersection. A dichotomy, therefore, may arise: The engineer may hold strongly to his judgmental belief, in which case he would insist that the accident is only a chance occurrence and the average accident rate remains *twice a year*, in spite of the most recent accident. However, if he only considers actual observed data, he would estimate the average accident rate to be *once a week*. Intuitively, it would seem that both types of information are relevant and ought to be used in determining the average accident rate. Within the classical method of statistical estimation, however, there is no formal basis for such analysis. Problems of this type is formally the subject of Bayesian estimation.

The *Bayesian method* approaches the estimation problem from another point of view. In this case, the unknown parameters of a distribution are assumed (or modeled) to be also random variables. In this way, uncertainty associated with the estimation of the parameters can be combined formally

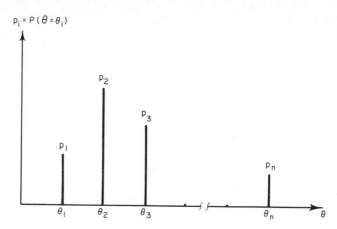

Figure 8.1   Prior PMF of parameter θ

(through Bayes' theorem) with the inherent variability of the basic random variable. With this approach, subjective judgments based on intuition, experience, or indirect information are incorporated systematically with observed data to obtain a balanced estimation. The Bayesian method is particularly helpful in cases where there is a strong basis for such judgments. We introduce the basic concepts of the Bayesian approach in the following sections.

8.2.   BASIC CONCEPTS—THE DISCRETE CASE

The *Bayesian approach* has special significance to engineering design, where available information is invariably limited and subjective judgment is often necessary. In the case of parameter estimation, the engineer often has some knowledge (perhaps inferred intuitively from experience) of the possible values, or range of values, of a parameter; moreover, he may also have some intuitive judgment on the values that are more likely to occur than others. For simplicity, suppose that the possible values of a parameter θ were assumed to be a set of discrete values θ_i, $i = 1, 2, \ldots, n$, with relative likelihoods $p_i = P(\Theta = \theta_i)$ as illustrated in Fig. 8.1 (Θ is the random variable whose values represent possible values of the parameter θ).

Then if additional information becomes available (such as the results of a series of tests or experiments), the prior assumptions on the parameter θ may be modified formally through Bayes' theorem as follows.

Let ϵ denote the observed outcome of the experiment. Then applying

Bayes' theorem of Eq. 2.18, we obtain the updated PMF for Θ as

$$P(\Theta = \theta_i \mid \epsilon) = \frac{P(\epsilon \mid \Theta = \theta_i)\, P(\Theta = \theta_i)}{\sum\limits_{i=1}^{n} P(\epsilon \mid \Theta = \theta_i)\, P(\Theta = \theta_i)} \qquad i = 1, 2, \ldots, n \quad (8.1)$$

The various terms in Eq. 8.1 can be interpreted as follows:

$P(\epsilon \mid \Theta = \theta_i)$ = the likelihood of the experimental outcome ϵ if $\Theta = \theta_i$; that is, the conditional probability of obtaining a particular experimental outcome assuming that the parameter is θ_i

$P(\Theta = \theta_i)$ = the *prior* probability of $\Theta = \theta_i$; that is, prior to the availability of the experimental information ϵ

$P(\Theta = \theta_i \mid \epsilon)$ = the *posterior* probability of $\Theta = \theta_i$; that is, the probability that has been revised in the light of the experimental outcome ϵ

Denoting the *prior* and *posterior* probabilities as $P'(\Theta = \theta_i)$ and $P''(\Theta = \theta_i)$, respectively, Eq. 8.1 becomes

$$P''(\Theta = \theta_i) = \frac{P(\epsilon \mid \Theta = \theta_i)P'(\Theta = \theta_i)}{\sum\limits_{i=1}^{n} P(\epsilon \mid \Theta = \theta_i)P'(\Theta = \theta_i)} \qquad (8.1a)$$

Equation 8.1a, therefore, gives the posterior probability mass function of Θ. (In general, we shall use $'$ and $''$ to denote the prior and posterior).

The *expected value* of Θ is then commonly used as the *Bayesian estimator** of the parameter; that is,

$$\hat{\theta}'' = E(\Theta \mid \epsilon) = \sum_{i=1}^{n} \theta_i P''(\Theta = \theta_i) \qquad (8.2)$$

We may point out that in Eq. 8.2 observational data and judgmental information are both used and combined in a systematic way to estimate the underlying parameter.

In the Bayesian framework, the significance of judgmental information is reflected also in the calculation of relevant probabilities. In the case above, where subjective judgments were used in the estimation of the parameter θ, such judgments would be reflected in the calculation of the probability, for example, $P(X \leq a)$, through the theorem of total proba-

---

* There are other Bayesian estimators depending on the assumed form of the "loss function" (discussed in Vol. II). Moreover, other parameters of the posterior distribution may serve as the estimator instead; for example, the *mode*.

bility using the posterior PMF of Eq. 8.1a. That is,

$$P(X \le a) = \sum_{i=1}^{n} P(X \le a \mid \Theta = \theta_i) \, P''(\Theta = \theta_i) \qquad (8.3)$$

This represents the up-to-date probability of the event $(X \le a)$ based on all available information. It may be emphasized that in Eq. 8.3 the uncertainty associated with the error of estimating the parameter [as reflected in $P''(\Theta = \theta_i)$] is combined with the inherent variability of the random variable X.

To clarify these general concepts, consider the following examples.

EXAMPLE 8.1

Piles for a building foundation were initially designed for 250-ton capacity each; however, this did not include the effect of high winds that occur only very rarely. On such rare occasions, it is estimated that some of the piles may be subjected to loads as high as 300 tons. In order to assess the safety of the initial design, the engineer in charge wishes to determine the probability of the piles failing under a maximum load of 300 tons.

Suppose that from the engineer's experience with this type of piles and the soil condition at the site, he estimated (judgmentally) that the probability p would range from 0.2 to 1.0 with 0.4 as the most likely value; more specifically, p is described by the prior PMF shown in Fig. E8.1a. The values of p are discretized at 0.2 intervals to simplify the presentation.

On the basis of this prior PMF, the estimated probability of a pile failing at a load of 300 tons would be (by virtue of the total probability theorem)

$$\hat{p}' = (0.2)(0.3) + (0.4)(0.4) + (0.6)(0.15) + (0.8)(0.10) + (1.0)(0.05)$$
$$= 0.44$$

In order to supplement his judgment, the engineer ordered a pile of the same type test-loaded at the site to a maximum load of 300 tons. The outcome of the test shows that the pile failed to carry the maximum load. Based on this single test result, the PMF of p would be revised according to Eq. 8.1a, obtaining the posterior PMF as follows:

$$P''(p = 0.2) = \frac{(0.2)(0.3)}{(0.2)(0.3) + (0.4)(0.4) + (0.6)(0.15) + (0.8)(0.1) + (1.0)(0.05)}$$
$$= 0.136$$

and, similarly,

$$P''(p = 0.4) = 0.364$$
$$P''(p = 0.6) = 0.204$$
$$P''(p = 0.8) = 0.182$$
$$P''(p = 1.0) = 0.114$$

which are shown graphically in Fig. E8.1b.
The Bayesian estimate for p, Eq. 8.2, therefore is

$$\hat{p}'' = E(p \mid \varepsilon) = 0.2(0.136) + 0.4(0.364) + 0.6(0.204) + 0.8(0.182) + 1.0(0.114)$$
$$= 0.55$$

$P'(p = p_i)$

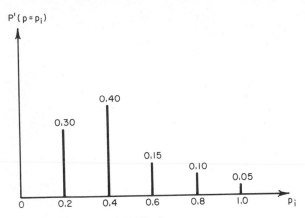

Figure E8.1a Prior PMF of p

In Fig. E8.1b, we see that as a result of the single unsuccessful load test, the probabilities for higher values of p_i are increased from those of the prior distribution, resulting in a higher estimate for p, namely, $\hat{p}'' = E(p \,|\, \varepsilon) = 0.55$, whereas the prior estimate was 0.44. Observe that the failure of one test pile does not imply the impossibility of such piles carrying the 300-ton load; instead, the test result merely serves to increase the estimated probability by 0.11 (from 0.44 to 0.55). Figure E8.1c illustrates how the PMF of p changes with increasing number of consecutive test pile failures; the distribution shifts toward $p = 1.0$ as $n \to \infty$.

Figure E8.1d shows the corresponding Bayesian estimate for p; observe that after a sequence of 6 consecutive failures the estimate for p is 0.90. If a long sequence of failures is observed, the Bayesian estimate of p approaches 1.0—a result that tends to the classical estimate; in such a case, there is overwhelming amount of observed data

$P''(p = p_i)$

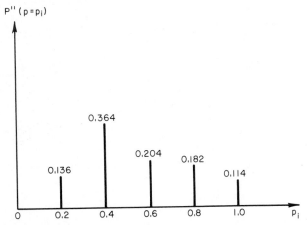

Figure E8.1b Posterior PMF of p

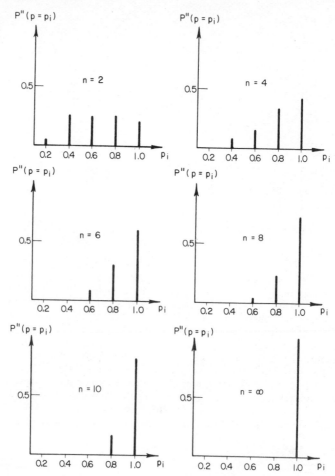

Figure E8.1c PMF of p for increasing number of test pile failures

to supersede any prior judgment. Ordinarily, however, where observational data are limited, judgment would be important and is reflected properly in the Bayesian estimation process.

Now suppose that each main column is supported on a group of three piles. If the piles carry equal loads and are statistically independent, the probability that none of the piles supporting a column will fail at a total column load of 900 tons (300 tons per pile) can be obtained by Eq. 8.3. Based on the posterior PMF of Fig. E8.1b, and denoting X as the number of piles failing, the required probability is

$$P(X = 0) = P(X = 0 \mid p = 0.2)P''(p = 0.2) + P(X = 0 \mid p = 0.4)P''(p = 0.4)$$
$$+ \cdots + P(X = 0 \mid p = 1.0)P''(p = 1.0)$$
$$= (0.8)^3(0.136) + (0.6)^3(0.364) + (0.4)^3(0.204) + (0.2)^3(0.182)$$
$$= 0.163$$

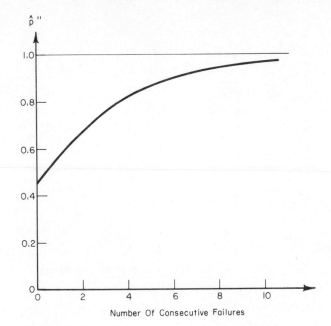

Figure E8.1d $\hat{p}''$ vs. no. of consecutive failures

EXAMPLE 8.2

A traffic engineer is interested in the average rate of accidents v at an improved road intersection. Suppose that from his previous experience with similar road and traffic conditions, he deduced that the expected accident rate would be between one and three per year, with an average of two, and the prior PMF shown in Fig. E8.2. Occurrence of accidents is assumed to be a Poisson process.

During the first month after completion of the intersection, one accident occurred.

(a) In the light of this observation, revise the estimate for v.

(b) Using the result of part (a), determine the probability of no accident in the next six months.

Solutions

(a) Let ε be the event that an accident occurred in one month. The posterior probabilities then are

$$P''(v = 1) = \frac{P(\varepsilon \mid v = 1)P'(v = 1)}{P(\varepsilon \mid v = 1)P'(v = 1) + P(\varepsilon \mid v = 2)P'(v = 2) + P(\varepsilon \mid v = 3)P'(v = 3)}$$

$$= \frac{e^{-1/12}(1/12)(0.3)}{e^{-1/12}(1/12)(0.3) + e^{-1/6}(1/6)(0.4) + e^{-1/4}(1/4)(0.3)}$$

$$= 0.166$$

Similarly,

$$P''(v = 2) = 0.411$$
$$P''(v = 3) = 0.423$$

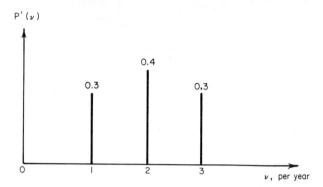

Figure E8.2   Prior distribution of ν

Hence the updated value of ν is

$$\hat{\nu}'' = E(\nu \mid \varepsilon) = (0.166)(1) + (0.411)(2) + (0.423)(3)$$
$$= 2.26 \text{ accidents per year}$$

(b) Let A be the event of no accidents in the next six months. Then

$$P(A) = P(A \mid \nu = 1)P''(\nu = 1) + P(A \mid \nu = 2)P''(\nu = 2) + P(A \mid \nu = 3)P''(\nu = 3)$$
$$= e^{-1/2}(0.166) + e^{-1}(0.411) + e^{-3/2}(0.423)$$
$$= 0.346$$

8.3. THE CONTINUOUS CASE

8.3.1. General formulation

In Section 8.2 the possible values of the parameter θ (such as p in Example 8.1 and ν in Example 8.2) were limited to a discrete set of values; this was purposely assumed to simplify the presentation of the concepts underlying

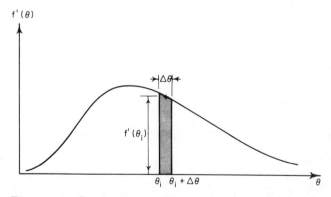

Figure 8.2   Continuous prior distribution of parameter θ

the Bayesian method of estimation. In many situations, however, the value of a parameter could be in a continuum of possible values. Thence, it would be appropriate to assume the parameter to be a continuous random variable in the Bayesian estimation. In this case we develop the corresponding results, analogous to Eqs. 8.1 through 8.3, as follows.

Let Θ be the random variable for the parameter of a distribution, with a prior density function $f'(\theta)$ shown in Fig. 8.2. The prior probability that θ will be between θ_i and $\theta_i + \Delta\theta$ then is $f'(\theta_i)\Delta\theta$. Then, if ϵ is an observed experimental outcome, the prior distribution $f'(\theta)$ can be revised in the light of ϵ using Bayes' theorem, obtaining the posterior probability that θ will be in $(\theta_i, \theta_i + \Delta\theta)$ as

$$f''(\theta_i)\Delta\theta = \frac{P(\epsilon \mid \theta_i)f'(\theta_i)\Delta\theta}{\sum_{i=1}^{n} P(\epsilon \mid \theta_i)f'(\theta_i)\Delta\theta}$$

where $P(\epsilon \mid \theta_i) = P(\epsilon \mid \theta_i < \theta \le \theta_i + \Delta\theta)$. In the limit, this yields

$$f''(\theta) = \frac{P(\epsilon \mid \theta)f'(\theta)}{\int_{-\infty}^{\infty} P(\epsilon \mid \theta)f'(\theta)\, d\theta} \qquad (8.4)$$

The term $P(\epsilon \mid \theta)$ is the conditional probability or likelihood of observing the experimental outcome ϵ assuming that the value of the parameter is θ. Hence $P(\epsilon \mid \theta)$ is a function of θ and is commonly referred to as the *likelihood function* of θ and denoted $L(\theta)$. The denominator is independent of θ; this is simply a normalizing constant required to make $f''(\theta)$ a proper density function. Equation 8.4 then can be expressed as

$$f''(\theta) = kL(\theta)f'(\theta) \qquad (8.5)$$

where the normalizing constant $k = \left[\int_{-\infty}^{\infty} L(\theta)f'(\theta)\, d\theta \right]^{-1}$; and

$L(\theta) =$ the likelihood of observing the experimental outcome ϵ assuming a given θ.

We observe from Eq. 8.5 that both the prior distribution and the likelihood function contribute to the posterior distribution of Θ. In this way, as in the discrete case, the significance of judgment and of observational data is combined properly and systematically; the former through $f'(\theta)$ and the latter in $L(\theta)$.

Analogous to the discrete case, Eq. 8.2, the expected value of Θ is commonly used as the point estimator of the parameter. Hence the updated estimate of the parameter θ, in the light of observational data ϵ, is given by

$$\hat{\theta}'' = E(\Theta \mid \epsilon) = \int_{-\infty}^{\infty} \theta f''(\theta) \, d\theta \tag{8.6}$$

The uncertainty in the estimation of the parameter can be included in the calculation of the probability associated with a value of the underlying random variable. For example, if X is a random variable

$$P(X \leq a) = \int_{-\infty}^{\infty} P(X \leq a \mid \theta) f''(\theta) \, d\theta \tag{8.7}$$

Physically, Eq. 8.7 is the average probability of $(X \leq a)$ weighted by the posterior probabilities of the parameter θ.

EXAMPLE 8.3

Consider again the problem of Example 8.1, in which the probability of pile failure at a load of 300 tons is of concern; this time, however, assume that the probability p is a continuous random variable. If there is no (prior) factual information on p, a uniform prior distribution may be assumed (known as the *diffuse prior*), namely,

$$f'(p) = 1.0 \qquad 0 \leq p \leq 1$$

On the basis of a single test, the likelihood function is simply the probability of the event $\varepsilon =$ capacity of test pile less than 300 tons, which is simply p. Hence the posterior distribution of p, according to Eq. 8.5, is

$$f''(p) = kp(1.0) \qquad 0 \leq p \leq 1$$

in which the constant

$$k = \left[\int_0^1 p \, dp \right]^{-1} = 2$$

Thus

$$f''(p) = 2p \qquad 0 \leq p \leq 1$$

The Bayesian estimate of p then is

$$\hat{p}'' = E(p \mid \varepsilon) = \int_0^1 p \cdot 2p \, dp$$

$$= 0.667$$

If a sequence of n piles were tested, out of which r piles failed at loads less than the maximum test load, then the likelihood function is the probability of observing r failures among the n piles tested. If the failure probability of each pile is p, and statistical independence is assumed between piles, the likelihood function would be

$$L(p) = \binom{n}{r} p^r (1 - p)^{n-r}$$

Then, with the diffuse prior, the posterior distribution of p becomes

$$f''(p) = k \binom{n}{r} p^r (1 - p)^{n-r} \qquad 0 \leq p \leq 1$$

where

$$k = \left[\int_0^1 \binom{n}{r} p^r (1 - p)^{n-r}\, dp \right]^{-1}$$

Thus the Bayesian estimator is

$$\hat{p}'' = E(p \mid \varepsilon) = \frac{\int_0^1 p \binom{n}{r} p^r (1 - p)^{n-r}\, dp}{\int_0^1 \binom{n}{r} p^r (1 - p)^{n-r}\, dp}$$

$$= \frac{\int_0^1 p^{r+1} (1 - p)^{n-r}\, dp}{\int_0^1 p^r (1 - p)^{n-r}\, dp}$$

Repeated integration-by-parts of the above integrals yields

$$\hat{p}'' = \frac{r + 1}{n} \frac{\int_0^1 (p^n - p^{n+1})\, dp}{\int_0^1 (p^{n-1} - p^n)\, dp}$$

$$= \frac{r + 1}{n + 2}$$

From this result, we may observe that as the number of tests n increases (with the ratio r/n remaining constant), the Bayesian estimate for p approaches that of the classical estimate; that is,

$$\frac{r + 1}{n + 2} \to \frac{r}{n} \qquad \text{for large } n$$

EXAMPLE 8.4

An engineer is designing a temporary structure subjected to wind load on a newly developed island in the Pacific. Of interest is the probability p that the annual maximum wind speed will not exceed 120 km/hr. Records for the annual maximum wind speed in the island are available only for the last five years; and among these, the 120 km/hr wind was exceeded only once. However, an adjacent island has a longer record of wind speeds. After a comparative study of the geographical condition in the two islands, the engineer inferred from this longer record that the average value of p for the newly developed island is 2/3 with a COV of 27%. Since p is bounded between 0 and 1.0, the following beta distribution (consistent with the above statistics) is also assumed for the prior distribution:

$$f'(p) = 20p^3(1 - p) \qquad 0 \le p \le 1$$

In this case, the likelihood that the annual maximum wind speed will exceed 120 kph in one out of five years is

$$L(p) = \binom{5}{4} p^4 (1 - p)$$

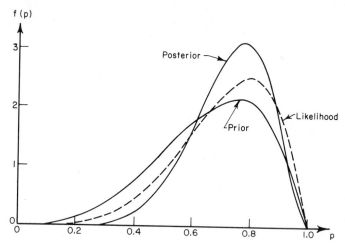

Figure E8.4 Prior, likelihood, and posterior functions

Hence the posterior density function of p is

$$f''(p) = kL(p)f'(p)$$

$$= k\left[\binom{5}{4} p^4(1-p) \right][20p^3(1-p)]$$

$$= 100kp^7(1-p)^2$$

where

$$k = \left[\int_0^1 100p^7(1-p)^2 \, dp \right]^{-1} = 3.6$$

Thus

$$f''(p) = 360p^7(1-p)^2 \qquad 0 \le p \le 1$$

In this case, the prior density function is equivalent to the assumption of one exceedance in four years, whereas the resulting posterior distribution is tantamount to two exceedances in nine years. In fact, the above posterior distribution is the same as that obtained for a case in which two exceedances were observed in nine years and a diffused prior distribution is assumed. This example should serve also to illustrate a property of the Bayesian approach—namely, that information from sources other than the observed data can be useful in the estimation process.

The relation between the likelihood function and the prior and posterior distributions of the parameter p is illustrated in Fig. E8.4. Observe that the posterior distribution is "sharper" than either the prior distribution or the likelihood function. This implies that more information is "contained" in the posterior distribution than in either the prior or the likelihood function.

EXAMPLE 8.5

The occurrences of earthquakes may be modeled as a Poisson process with mean occurrence rate ν (Benjamin, 1968). Suppose that historical record for a region A

shows that n_0 earthquakes have occurred in the past t_0 years. The corresponding likelihood function is then given by

$$L(v) = P(n_0 \text{ quakes in } t_0 \text{ years} \mid v)$$

$$= \frac{(vt_0)^{n_0}}{n_0!} e^{-vt_0} \qquad v \geq 0$$

If there is no other information for estimating v, a uniform diffuse prior may be assumed; this implies that $f'(v)$ is independent of the values of v and thus can be absorbed into the normalizing constant k. Then the posterior distribution of v becomes

$$f''(v) = kL(v)$$

$$= k \frac{(vt_0)^{n_0}}{n_0!} e^{-vt_0} \qquad v \geq 0$$

Upon normalization, $k = t_0$; this result may also be obtained by comparing the foregoing $f''(v)$ with the gamma density function of Eq. 3.44b (for the random variable v).

The probability of the event ($E = n$ earthquakes in the next t years in region A) is then given by Eq. 8.7 as follows:

$$P(E) = \int_0^\infty P(E \mid v) f''(v) \, dv$$

$$= \int_0^\infty \frac{(vt)^n}{n!} e^{-vt} \cdot \frac{t_0(vt_0)^{n_0}}{n_0!} e^{-vt_0} \, dv$$

$$= \left(\int_0^\infty \frac{(t + t_0)[v(t + t_0)]^{n+n_0}}{(n + n_0)!} e^{-v(t+t_0)} \, dv \right) \frac{(n + n_0)!}{n! \, n_0!} \frac{t^n t_0^{(n_0+1)}}{(t + t_0)^{n+n_0+1}}$$

Since the integrand inside the parentheses is a gamma density function, the integral is equal to 1.0. Hence

$$P(E) = \frac{(n + n_0)!}{n! \, n_0!} \frac{t^n t_0^{(n_0+1)}}{(t + t_0)^{n+n_0+1}} = \frac{(n + n_0)!}{n! \, n_0!} \frac{(t/t_0)^n}{(1 + t/t_0)^{n+n_0+1}}$$

a result that was first derived by Benjamin (1968).

As an illustration, suppose that historical records in region A show that two earthquakes with intensity exceeding VI (MM scale) had occurred in the last 60 years. The probability that there will be no earthquakes with this intensity in the next 20 years, therefore, is

$$P(E) = \frac{(0 + 2)!}{0! \, 2!} \frac{(20/60)^0}{(1 + 20/60)^3}$$

$$= 0.42$$

8.3.2. A special application of Bayesian up-dating process

An interesting application of the Bayesian updating process is in the inspection and detection of material defects (Tang, 1973). Fatigue and fracture failures in metal structures are frequently the result of unchecked propagation of flaws or cracks in the joints (welds) or base metals. Periodic inspection and repair can be used to minimize the risk of fracture

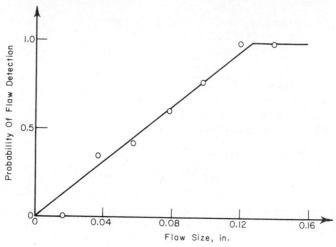

Figure 8.3 Detectability versus actual flaw depth (data from Packman et al., 1968)

failure by limiting the existing flaw sizes. Methods of detecting flaws, such as nondestructive testing (NDT), however, are invariably imperfect; consequently, not all flaws may be detected during an inspection.

The probability of detecting a flaw generally increases with the flaw size and the detection power of the device. An example of a detectability curve for ultrasonics method is shown in Fig. 8.3. Hence, even when a structure is inspected and all detected flaws are repaired, it is difficult to ensure that there are no flaws larger than some specified size.

Suppose that an NDT device is used to inspect a set of welds in a structure and all detected flaws are fully repaired. On the basis of this assumption, the flaws that remain in the weld would be those that were not detected. Let X be the flaw size and D the event that a flaw is detected. The probability that a flaw size (for example, depth) will be between x and $(x + dx)$ given that the flaw was not detected is, therefore,

$$P(x < X \leq x + dx \mid \bar{D}) = \frac{P(\bar{D} \mid x < X \leq x + dx)P(x < X \leq x + dx)}{P(\bar{D})}$$

This can be expressed also in terms of density functions as

$$f_X(x \mid \bar{D}) = kP(\bar{D} \mid x)f_X(x) \tag{8.8}$$

in which $f_X(x)$ is the distribution of the flaw size prior to inspection and repair, whereas $f_X(x \mid \bar{D})$ is the corresponding distribution after inspection and repair. Also $P(\bar{D} \mid x) = 1 - P(D \mid x)$, where $P(D \mid x)$ is simply the probability of detecting a flaw with depth x, which is the function defined by the detectability curve, such as that shown in Fig. 8.3. Comparing Eq. 8.8 with Eq. 8.5, we observe that Eq. 8.8 is of the same form as Eq. 8.5, with the following equivalences:

$$f_X(x \mid \bar{D}) \sim \text{the posterior distribution}$$

$$P(\bar{D} \mid x) \sim \text{the likelihood function}$$

$$f_X(x) \sim \text{the prior distribution}$$

EXAMPLE 8.6

As an illustration, suppose the initial (prior) distribution of flaw depths X in a series of welds has a triangular shape described as follows (see Fig. E8.6):

$$f_X(x) = \begin{cases} 208.3x & 0 < x \le 0.06 \\ 20 - 125x & 0.06 < x \le 0.16 \\ 0 & x > 0.16 \end{cases}$$

Assume also that the NDT device used in the inspection has the detectability curve shown in Fig. 8.3; mathematically, this curve is given by

$$P(D \mid x) = \begin{cases} 0 & x \le 0 \\ 8x & 0 < x \le 0.125 \\ 1.0 & x > 0.125 \end{cases}$$

Substituting the appropriate expressions for each interval of X into Eq. 8.8, we

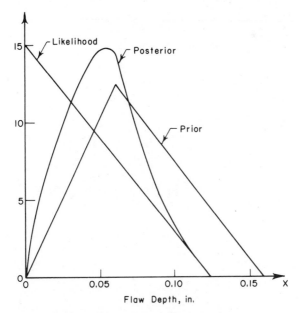

Figure E8.6   Distribution of flaw depth

obtain the updated density function of flaw depths

$$f_X(x \mid \bar{D}) = \begin{cases} 0 & x \leq 0 \\ k(1 - 8x)(208.3x) & 0 < x \leq 0.06 \\ k(1 - 8x)(20 - 125x) & 0.06 < x \leq 0.125 \\ 0 & x > 0.125 \end{cases}$$

which, after normalization, becomes

$$f_X(x \mid \bar{D}) = \begin{cases} 0 & x \leq 0 \\ 495x - 3964x^2 & 0 < x \leq 0.06 \\ 47.6 - 678x + 2379x^2 & 0.06 < x \leq 0.125 \\ 0 & x > 0.125 \end{cases}$$

The above prior, likelihood, and posterior functions are plotted in Fig. E8.6. It can be observed that the likelihood function, which is the "complementary function" of Fig. 8.3, behaves as a filter; it cuts off flaws larger than 0.125 in. and also eliminates many of the remaining larger flaws. Thus, after the inspection and repair program, the distribution of flaw depth is shifted toward smaller values.

8.4.   BAYESIAN CONCEPTS IN SAMPLING THEORY

8.4.1.   General formulation

If the experimental outcome ϵ in Eq. 8.4 is a set of observed values $x_1, x_2, \ldots, x_n$, representing a random sample (see Section 5.2.1) from a population X with underlying density function $f_X(x)$, the probability of observing this particular set of values, assuming that the parameter of the distribution is θ, is

$$P(\epsilon \mid \theta) = \prod_{i=1}^{n} f_X(x_i \mid \theta) \, dx$$

Then, if the prior density function of Θ is $f'(\theta)$, the corresponding posterior density function becomes, according to Eq. 8.4,

$$f''(\theta) = \frac{\left[\displaystyle\prod_{i=1}^{n} f_X(x_i \mid \theta) \, dx \right] f'(\theta)}{\displaystyle\int_{-\infty}^{\infty} \left[\prod_{i=1}^{n} f_X(x_i \mid \theta) \, dx \right] f'(\theta) \, d\theta}$$

$$= kL(\theta)f'(\theta) \tag{8.9}$$

in which the normalizing constant is

$$k = \left[\int_{-\infty}^{\infty} \left(\prod_{i=1}^{n} f_X(x_i \mid \theta) \right) f'(\theta) \, d\theta \right]^{-1}$$

whereas the likelihood function $L(\theta)$ is the product of the density function of X evaluated at $x_1, x_2, \ldots, x_n$, or

$$L(\theta) = \prod_{i=1}^{n} f_X(x_i \mid \theta) \tag{8.10}$$

Using the posterior density function for Θ of Eq. 8.9 in Eq. 8.6, we therefore obtain the Bayesian estimator of the parameter θ. It is interesting to observe that the likelihood function of Eq. 8.10 is the same as that given earlier in Eq. 5.4 in connection with the classical method of maximum likelihood estimation. Furthermore, if a diffuse prior distribution is assumed (for example, as in Eq. 8.13), then the mode of the posterior distribution, Eq. 8.9, gives the maximum likelihood estimator.

8.4.2. Sampling from normal population

In the case of a Gaussian population with known standard deviation σ, the likelihood function for the parameter μ, according to Eq. 8.10, is

$$L(\mu) = \prod_{i=1}^{n} \frac{1}{\sqrt{2\pi}\sigma} \exp\left[-\frac{1}{2}\left(\frac{x_i - \mu}{\sigma}\right)^2\right] = \prod_{i=1}^{n} N_\mu(x_i, \sigma)$$

where $N_\mu(x_i, \sigma)$ denotes the density function of μ with mean value x_i and standard deviation σ. It can be shown (for instance, Tang, 1971) that the product of m normal density functions with respective means μ_i and standard deviations σ_i is also a normal density function with mean and variance

$$\mu^* = \frac{\sum\limits_{i=1}^{m}(\mu_i/\sigma_i^2)}{\sum\limits_{i=1}^{m} 1/\sigma_i^2} \quad \text{and} \quad (\sigma^*)^2 = \frac{1}{\sum\limits_{i=1}^{m} 1/\sigma_i^2} \tag{8.11}$$

Therefore the likelihood function $L(\mu)$ becomes

$$L(\mu) = N_\mu\left(\frac{\sum\limits_{i=1}^{n}(x_i/\sigma^2)}{\sum\limits_{i=1}^{n}(1/\sigma^2)}, \frac{1}{\sqrt{\sum\limits_{i=1}^{n}(1/\sigma^2)}}\right) = N_\mu\left(\frac{(1/\sigma^2)\sum\limits_{i=1}^{n} x_i}{n/\sigma^2}, \frac{1}{\sqrt{n/\sigma^2}}\right)$$

$$= N_\mu\left(\bar{x}, \frac{\sigma}{\sqrt{n}}\right) \tag{8.12}$$

where $\bar{x}$ is the sample mean of Eq. 5.1.

Without prior information. In the absence of prior information on μ, a diffuse prior distribution may be assumed. In such a case we obtain the posterior distribution for μ, as

$$f''(\mu) = kL(\mu)$$

$$= kN_\mu\left(\bar{x}, \frac{\sigma}{\sqrt{n}}\right)$$

$$= N_\mu\left(\bar{x}, \frac{\sigma}{\sqrt{n}}\right) \tag{8.13}$$

where k is necessarily equal to 1.0 upon normalization. Therefore, without prior information, the posterior distribution of μ is Gaussian with a mean value equal to the sample mean $\bar{x}$ and standard deviation $\sigma/\sqrt{n}$.

Using the expected value of μ as the Bayesian estimator we obtain, in accordance with Eq. 8.6,

$$\hat{\mu}'' = E(\mu \mid \epsilon) = \bar{x}$$

That is, the sample mean $\bar{x}$ is the point estimate of the population mean. We recognize that this is the same as the classical estimate of Eq. 5.1. Therefore, in the absence of prior information, the Bayesian and classical methods give the same estimates for the population mean. Conceptually, however, the Bayesian basis for this estimate differs from that of the classical approach. Whereas Eq. 8.13 says that the posterior distribution of μ is Gaussian with mean $\bar{x}$ and standard deviation $\sigma/\sqrt{n}$, the classical approach (of Sect. 5.2) says the sample mean $\bar{X}$ is a Gaussian random variable with mean μ and standard deviation $\sigma/\sqrt{n}$.

Significance of prior information. In contrast to the classical approach, however, prior information can be included in the estimation of the parameter μ. This is accomplished explicitly through the prior distribution $f'(\theta)$; we demonstrate this for the case of a Gaussian population as follows.

In the case where X is Gaussian with known variance, it is mathematically convenient to assume also a Gaussian prior (see Sect. 8.4.4). Suppose that $f'(\mu)$ is $N(\mu', \sigma')$. Then, with the likelihood function of Eq. 8.12, the posterior distribution of μ becomes

$$f''(\mu) = kL(\mu)f'(\mu)$$

$$= kN_\mu\left(\bar{x}, \frac{\sigma}{\sqrt{n}}\right) N_\mu(\mu', \sigma')$$

which is a product of two normal density functions. Again, it can be shown that $f''(\mu)$ is also Gaussian with mean

$$\mu'' = \frac{[\bar{x}/(\sigma/\sqrt{n})^2] + [\mu'/(\sigma')^2]}{[1/(\sigma/\sqrt{n})^2] + [1/(\sigma')^2]} = \frac{\bar{x}(\sigma')^2 + \mu'(\sigma^2/n)}{(\sigma')^2 + (\sigma^2/n)} \tag{8.14}$$

and standard deviation

$$\sigma'' = \sqrt{\frac{(\sigma')^2(\sigma^2/n)}{(\sigma')^2 + (\sigma^2/n)}} \qquad (8.15)$$

In this case the Bayesian estimator of μ, Eq. 8.6, yields

$$\hat{\mu}'' = \mu''$$

That is, the Bayesian estimate of the mean value is an average of the prior mean μ' and the sample mean $\bar{x}$, weighted inversely by the respective variances.

Equation 8.14 is an example of how prior information is combined systematically with observed data—in the present case, to estimate the mean value μ.

It is important to observe that the posterior variance of μ, as given by Eq. 8.15, is always less than* $(\sigma')^2$ or (σ^2/n); that is, the variance of the posterior distribution is always less than that of the prior distribution or of the likelihood function.

On the basis of the posterior distribution of μ, that is, $N_\mu(\bar{x}, \sigma/\sqrt{n})$ of Eq. 8.13 or $N_\mu(\mu'', \sigma'')$ with Eqs. 8.14 and 8.15, we may also determine the interval for μ corresponding to a specified probability. For example, the probability that μ is between a and b is given by

$$P(a < \mu \leq b) = \int_a^b f''(\mu)\, d\mu$$

8.4.3. Error in estimation

Any error in the estimation of a parameter θ can be combined with the inherent variability of the underlying random variable, for example X, to obtain the total uncertainty associated with X. Accounting for the error in the estimation of θ, the density function of X becomes (by virtue of the

* Since $(\sigma')^2 \geq 0$, and $\sigma^2/n \geq 0$

$$(\sigma'^2)^2 + (\sigma'^2)\left(\frac{\sigma^2}{n}\right) \geq (\sigma'^2)\left(\frac{\sigma^2}{n}\right)$$

$$(\sigma')^2\left(\sigma'^2 + \frac{\sigma^2}{n}\right) \geq (\sigma'^2)\left(\frac{\sigma^2}{n}\right)$$

or

$$(\sigma')^2 \geq \frac{(\sigma'^2)(\sigma^2/n)}{(\sigma')^2 + \sigma^2/n} = \sigma''^2$$

Similarly, it can be shown that $(\sigma'')^2 \leq \sigma^2/n$.

total probability theorem)

$$f_X(x) = \int_{-\infty}^{\infty} f_X(x \mid \theta) f''(\theta) \, d\theta \qquad (8.16)$$

In the case of a Gaussian variate X, with known σ, and μ estimated from sample data,

$$f_X(x) = \int_{-\infty}^{\infty} f_X(x \mid \mu) f''(\mu) \, d\mu$$

where $f_X(x \mid \mu) = N_X(\mu, \sigma)$, and $f''(\mu)$ is given by Eq. 8.13. Again it can be shown (for instance, Tang, 1971) that this last integral yields the normal density function $N_X(\bar{x}, \sqrt{\sigma^2 + \sigma^2/n})$; that is,

$$f_X(x) = N(\bar{x}, \sqrt{\sigma^2 + \sigma^2/n}) \qquad (8.17)$$

The overall uncertainty in X here is reflected in its variance, $\sigma^2 + \sigma^2/n$, which is composed of the variance of the basic random variable X and that of the parameter μ. Effectively, the error in the estimation of μ serves to increase the total uncertainty in X, by an amount that decreases with the sample size n.

EXAMPLE 8.7

A toll bridge was recently opened to traffic. For the past two weeks, records on rush-hour traffic during the last 10 workdays showed a sample mean of 1535 vehicles per hour (vph). Suppose that rush-hour traffic has a normal distribution with a standard deviation of 164 vph. Based on this observational information, the posterior distribution of the mean rush-hour traffic μ is, according to Eq. 8.13, $N(1535, 164/\sqrt{10})$ or $N(1535, 51.9)$ vph. The point estimate of μ, therefore, is 1535 vph.

The probability that μ will be between 1500 and 1600 vph is given by

$$P(1500 < \mu \leq 1600) = \Phi\left(\frac{1600 - 1535}{51.9}\right) - \Phi\left(\frac{1500 - 1535}{51.9}\right)$$
$$= \Phi(1.253) - \Phi(-0.674)$$
$$= 0.6445$$

Of greater interest are probabilities associated with the rush-hour traffic (rather than its mean) on a given workday. Suppose that for the present toll collection procedure, serious problems would arise if the rush-hour traffic exceeds 1700 vph on a given day. Then the probability that this will occur on any given day, based on Eq. 8.17, is given by

$$P(X > 1700) = 1 - \Phi\left(\frac{1700 - 1535}{\sqrt{(164)^2 + (51.9)^2}}\right)$$
$$= 1 - \Phi(0.958)$$
$$= 0.169$$

In other words, in about 17% of the working days, the present toll collection system will be inadequate during rush hours. Observe that the error in the estimation of μ has been included in computing this probability.

Now suppose that before the toll bridge was opened for traffic, simulation was performed to predict the rush-hour traffic on the bridge. Based on the simulation results alone, it was estimated that the mean rush-hour traffic on a workday would be 1500 ± 100 with 90% confidence. How can this information be used with the observed traffic flow in the estimation of μ?

Assuming a Gaussian prior and with the foregoing simulation results, we obtain the prior distribution of the mean rush-hour traffic μ to be N (1500, 60.8) vph. Then, applying Eqs. 8.14 and 8.15, the posterior distribution of μ is Gaussian with

$$\mu'' = \frac{1535(60.8)^2 + 1500(51.9)^2}{(60.8)^2 + (51.9)^2} = 1520 \text{ vph}$$

and

$$\sigma'' = \sqrt{\frac{(60.8)^2(51.9)^2}{(60.8)^2 + (51.9)^2}} = 39.5 \text{ vph}$$

Therefore, by incorporating the result of simulation, the estimated mean rush-hour traffic is 1520 vph and corresponding standard deviation is 39.5 vph.

EXAMPLE 8.8

Five repeated measurements of the elevation (relative to a fixed datum) of a bridge pier under construction were made as follows:

$$20.45 \text{ m}$$
$$20.38 \text{ m}$$
$$20.51 \text{ m}$$
$$20.42 \text{ m}$$
$$20.46 \text{ m}$$

Assume that the measurement error is Gaussian with zero mean and standard deviation 0.08 m.

(a) Estimate the actual elevation of the pier based on the given measurements.

(b) Suppose that the elevation of the pier was previously measured by another surveying crew; the elevation was estimated to be 20.42 ± 0.02 m (that is, the mean measurement was 20.42 m with a standard error of 0.02 m). Estimate the elevation of the pier taking advantage of this prior information.

Solution

The estimation of an actual dimension δ in surveying and photogrammetry is equivalent to the estimation of the mean value of a random variable (see Section 5.2.3). Measurement error is invariably assumed to be Gaussian with zero mean; this means tacitly that a set of measurements constitute a sample from a normal population. Therefore the results derived in Section 8.4.2 are applicable to the estimation of geometric quantities in surveying and photogrammetry.

(a) The sample mean of the five measurements is

$$\bar{d} = \tfrac{1}{5}(20.45 + 20.38 + 20.51 + 20.42 + 20.46)$$
$$= 20.444 \text{ m}$$

Hence, on the basis of the five observations, the actual elevation of the pier has a

Gaussian distribution $N (20.444, 0.08/\sqrt{5})$ or $N (20.444, 0.036)$ m. In the convention of surveying and photogrammetry, the elevation of the pier would be given as 20.444 ± 0.036 m.

(b) In the case where prior information is available, such information can be incorporated through the prior distribution of δ. In the present case, using the pier elevation estimated earlier by another crew, the prior distribution of δ can be modeled as $N (20.420, 0.020)$ m. Then applying Eqs. 8.14 and 8.15, the Bayesian estimate of the elevation is

$$\hat{d}'' = \frac{(20.420)(0.036)^2 + (20.444)(0.020)^2}{(0.036)^2 + (0.020)^2}$$

$$= 20.426 \text{ m}$$

and the corresponding standard error is

$$\sigma_d'' = \sqrt{\frac{(0.036)^2(0.020)^2}{(0.036)^2 + (0.020)^2}}$$

$$= 0.017 \text{ m}$$

EXAMPLE 8.9

The annual maximum flow of a stream has been recorded for the last five years as follows:

$$21.5, 19.2, 23.4, 20.1, 18.1 \ (100 \text{ m}^3/\text{sec})$$

Based on extensive data from adjacent streams, the annual maximum stream flow may be modeled by a log-normal distribution. Assume that the parameter ζ in the log-normal distribution is equal to the value obtained from the five sample values. The problem here is to estimate the parameter λ.

In Chapter 4 (Example 4.2) it is shown that if a random variable Y is log-normal, then $X = \ln Y$ is normal. Hence the logarithm of the stream flow will be Gaussian with mean λ and known standard deviation ζ.

The natural logarithm of the above data values are, respectively,

$$3.07, 2.96, 3.15, 3.00, 2.90$$

from which we obtain the sample mean $\bar{x} = 3.016$, and sample standard deviation $\zeta = 0.097$.

Without any prior information, the posterior distribution of λ, according to Eq. 8.13, is $N (\bar{x}, \zeta/\sqrt{5})$ or $N (3.016, 0.097/\sqrt{5}) = N (3.016, 0.043)$.

If prior information is available, it can be incorporated through the prior distribution of λ. For example, suppose that $f'(\lambda)$ is assumed to be $N (2.9, 0.06)$; then from Eqs. 8.14 and 8.15 the posterior distribution $f''(\lambda)$ will be normal with

$$\mu_\lambda'' = \frac{3.016(0.06)^2 + 2.9(0.0435)^2}{(0.06)^2 + (0.0435)^2} = 2.98$$

and

$$\sigma_\lambda'' = \sqrt{\frac{(0.06)^2(0.0435)^2}{(0.06)^2 + (0.0435)^2}} = 0.035$$

That is, in this latter case, the posterior distribution of λ is $N (2.98, 0.035)$.

8.4.4. Use of conjugate distributions

In deriving the posterior distribution of a parameter by Eq. 8.5 or 8.9, considerable mathematical simplification can be achieved if the distribution of the parameter is appropriately chosen with respect to that of the underlying random variable. We saw this in Sect. 8.4.2 in the case of the Gaussian random variable X with known σ; by assuming the prior distribution of μ to be also Gaussian, the posterior distribution of μ remains Gaussian. This was similarly demonstrated for the discrete case in Example 8.4, in which the random variable has a binomial distribution and the prior distribution for p was assumed to be a beta distribution (with parameters $q' = 4$ and $r' = 2$). The resulting posterior distribution for p is also a beta distribution, with updated parameters $q'' = 8$ and $r'' = 3$.

Such pairs of distributions are known in the Bayesian terminology as *conjugate pairs* or *conjugate distributions*. By choosing a prior distribution that is a conjugate of the distribution of the underlying random variable, convenient posterior distribution, which is usually of the same mathematical form as the prior, is obtained. This has been illustrated earlier in the case of the normal-normal and the binomial-beta distributions. Other pairs of conjugate distributions may be developed; Table 8.1 summarizes some of these involving certain common distributions.

It should be emphasized that conjugate distributions are chosen solely for mathematical convenience and simplicity. For a random variable with a specified distribution, its conjugate prior distribution may be adopted if there is no other basis for the choice of the prior distribution. However, if there is evidence to support a particular prior distribution, then such a distribution ought to be used, mathematical complications notwithstanding.

EXAMPLE 8.10

The occurrence of flaws in a weld joint may be modeled by a Poisson process with a mean occurrence rate of μ flaws per meter of weld. Actual observation with a powerful device (assume it would not miss detecting any significant flaw) detected 5 flaws in a weld of 9.2 meters. However, from previous experience with the same type of weld and quality of workmanship, the mean flaw rate is believed to be 0.5 flaw/m with a COV of 40%. Determine the mean and COV of μ for this type of weld, using the observed data as well as the information from prior experience.

Since the number of flaws in a given weld length is described by the Poisson distribution, it is convenient, according to Table 8.1, to prescribe its conjugate gamma distribution as the prior distribution for the parameter μ. From the information given above, and observing from Section 3.2.8 the mean and variance of the gamma distribution, we have

$$E'(\mu) = \frac{k'}{v'} = 0.5$$

Table 8.1 Conjugate Distributions

Basic random variable	Parameter	Prior and posterior distributions of parameter
Binomial		Beta
$p_X(x) = \binom{n}{x} \theta^x (1-\theta)^{n-x}$	θ	$f_\Theta(\theta) = \dfrac{\Gamma(q+r)}{\Gamma(q)\Gamma(r)} \theta^{q-1}(1-\theta)^{r-1}$
Exponential		Gamma
$f_X(x) = \lambda e^{-\lambda x}$	λ	$f_\Lambda(\lambda) = \dfrac{\nu(\nu\lambda)^{k-1}e^{-\nu\lambda}}{\Gamma(k)}$
Normal		Normal
$f_X(x) = \dfrac{1}{\sqrt{2\pi}\sigma} \exp\left[-\dfrac{1}{2}\left(\dfrac{x-\mu}{\sigma}\right)^2\right]$ (with known σ)	μ	$f_M(\mu) = \dfrac{1}{\sqrt{2\pi}\sigma_\mu} \exp\left[-\dfrac{1}{2}\left(\dfrac{\mu-\mu_\mu}{\sigma_\mu}\right)^2\right]$
Normal		Gamma-Normal
$f_X(x) = \dfrac{1}{\sqrt{2\pi}\sigma} \exp\left[-\dfrac{1}{2}\left(\dfrac{x-\mu}{\sigma}\right)^2\right]$	μ, σ	$f(\mu, \sigma)$ $= \left\{\dfrac{1}{\sqrt{2\pi}\sigma/n} \exp\left[-\dfrac{1}{2}\left(\dfrac{\mu-\bar{x}}{\sigma/\sqrt{n}}\right)^2\right]\right\}$ $\cdot \left\{\dfrac{[(n-1)/2]^{(n+1)/2}}{\Gamma[(n+1)/2]}\left(\dfrac{s^2}{\sigma^2}\right)^{(n-1)/2}\right.$ $\left.\cdot\exp\left(-\dfrac{n-1}{2}\dfrac{s^2}{\sigma^2}\right)\right\}$
Poisson		Gamma
$p_X(x) = \dfrac{(\mu t)^x}{x!} e^{-\mu t}$	μ	$f_M(\mu) = \dfrac{\nu(\nu\mu)^{k-1}e^{-\nu\mu}}{\Gamma(k)}$
Lognormal		Normal
$f_X(x) = \dfrac{1}{\sqrt{2\pi}\zeta x} \cdot$ $\cdot \exp\left[-\dfrac{1}{2}\left(\dfrac{\ln x - \lambda}{\zeta}\right)^2\right]$ (with known ζ)	λ	$f_\Lambda(\lambda) = \dfrac{1}{\sqrt{2\pi}\sigma} \exp\left[-\dfrac{1}{2}\left(\dfrac{\lambda-\mu}{\sigma}\right)^2\right]$

Mean and Variance of Parameter	Posterior Statistics
$E(\Theta) = \dfrac{q}{q+r}$	$q'' = q' + x$
$\mathrm{Var}(\Theta) = \dfrac{qr}{(q+r)^2(q+r+1)}$	$r'' = r' + n - x$
$E(\lambda) = \dfrac{k}{\nu}$	$\nu'' = \nu' + \sum_i x_i$
$\mathrm{Var}(\lambda) = \dfrac{k}{\nu^2}$	$k'' = k' + n$
$E(\mu) = \mu_\mu$	$\mu_\mu'' = \dfrac{\mu_\mu'(\sigma^2/n) + \bar{x}\sigma_\mu'^2}{\sigma^2/n + (\sigma_\mu')^2}$
$\mathrm{Var}(\mu) = \sigma_\mu^2$	$\sigma_\mu'' = \sqrt{\dfrac{(\sigma_\mu')^2(\sigma^2/n)}{(\sigma_\mu')^2 + \sigma^2/n}}$
$E(\mu) = \bar{x}$	$n'' = n' + n$
$\mathrm{Var}(\mu) = s^2\left[\dfrac{n-1}{n(n-3)}\right]$	$n''\bar{x}'' = n'\bar{x}' + n\bar{x}$
$E(\sigma) = s\sqrt{\dfrac{n-1}{2}}\dfrac{\Gamma[(n-2)/2]}{\Gamma[(n-1)/2]}$	$(n''-1)s''^2 + n''\bar{x}''^2$ $= [(n'-1)s'^2 + n'\bar{x}'^2]$
$\mathrm{Var}(\sigma) = s^2\left(\dfrac{n-1}{n-3}\right) - E^2(\sigma)$	$+ [(n-1)s^2 + n\bar{x}^2]$
$E(\mu) = \dfrac{k}{\nu}$	$\nu'' = \nu' + t$
$\mathrm{Var}(\mu) = \dfrac{k}{\nu^2}$	$k'' = k' + x$
$E(\lambda) = \mu$	$\mu'' = \dfrac{\mu'(\zeta^2/n) + \sigma^2\overline{\ln x}}{\zeta^2/n + \sigma^2}$
$\mathrm{Var}(\lambda) = \sigma^2$	$\sigma'' = \sqrt{\dfrac{\sigma^2(\zeta^2/n)}{\sigma^2 + \zeta^2/n}}$

and

$$\delta'(\mu) = \frac{\sqrt{\overline{k'/\nu'^2}}}{k'/\nu'} = \frac{1}{\sqrt{k'}} = 0.4$$

Thus the prior parameters of the gamma distribution are $k' = 6.25$; and $\nu' = 12.5$.

It follows then that the posterior distribution of μ is also gamma. From the relationships given in Table 8.1 between the prior and posterior statistics, and the sample data, we evaluate the parameters k'' and ν'' of the posterior gamma distribution as follows:

$$k'' = k' + x = 6.25 + 5 = 11.25$$
$$\nu'' = \nu' + t = 12.5 + 9.2 = 21.7$$

Hence the updated mean and COV of the average flaw rate μ are

$$E''(\mu) = \frac{k''}{\nu''} = \frac{11.25}{21.7} = 0.52 \text{ flaw/m}$$

$$\delta''(\mu) = \frac{1}{\sqrt{k''}} = \frac{1}{\sqrt{11.25}} = 0.30$$

8.5. CONCLUDING REMARKS

In the process of engineering planning and design, judgmental assumptions and inferential information are often useful and necessary. The significance of such prior information and its role (in combination with observational data) in the process of estimation are formally the subject of Bayesian statistics. The basic concepts of the Bayesian approach have been introduced here with special reference to sampling and estimation. Applications of these concepts in Bayesian statistical decision will be covered in Vol. II.

Philosophically, there are fundamental differences between the Bayesian and classical statistics. Within the Bayesian context, a probability or a probability statement is an expression of the *degree-of-belief*, whereas in the classical sense, probability is a verifiable measure of *relative frequency*. Furthermore, in estimation, the Bayesian approach assumes that a parameter is a random variable, whereas in the classical approach it is an unknown constant.

Relative to engineering planning and design, the Bayesian approach offers the following advantages:

1. It provides the formal framework for incorporating engineering judgment (expressed in probability terms) with observational data.
2. It systematically combines uncertainties associated with randomness and those arising from errors of estimation and prediction (see Vol. II).
3. It provides a formal procedure for systematic updating of information.

PROBLEMS

8.1 A new structure is subjected to proof testing. Assume that the maximum proof load is specified at a reasonably high level so that the calculated probability of the structure surviving the maximum proof load is 0.90. However, it is felt that this calculation is only 70% reliable, and there is a 25% chance that the true probability may be 0.50; moreover, there is even a 5% chance that it may be only 0.10.

(a) What is the expected probability of survival before the proof test?

(b) If only one structure is proof-tested, and it survives the maximum proof load, determine the updated distribution of the survival probability.

(c) What is the expected probability of survival after the proof test?

(d) If three structures were proof-tested, and two of the structures survived whereas one failed under the maximum proof load, determine the updated expected probability of survival.

8.2 A new waste-treatment process has been developed. In order to evaluate its effectiveness, the treatment process is installed for a trial period. Each day the output from the treatment process is inspected to see if it satisfies the specified standard. Suppose that the outputs between days are statistically independent, and there is a probability p that the daily output will be acceptable. If the prior PMF is as shown in Fig. P8.2, determine the posterior distribution of p with each of the following observations.

(a) The output on the first day of the trial period is of unacceptable quality.

(b) For a three-day trial period, the quality is unacceptable in only one day.

(c) For a three-day trial period, the first two days are satisfactory whereas the quality is unacceptable on the third day.

In each case, determine also the Bayesian estimate for p. *Ans. 0.536, 0.617, 0.617.*

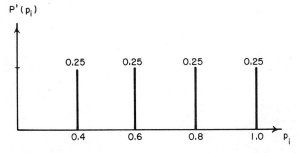

Figure P8.2

8.3 A hazardous street intersection has been improved by changing the geometric design to reduce the accident and fatality rates. For simplicity, assume that accident and fatality rates can be classified as high H or low L, leading to the following possible conditions: $H_A H_F$ (high accident rate, high fatality rate), $H_A L_F$, $L_A H_F$, and $L_A L_F$. Preliminary evaluation revealed that the relative likelihoods for these four conditions are $3:3:2:2$.

An accident rate prediction model, for example, the Tharp model, was used to obtain a better evaluation of the accident potential at this (improved) intersection. Because of possible inaccuracies in the prediction model, a predicted condition may not be actually realized. Furthermore, the probability of a correct prediction depends on the underlying actual condition, as indicated in the following table of conditional probabilities.

Predicted \ Actual	$H_A H_F$	$H_A L_F$	$L_A H_F$	$L_A L_F$
$H_A' \, H_F'$	0.30	0.40	0.20	0.25
$H_A' \, L_F'$	0.30	0.30	0.20	0.25
$L_A' \, H_F'$	0.20	0.20	0.50	0.25
$L_A' \, L_F'$	0.20	0.10	0.10	0.25

(a) What is the probability that the model will indicate $H_A' H_F'$?

(b) Suppose that the model predicted $H_A' H_F'$; what is the probability that the condition of the improved intersection will actually be $H_A H_F$?

(c) If the model predicted $L_A' L_F'$; what is the updated relative likelihoods of the four possible conditions?

8.4 An instrument is used to check the accuracy of a set of measurements. However, it can only record three readings, namely $x = 1, 2,$ or 3. The reading $x = 2$ implies that the previously measured value is within a tolerable error, whereas $x = 1$ and $x = 3$ denote that the measurement is on the low and high side, respectively. Suppose the distribution of X is given as follows:

$$p_X(x_i) = \begin{cases} \dfrac{1-m}{2} & x_i = 1 \\[2mm] m & x_i = 2 \\[2mm] \dfrac{1-m}{2} & x_i = 3 \end{cases}$$

where m is the parameter. For a particular set of measurements, the engineer estimated that the value of m would be 0.4 or 0.8 with equal likelihood. However, on checking a set of measurements, the first one indicates $x = 2$.

(a) What should be the engineer's revised distribution of m?

(b) Estimate the probability that at least two out of the next three measurements will be accurate.

8.5 An engineer plans to build a log cabin in the middle of a forest where logs of similar size are available. He assumes that the bending capacity M of each log follows a Rayleigh distribution

$$f_M(m) = \frac{m}{\lambda^2} e^{-(1/2)(m/\lambda)^2} \qquad m \geq 0$$

where the parameter λ is the modal value of the distribution.

From previous experience with similar logs, he feels that λ would be 4(kip-ft) with probability 0.4 or 5 (kip-ft) with probability 0.6. Not entirely satisfied with these subjective probabilities, he decided to get a better measure of the parameter λ. Being pressed for time and with limited supply of logs, he can only afford to test the bending capacity of two logs by simple load test on the site. The test results yielded 4.5 kip-ft and 5.2 kip-ft for the two tests.

(a) Determine the posterior distribution (discrete) for the parameter λ.
(b) Derive the distribution of the bending capacity of the logs M, based on the posterior distribution of λ.
(c) What is the probability that M is less than 2 kip-ft?

8.6 The absolute error E (in cm) of each measurement from a surveying instrument is governed by the triangular distribution shown in Fig. P8.6, where α denotes the upper limit of the error.

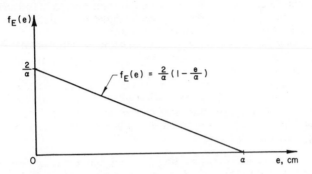

Figure P8.6

Two measurements were made and the errors are 1 and 2 cm, respectively.
(a) Suppose that α is assumed to be 2 or 3 cm with equal likelihood prior to the two measurements; determine the updated distribution of α. Estimate the value of α based on this updated distribution.
(b) Now suppose that the prior density function of α is uniform between 2 and 3; determine and plot the updated distribution of α, and evaluate the corresponding Bayesian estimate for α.

8.7 Suppose that the prior density function of the mean accident rate ν in Example 8.2 is

$$f'(\nu) = \frac{0.271}{\nu} \qquad 0.5 \leq \nu < 20$$

$$= 0 \qquad \text{elsewhere}$$

Determine the posterior density function of ν based on the observation that an accident was recorded during the first month of operation.

8.8 In Problem 8.1 suppose that the survival probability has a prior density function as follows.
(i) Uniform between $p = 0$ to 0.9.
(ii) Uniform between $p = 0.9$ to 1.0.
(iii) It is more likely that p will exceed 0.9 than be less than 0.9; the relative likelihood between these two possibilities is 7 to 3.
(a) Determine the prior density function of p.
(b) If three structures were proof tested, and all three survived the maximum proof load, determine and plot the posterior density function of p.
(c) What is the estimated value of p in light of the results in part (b)?

8.9 The occurrence of fire in a city may be modeled by a Poisson process. Suppose the average occurrences of fires v is assumed to be 15 or 20 times a year; the likelihood of 20 fires a year (on the average) is twice that of 15 fires a year.

(a) Determine the probability that there will be 20 occurrences of fire in the next year.

(b) If there are actually 20 fires in the next year, what will be the updated PMF of v?

(c) What is the probability that there will be 20 occurrences of fire in the year after next, in the light of part (b)?

8.10 Consider a case where the mean compression index of a soil stratum is to be estimated. Assume that the compression index of a soil sample is $N(\mu, \sigma)$ and σ is assumed to be equal to 0.16. Laboratory tests on four speciments show the following compression index values: 0.75, 0.89, 0.91, and 0.81.

(a) What is the posterior distribution of μ if there is no other information except the observed data?

(b) Suppose there is prior information to indicate that μ is Gaussian with mean 0.8 and COV 25%. What will be the posterior distribution of μ if this prior information is taken into account?

(c) What is the probability that μ will be less than 0.95, using the data from part (b)? *Ans. 0.938.*

8.11 An air passenger is commuting between San Francisco and Los Angeles regularly. Lately, he started recording the time of each flight. He computed the average flight time from his five previous trips to be 65 minutes. Suppose the flight time T is a Gaussian random variable with known standard deviation of 10 minutes.

(a) Based purely on the data, what is the posterior distribution of μ_T?

(b) The passenger is now on a plane from San Francisco to Los Angeles. By coincidence, the passenger sitting next to him also has been keeping track of the flight time. From his record of 10 previous trips, he obtained an average of 60 minutes. Assume that these two passengers have never taken a plane together before. With this additional information, what would be the updated distribution of μ_T?

(c) What is the probability that their flight will take more than 80 minutes? *Ans. 0.038.*

8.12 Six measurements were made of an angle as follows:

$$32°04' \qquad 32°05'$$
$$31°59' \qquad 31°57'$$
$$32°01' \qquad 32°00'$$

Assume that the measurement error is Gaussian with zero mean; and the standard deviation of each measurement can be represented by the sample standard deviation of the six measurements above.

(a) Estimate the angle.

(b) Subsequently, the engineer discovered that the angle has been measured before, and recorded as $32°00' \pm 2'$. Estimate the actual angle using both sets of measurements.

8.13 A distance L is measured independently by three surveyors with three sets of instruments. The respective measurements are 2.15, 2.20, 2.18 km. Suppose the ratio of the standard error of the three measurements is $1:2:3$. Estimate the actual distance L on the basis of the three sets of measurements. Assume that measurement error is Gaussian with zero mean.

8.14 In designing reinforced concrete structural members to resist ultimate load, a capacity reduction factor ϕ is often used. Suppose that the structural member is a beam element and it is designed for pure flexure. The conventional value of ϕ is 0.9. However, a committee is investigating the effect on the probability of failure of beams against ultimate load if ϕ is increased to 0.95. Twelve beams are designed using $\phi = 0.95$, and each of them is subjected to the designed ultimate load in the laboratory. It is desired to estimate p, the probability of failure of such beams against ultimate load based on the prior judgmental information as well as experimental outcome.

Suppose that the prior distribution of p has a mean of 0.1 and standard deviation of 0.06; and one out of 12 beams tested failed the ultimate load.

Suggest a suitable prior distribution and determine the mean and variance of p from these data.

8.15 The time between breakdowns of a certain type of construction equipment follows an exponential distribution

$$f_X(x) = \lambda e^{-\lambda x} \qquad x \geq 0$$

where the mean rate of failure λ was rated by the manufacturer to have a mean of 0.5 per year and a COV of 20%. A contractor owns two pieces of this construction equipment. The operational times until breakdown of the equipments were subsequently observed to be 12 and 18 months, respectively. Using conjugate distributions, determine the updated mean and COV of the parameter λ.

9. Elements of Quality Assurance and Acceptance Sampling

The assurance of product quality in manufacturing has long been a problem of industrial and production engineers. Quality assurance, however, is of concern to all engineers. Compliance with minimum standards of construction and fabrication, and of quality in materials and workmanship, is necessary to ensure the design performance capability of an engineering system. For these purposes, acceptance criteria and acceptance sampling programs are necessary. For example, in the construction of highway pavements, acceptance criteria are necessary to ensure compliance with construction specifications; similarly, in stream pollution control, inspection plans are necessary to enforce water quality standards.

Probability concepts and statistical techniques are pertinent and useful to a variety of quality assurance problems. In this chapter we present and develop those statistical concepts that form the bases of some commonly used acceptance sampling programs. Such programs are of two types—*sampling by attributes* and *sampling by variables*.

9.1. ACCEPTANCE SAMPLING BY ATTRIBUTES

When a *lot* of material of size N is submitted for inspection, a sample of n items may be selected at random from the lot and subjected to inspection and testing. In acceptance sampling by attributes, each of these n items is classified as *good* (acceptable) or *bad* (defective) after the test. It is a common criterion that if more than r defective items are found from the sample of n, the lot will be rejected. Conversely, the lot will be accepted if there are r or less defectives. If among the lot of size N, the actual fraction of defectives is p, then the total number of defective items in the sample of size n is described by the hypergeometric distribution (see Section 3.2.9), and the *probability of accepting the lot* is accordingly given by

$$g(p) = \sum_{x=0}^{r} \frac{\binom{Np}{x}\binom{Nq}{n-x}}{\binom{N}{n}} \tag{9.1}$$

where $q = 1 - p$. If n is small relative to N, it can be shown (Hald, 1952) that $g(p)$ in Eq. 9.1 can be approximated by

$$g(p) \simeq \sum_{x=0}^{r} \binom{n}{x} p^x q^{n-x} \tag{9.2}$$

which involves the binomial distribution. Equation 9.2 can also be written

$$g(p) = 1 - \sum_{x=r+1}^{n} \binom{n}{x} p^x q^{n-x} \tag{9.2a}$$

In this latter form, the term $\sum_{x=r+1}^{n} \binom{n}{x} p^x q^{n-x}$ can be found in tables of binomial probabilities, for example, Eisenhart (1950) or Aiken (1955).

9.1.1. The operating characteristic (OC) curve

The function $g(p)$ of Eq. 9.2 is referred to as the OC curve (*operating characteristic curve*). Examples of the OC curve are shown in Fig. 9.1 for various sampling plans (with different combinations of n and r). It can be observed from each of the OC curves in Fig. 9.1 that as the fraction of defective items increases, the probability that the lot will be accepted decreases. For example, according to the sampling plan of Fig. 9.1b, with $n = 15$ and $r = 1$, there is less than 5% probability that a lot with 24% defective items will be accepted, whereas the probability of acceptance increases to 88% for a lot with 4% defective. From the appropriate OC curve, therefore, we can read off the probabilities of accepting and rejecting lots containing various percentages of defective items.

Generally, in determining the optimal inspection plan, it should be kept in mind that the plan has to be accepted by both the *supplier* and *receiver* of the lot.

- For the supplier, it is desirable that the plan have a low probability of rejecting a lot in which the actual fraction of defective items p is less than p_1, the maximum fraction of defective units permitted in good quality lots.
- For the receiver, it is desirable that there be a low probability of accepting a lot if p exceeds p_2, the minimum fraction of defective units sufficient to define poor quality lots.

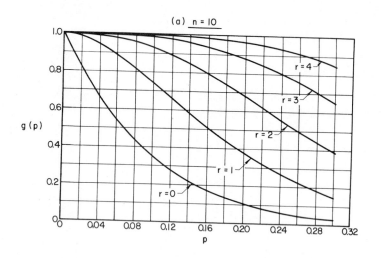

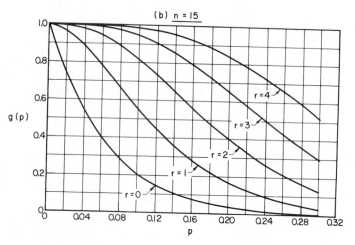

Figure 9.1a and b   Operating characteristic (OC) Curves. (*a*) $n = 10$. (*b*) $n = 15$.

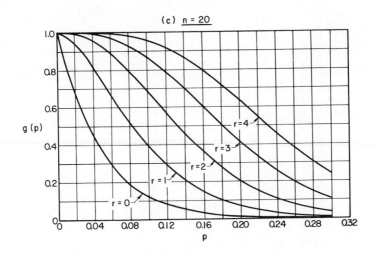

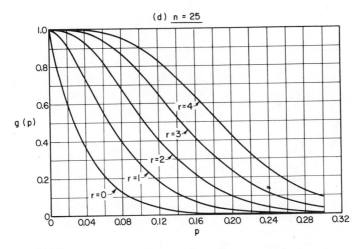

Figure 9.1c and d Operating characteristic (OC) Curves. (c) $n = 20$. (d) $n = 25$.

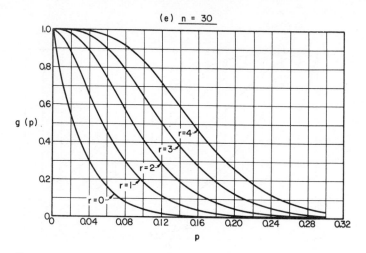

Figure 9.1e   Operating-characteristic (OC) curve $n = 30$.

The supplier's risk of rejecting good-quality lots and the receiver's risk of accepting inferior-quality lots may be referred to as the *producer's risk* and the *consumer's risk*, respectively. The optimal inspection plan should have values of n and r such that the corresponding OC curve will mutually satisfy these risk levels.

EXAMPLE 9.1

In the construction of an earth embankment, the fill material is compacted to a specified CBR (California Bearing Ratio). Suppose that unsatisfactory performance will result if more than 15% of the fill falls below the specified CBR limit; however, the best-quality compaction that can be expected at the contract price would contain 1% of the embankment falling below the specified CBR limit. Assume a 5% risk for both the producer and consumer; what values of n and r should be used in the quality control program?

Using the approximation of Eq. 9.2, the conditions to be satisfied are

$$g(0.01) = \sum_{x=0}^{r} \binom{n}{x} (0.01)^x (0.99)^{n-x} = 0.95$$

and

$$g(0.15) = \sum_{x=0}^{r} \binom{n}{x} (0.15)^x (0.85)^{n-x} = 0.05$$

The required values of n and r may be determined by trial and error from the solutions to the two simultaneous equations given above. Alternatively, we can examine the available OC curves (such as Fig. 9.1) and select the appropriate one. It may be observed from Fig. 9.1e that $n = 30$ and $r = 1$ will suffice.

EXAMPLE 9.2

The density of asphalt concrete in a roadway is to be inspected. Twenty specimens (15-in. square each) were cored at random locations over a 5-mile stretch of the roadway. Laboratory tests show that only one of the specimens has a density below the specified limit. Suppose that the maximum permissible fraction of defective asphalt concrete is 15% and it is desired to limit the risk of accepting inferior quality material to 4%. Should the asphalt concrete be rejected?

The acceptance function $g(p)$ in Eq. 9.2 can be applied. In this case the criterion for accepting the asphalt concrete is

$$g(0.15) \leq 0.04$$

For $n = 20$ and $r = 1$, Eq. 9.2 gives

$$g(0.15) \simeq \binom{20}{0}(0.15)^0(0.85)^{20} + \binom{20}{1}(0.15)^1(0.85)^{19}$$
$$= 0.0388 + 0.1374$$
$$= 0.1762$$

Since $g(0.15)$ exceeds 0.04, the asphalt concrete should be rejected. In this case, for the roadway to be acceptable, all the 20 cored specimens must have at least the minimum specified density; because with $r = 0$,

$$g(0.15) \simeq \binom{20}{0}(0.15)^0(0.85)^{20}$$
$$= 0.0388 < 0.04$$

EXAMPLE 9.3

Part of the probabilistic stream standard proposed by Loucks and Lynn (1966) requires that the probability of dissolved oxygen (DO) in a stream falling below 4 mg/l in any one day should be less than 0.2. Suppose that the DO concentrations in the stream between days are independent and identically distributed; how many days of measurements are required to achieve a 95% confidence that this standard is met, that is, $P(\text{DO} < 4 \text{ mg/l}) < 0.2$, for each of the following cases.

(i) The daily DO cannot be less than 4 mg/l during the period of measurement.
(ii) The daily DO is allowed to be less than 4 mg/l at most once during the period of measurement.

Let p denote the probability that the daily DO is less than 4 mg/l. The problem here is to determine n so that the probability of rejecting a stream quality with p exceeding 0.2 is 0.95, or

$$g(0.2) = 1 - 0.95 = 0.05$$

Since the lot size here is conceivably unlimited, we may apply Eq. 9.2. For case (i), $r = 0$; hence

$$g(0.2) = (0.8)^n = 0.05$$

obtaining

$$n = \frac{\ln(0.05)}{\ln(0.8)} = 13.4 \simeq 14 \text{ days}$$

For case (ii), $r = 1$; hence

$$g(0.2) = (0.8)^n + n(0.2)(0.8)^{n-1} = 0.05$$

By trial-and-error, we obtain the required period of measurement to be 22 days.

It may be observed, from the above results, that a two-stage sequential sampling plan may be devised to achieve the same degree of confidence in meeting the required stream quality standard, as follows.

Step 1

Take 14 days of measurements; if the daily DO concentration during all the 14 days exceeds 4 mg/l, the standard is met; if the DO concentration falls below 4 mg/l for two or more days, the standard is not met, whereas if the DO concentration is below 4 mg/l on only one day, go to the second step.

Step 2

Continue taking measurements for another eight days; if the DO concentration exceeds 4 mg/l in each of the eight days, the standard is met; otherwise, the standard is not met.

9.1.2. The success run

A type of quality control problem commonly encountered is as follows.

A "batch" of material or manufactured products is submitted for inspection. Suppose that a sample of n specimens are picked at random from the batch and each one is tested for compliance with a standard specification. If acceptance of the batch requires no failure out of the n specimens tested, what should the value of n be in order to ensure a *reliability* R for the manufactured products with a *confidence level C*?

This problem can be solved by applying Eq. 9.2 with r equal to zero. A batch with reliability R implies that the fraction defective is $p = 1 - R$; hence the probability of accepting a batch, whose reliability is R, is given by

$$g(1 - R) = R^n \qquad (9.3)$$

Since a confidence level C is desired, the probability of acceptance should be limited to $1 - C$. Therefore,

$$R^n = 1 - C \qquad (9.4)$$

from which we obtain

$$n = \frac{\ln(1 - C)}{\ln R} \qquad (9.5)$$

It may be emphasized that this is the minimum number of specimens that must be tested without any failures (that is, a *success run* of n), in order to assure that the reliability of the product is R with confidence C.

EXAMPLE 9.4

In a prestressed concrete reactor containment structure, 900 tendons are used in prestressing the concrete wall. After operating for a period of time, the level of prestressing in these tendons may decrease. Suppose that to assure proper performance of the structure, no more than 4% of the prestressing tendons can be allowed to have less than the specified minimum prestressing force. Then, if a 95% confidence is desired, how many tendons must be tested without failures?

In this case we have $R = 0.96$, and $C = 0.95$; thus Eq. 9.5 yields

$$n = \frac{\ln (1 - 0.95)}{\ln 0.96} \simeq 74$$

Hence, in order to assure proper structural performance with a 95% confidence, 74 tendons must be tested and all must have prestressing forces equal to or above the specified minimum.

9.1.3.   The average outgoing quality curve

Another commonly used quality control sampling plan involves the AOQ (*average outgoing quality*) curve; this is a plot of the expected fraction of defective units in the accepted product (after inspection) as a function of the assumed fraction of defective units in the as-submitted lot. In other words, the AOQ curve indicates the degree of protection offered by the inspection program by providing information on the average quality of the accepted product.

Consider an inspection plan where each defective unit is replaced by an acceptable one. If the lots submitted for inspection consist of $(100\,p)\%$ defective units, the average fraction of these lots that will be accepted is given by $g(p)$ of Eq. 9.2, whereas the average fraction that will be rejected is $[1 - g(p)]$. Moreover, among the accepted lots, the fraction of defective units is p, whereas the lots that were rejected will presumably be screened and returned as perfect products and thus will contain no defective units. Hence the average fraction of defective units in the final product is given by

$$\mathrm{AOQ} = p\,g(p) + 0[1 - g(p)]$$

$$= p\,g(p) \tag{9.6}$$

The AOQ curve corresponding to the OC curve of Fig. 9.1c (with $n = 20$, $r = 1$) is plotted in Fig. 9.2. It can be observed that when p is small, the AOQ is also low as expected. However, the AOQ is also low for high values of p; this is because, in this case, the lots will have a high likelihood of being rejected and the defective units replaced with good ones before resubmission. Therefore, at some moderate value of p (8% in the case of Fig. 9.2), the AOQ will attain its maximum value; this value of AOQ is referred to as the AOQL (*average outgoing quality limit*), denoting the maximum possible fraction of defective units in the resultant product associated with

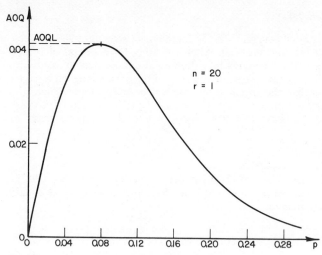

Figure 9.2 Average outgoing quality (AOQ) curve

this quality control plan. In other words, using this acceptance inspection procedure, we can ensure that the fraction of defectives in the overall quality of the product will be less than the AOQL.

EXAMPLE 9.5

In Example 9.1 where the quality of compacted fill is being inspected, suppose that the sampling plan on a section of the embankment requires CBR test at 30 random locations, and the section will be accepted if no more than one of these 30 tests show substandard compaction. On the other hand, if there is more than one test showing substandard compaction, the entire section will be recompacted (assume that recompacting will correct any substandard quality in the original fill). What is

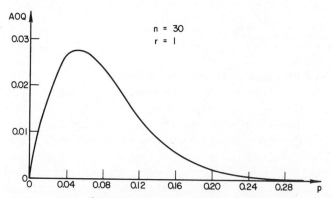

Figure E9.5 AOQ curve for compacted fill of Example 9.5

the resulting average quality of compacted fill as a function of the percentage of substandard compaction in the original fill?

Let p denote the percentage of substandard compacted fill. Apply Eq. 9.6 to the OC curve of Fig. 9.1e for $n = 30$ and $r = 1$. The resulting AOQ curve is plotted in Fig. E9.5. This curve gives the resulting average percentage of substandard compaction as a percentage of the substandard portion of the original fill. According to this sampling plan, the worst quality of compaction would have an average of 2.7% of substandard compaction in the final embankment. This will occur if the original embankment contains approximately 5% substandard compaction.

9.2. ACCEPTANCE SAMPLING BY VARIABLES

In sampling by attributes, as presented in Section 9.1, the quality of each item is classified simply as *good* or *bad*. With this procedure, one item could have better quality than another even though both of them are "good." It would seem that the actual measurements of the items tested should have some bearing on the quality of a lot. For this reason, a sampling plan that is based simply on a good-or-bad classification for each item tested would not fully utilize the information from the test results. An alternative procedure is *sampling by variables*, in which the values of the variable (a quality indicator) for each specimen in a lot are measured. By comparing certain statistic(s) of the observed data, such as the sample mean, with some standard value, the quality of the lot may be determined. Since the measured data are more fully utilized here than in sampling by attributes, the sample size required to achieve the same degree of quality control can be significantly smaller than that of sampling by attributes.

In sampling by variables, the distribution of the sample statistic is required, and this depends on the distribution of the underlying random variable. In practice, the Gaussian distribution is usually assumed. In many cases, such as the degree of compaction and material density, the variable appears to be Gaussian; in other cases, when the sample size is large, the distribution of the sample statistic (such as the mean value) is Gaussian by virtue of the central limit theorem. The following are some of the criteria used in acceptance sampling by variables.

9.2.1. Average quality criterion, σ known

In many quality control programs, such as inspection of compaction level, moisture content, and bulk density of construction materials, the assurance of quality may be based on the average quality of the lot. Suppose that a mean-value μ_a for a lot constitutes good or acceptable quality (from the producer's standpoint), whereas a mean value of μ_t for the lot corresponds to bad or unacceptable quality (from the consumer's position). A common acceptance plan would be to minimize the probability that a good lot (that is, with μ_a) will be rejected, and also minimize the probability that

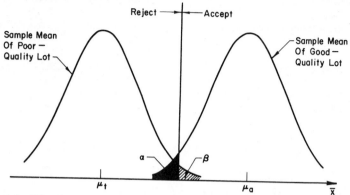

Figure 9.3   Distribution of sample mean and corresponding risk measures

a bad lot (with μ_t) will be accepted. Let α and β denote the producer's and consumer's risks, respectively; then the required sample size n and the standard mean-value L can be determined to satisfy these risks.

Consider first the case in which the standard deviation σ of the variable is known in advance (perhaps from experience) and the variable is Gaussian. For a sample of size n, the sample mean will be Gaussian with standard deviation $\sigma/\sqrt{n}$. Then if a lot is of acceptable quality (its mean value is μ_a), the sample mean will have a distribution $N(\mu_a, \sigma/\sqrt{n})$. Now suppose that the lot will be rejected if the sample mean $\bar{x}$ is less than L; then to limit the producer's risk to α, we should have

$$P(\bar{X} < L \mid \mu_a) = \Phi\left(\frac{L - \mu_a}{\sigma/\sqrt{n}}\right) = \alpha \tag{9.7}$$

Similarly, if the lot is of poor quality (that is, with mean-value μ_t), the sample mean will be $N(\mu_t, \sigma/\sqrt{n})$; then to satisfy the consumer's risk, we should require

$$P(\bar{X} > L \mid \mu_t) = 1 - \Phi\left(\frac{L - \mu_t}{\sigma/\sqrt{n}}\right) = \beta \tag{9.8}$$

Equations 9.7 and 9.8 are portrayed graphically in Fig. 9.3; solution of these two equations then yield L and n for given values of μ_a, μ_t, α, and β.

EXAMPLE 9.6

The density of a newly completed airport pavement is to be inspected by coring block specimens from the pavement. Suppose that from past experience, the density of such pavements has a standard deviation of 3%. The mean densities of good and poor quality lots are 96% and 92%, respectively. Determine the sampling plan if both consumer's and producer's risks are taken at 5%.

With the following substitutions:

$$\mu_a = 0.96, \ \mu_t = 0.92, \ \alpha = 0.05, \ \beta = 0.05, \ \sigma = 0.03$$

Eqs. 9.7 and 9.8 become, respectively,

$$\Phi\left(\frac{L - 0.96}{0.03/\sqrt{n}}\right) = 0.05$$

and

$$1 - \Phi\left(\frac{L - 0.92}{0.03/\sqrt{n}}\right) = 0.05$$

or

$$L - 0.96 = \frac{0.03}{\sqrt{n}} \Phi^{-1}(0.05) = \frac{-0.049}{\sqrt{n}}$$

and

$$L - 0.92 = \frac{0.03}{\sqrt{n}} \Phi^{-1}(0.95) = \frac{0.049}{\sqrt{n}}$$

From which we obtain $L = 0.94$ and $n \simeq 6$. Therefore the sampling plan here requires that six core specimens should be tested and the average density of the six specimens should be at least 94%, to ensure the density of the pavement.

EXAMPLE 9.7 (Excerpted from Grandage, 1966)

Cement is supplied in bags to a contractor. He is willing to accept cement with 0.3% moisture (considered to be excessively moist for construction purposes) only 10% of the time. However, the cement supplier demands that cement with 0.1% moisture (considered to be excellent quality cement) should have a low probability of rejection, no more than, for example, 5%.

Assuming that these moisture contents are average values, we have

$$\mu_a = 0.1\%; \qquad \mu_t = 0.3\%$$

and

$$\alpha = 0.05; \qquad \beta = 0.1$$

Assume further that $\sigma = 0.1\%$.

In this case, a good cement would be rejected if its mean moisture from a sample exceeds the specified standard L; therefore we require

$$P(\bar{X} > L \mid \mu_a) = 1 - \Phi\left(\frac{L - 0.001}{0.001/\sqrt{n}}\right) = 0.05$$

whereas a poor cement would be accepted if its sample mean moisture is less than L; hence we also require

$$P(\bar{X} < L \mid \mu_t) = \Phi\left(\frac{L - 0.003}{0.001/\sqrt{n}}\right) = 0.1$$

After simplification, we obtain

$$\frac{\sqrt{n}(L - 0.001)}{0.001} = \Phi^{-1}(0.95) = 1.64$$

and

$$\frac{\sqrt{n}(L - 0.003)}{0.001} = \Phi^{-1}(0.1) = -1.28$$

The nearest integer value of n that satisfies these equations is 3, and the corresponding value of L is 0.21%. Therefore, if the mean moisture of three bags of cement exceeds 0.21%, the lot should be rejected; otherwise, the lot should be accepted.

9.2.2. Average quality criterion, σ unknown

Quite often σ is not known in advance but has to be estimated from the measured data; in such cases the sample mean will no longer be Gaussian. Instead, the Student's t-distribution will apply (see Section 5.2.2). Then, in place of Eqs. 9.7 and 9.8, the two simultaneous equations needed to determine the required sample size n and specified standard L are

$$\frac{\sqrt{n}(L - \mu_a)}{s} = -t_{\alpha,n-1} \qquad (9.9)$$

and

$$\frac{\sqrt{n}(L - \mu_t)}{s} = t_{\beta,n-1} \qquad (9.10)$$

where $t_{\alpha,n-1}$ and $t_{\beta,n-1}$ are the $100(1 - \alpha)$ and $100(1 - \beta)$ percentile values of the t-distribution with $(n - 1)$ degrees of freedom; and s is the sample standard deviation. Here $t_{\alpha,n-1}$ and $t_{\beta,n-1}$ depend on n; hence the solutions for L and n from Eqs. 9.9 and 9.10 would require trial-and-error procedure.

EXAMPLE 9.8

In Example 9.7, if the standard deviation of the moisture content σ is unknown, the sample standard deviation s must be used. Suppose that the sample data yielded $s = 0.1\%$; then the constraint equations become

$$\frac{\sqrt{n}(L - 0.01)}{0.001} = t_{0.05,n-1}$$

$$\frac{\sqrt{n}(L - 0.003)}{0.001} = -t_{0.1,n-1}$$

After a trial-and-error procedure, the required sample size n is found to be 4 and the corresponding value of L becomes 0.22%.

9.2.3. Fraction defective criterion

Sometimes the consumer is not so much concerned with the mean quality and variability of a lot, but rather with the fraction of the lot that is defective.

In a sampling plan to control the *fraction defective* in a lot, a sample of size n is selected at random and measured. The fraction defective p may then be estimated from the sample mean and standard deviation. The criterion is that if the estimated fraction $\hat{p}$ is greater than a maximum allowable fraction M, the lot will be rejected.

Consider the case in which the required minimum value of a variable X is x_0; then the fraction defective is simply the probability $P(X < x_0)$. Suppose that the distribution of X is normal with standard deviation σ; then $\hat{p}$ may be estimated as

$$\hat{p} = P(X < x_0)$$

$$= \Phi\left(\frac{x_0 - \bar{X}}{\sigma}\right) \tag{9.11}$$

where $\bar{X}$ is the sample mean.

Again, we define p_a and p_t, respectively, as the acceptable and unacceptable fractions of defectives in a lot. If the fraction defective in a lot is p_a, the probability that an item will be defective is

$$P(X < x_0) = p_a$$

or

$$\Phi\left(\frac{x_0 - \mu_a}{\sigma}\right) = p_a$$

That is, the mean value of X for a lot with p_a fraction defective is given by

$$\mu_a = x_0 - \sigma\Phi^{-1}(p_a)$$

The event that this acceptable lot will be rejected is the same as $\hat{p}$ exceeding M. From Eq. 9.11, this probability is

$$P(\hat{p} > M) = P\left[\Phi\left(\frac{x_0 - \bar{X}}{\sigma}\right) > M\right]$$

$$= P[\bar{X} < x_0 - \sigma\Phi^{-1}(M)]$$

The sample mean $\bar{X}$ is Gaussian $N(\mu_a, \sigma/\sqrt{n})$ or $N[x_0 - \sigma\Phi^{-1}(p_a), \sigma/\sqrt{n}]$; hence, limiting the risk of rejecting a good lot to α, we have

$$P(\hat{p} > M) = \Phi\left[\frac{x_0 - \sigma\Phi^{-1}(M) - x_0 + \sigma\Phi^{-1}(p_a)}{\sigma/\sqrt{n}}\right]$$

$$= \Phi[\sqrt{n}\{\Phi^{-1}(p_a) - \Phi^{-1}(M)\}] = \alpha$$

From which

$$\sqrt{n}\{\Phi^{-1}(p_a) - \Phi^{-1}(M)\} = \Phi^{-1}(\alpha) \tag{9.12}$$

Similarly, if the fraction defective in a lot is p_t, limiting the risk of accepting a bad lot to β, we obtain

$$\sqrt{n}\{\Phi^{-1}(p_t) - \Phi^{-1}(M)\} = \Phi^{-1}(1 - \beta) \tag{9.13}$$

Solution of Eqs. 9.12 and 9.13 yields

$$n = \left[\frac{\Phi^{-1}(1 - \beta) - \Phi^{-1}(\alpha)}{\Phi^{-1}(p_t) - \Phi^{-1}(p_a)}\right]^2 \tag{9.14}$$

and

$$M = \Phi\left[\Phi^{-1}(p_a) - \frac{\Phi^{-1}(\alpha)}{\sqrt{n}}\right] \tag{9.15}$$

Therefore, in a sampling plan to control the fraction defective, the sample size n and tolerable fraction defective M are given by Eqs. 9.14 and 9.15.

If σ is unknown, acceptance sampling plans would be more difficult to derive. In such cases, the required sampling plans may be developed using standard tables, such as those provided in MIL-STD-414 (1957).

EXAMPLE 9.9

One of the principal items under control for a large earth dam embankment project is D, the ratio of fill dry density to laboratory standard maximum dry density. Suppose material of good-quality compaction requires that D exceeds 0.99; otherwise, the fill will be rated as poorly compacted. The engineer estimates that for satisfactory performance of the embankment, the maximum tolerable fraction of poorly compacted fill should not exceed 8%, whereas a 3% poorly compacted fill is acceptable. Assume that the producer's and consumer's risks are both 5%. Devise a sampling plan assuming that D has a Gaussian distribution.

With

$$\alpha = 0.05 \qquad \beta = 0.05$$

and

$$p_a = 0.03 \qquad p_t = 0.08$$

we obtain the required sample size n from Eq. 9.14 as

$$n = \left[\frac{\Phi^{-1}(0.95) - \Phi^{-1}(0.05)}{\Phi^{-1}(0.08) - \Phi^{-1}(0.03)}\right]^2$$

$$= \left[\frac{1.64 - (-1.64)}{(-1.40) - (-1.88)}\right]^2$$

$$= 47$$

The corresponding tolerable fraction defective M is obtained from Eq. 9.15 as

$$M = \Phi\left[\Phi^{-1}(0.03) - \frac{\Phi^{-1}(0.05)}{\sqrt{47}}\right]$$

$$= \Phi\left[(-1.88) - \frac{(-1.64)}{\sqrt{47}}\right]$$

$$= \Phi(-1.64)$$

$$= 0.05$$

Suppose that the sample mean of D from 47 specimens is computed to be 1.005; would this section of embankment be acceptable? Assume that the standard deviation of D, namely σ, is 0.01.

With the mean value of D estimated to be 1.005, the probability that a fill specimen will be of poor quality is given by

$$P(D < 0.99) = \Phi\left(\frac{0.99 - 1.005}{0.01}\right)$$
$$= \Phi(-1.5)$$
$$= 0.036$$

Hence the estimated fraction defective $\hat{p}$ is 0.036 based on the observed data. Since this is less than 0.05, this section of the embankment is acceptable.

EXAMPLE 9.10

In Example 9.3 it is specified that the probability p of the daily DO concentration in the stream falling below 4 mg/l is 0.2 with a 95% confidence. In other words, if $p = 0.2$, the probability that the stream quality will be unacceptable is 95%, or the risk of accepting poor-quality stream is 5%. Hence $\beta = 0.05$ and $p_t = 0.2$. Since only consumer's risk is involved in this case, Eq. 9.14 may be used (assuming the daily DO is Gaussian) to determine the required period of measurement (in days). Suppose that the daily DO is allowed to fall below 4 mg/l at most once during the measurement period; then the stream quality will not meet the standard if $\hat{p}$, estimated from n observed measurements according to Eq. 9.11, exceeds $1/n$. Thus $M = 1/n$ in this case. Equation 9.14 then yields

$$\sqrt{n} = \frac{\Phi^{-1}(0.95)}{\Phi^{-1}(0.2) - \Phi^{-1}(1/n)}$$
$$= \frac{1.64}{-0.84 - \Phi^{-1}(1/n)}$$

By trial and error, the solution to the foregoing equation gives $n = 11$. Therefore a period of 11 days of measurement is required; if the measured daily DO does not fall below 4 mg/l more than once during this period, the stream quality satisfies the DO standard with 95% confidence.

This result may be compared with the period of 22 days based on sampling by attributes in Example 9.3. Assuming that DO has a normal distribution, the sampling by variables therefore reduces the required period of observations to half that of sampling by attributes.

9.3. MULTIPLE-STAGE SAMPLING

In addition to the single-stage sampling plan discussed above, *multiple-stage sampling* is sometimes used in inspection programs. An example of this has been discussed in Example 9.3. The objective is to accept the unquestionably high-quality products and to reject the unquestionably low-quality products based on relatively small samples taken at the first stage. The lots with questionable quality are subjected to a second-stage sampling. Based on the combined samples from the two stages, these lots will be accepted or rejected, or be subjected to a third-stage sampling. This process may be continued to several sampling stages. The risks (producer's and consumer's) involved in the multiple-stage sampling can be made equivalent to that in the single-stage sampling, provided the OC curves are the same in each sampling plan through appropriate choice of sample sizes and acceptance-rejection criteria. The reader is referred to Lipson and Sheth (1973) and Hald (1952) for a further treatment of this subject.

9.4. CONCLUDING REMARKS

In this concluding chapter, we have introduced the elementary concepts pertinent to the development of programs for quality assurance and acceptance sampling. Acceptance sampling programs or plans are of two general types—*sampling by attributes* and *sampling by variables*. In all cases, an important consideration in the development of a program is the resolution of the conflicting interests of the producer and consumer.

The application of the concepts introduced here is facilitated greatly by the availability of pertinent tables and charts, such as the tables of binomial probabilities (Eisenhart, 1950; Aiken, 1955). Moreover, tables corresponding to standard sampling plans have been developed for both acceptance sampling by attributes and sampling by variables. For example, the military standard MIL-STD-414 (1957) and MIL-STD-105D (1963) provide charts giving the required sample size and the acceptance-rejection criteria for both single and multiple sampling plans once the lot size, the tightness of the inspection level (as defined by the level of producer's and consumer's risks), and the tolerable fraction of defectives have been specified.

PROBLEMS

9.1 (a) A large set of welds is submitted for inspection. One out of 20 welds inspected was found to contain flaws. Suppose that the workmanship would be unacceptable if more than 10% of the welds contain flaws. Should the welds be accepted if the risk of accepting welds with unsatisfactory workmanship is limited to 5%?

  (b) Suppose further that good workmanship is defined as no more than 3% of the welds contain flaws. If the risk of rejecting welds with good workmanship is limited to 10%, devise a sampling plan that will satisfy the producer's and consumer's risk requirements. Use any pertinent data of part (a) as necessary.

  (c) Consider a sampling plan that requires inspection of 25 welds and at least 24 welds must be flawless for acceptance. Those welds that are rejected are required to be repaired. Determine the AOQ curve and AOQL corresponding to this sampling plan.

9.2 Sulphur dioxide is a main source of air pollution in a major city. Suppose that a concentration of SO_2 less that 0.1 unit (for example, parts per million parts of air) is harmless to human beings. If it is desired to maintain this condition for at least 90% of the time, what minimum number of daily measurements of SO_2 concentration is required to assure the desired air quality with 95% confidence? Assume that SO_2 concentration between days are statistically independent. *Ans. 29.*

9.3 (a) Soft lenses of sand deposits are hazardous to foundation safety during earthquakes. Soil borings is one way of detecting the presence of such lenses. Record of ten borings made at random locations over a large building site shows no signs of soft lenses of sand deposit. What is the confidence (probability) that sand lenses would not be found in more than 15% of the area beneath the site? *Ans. 0.803.*

   (b) Suppose that the engineer would like to have a 99% confidence that sand lenses would occupy less than 15% of the site area. How many additional borings should be made? *Ans. 19.*

9.4 Individual diesel engines used for generating emergency electrical power must have a high reliability of starting during an emergency. If a reliability (to start) of 99% is required, how many consecutive successful starts would be necessary to ensure this reliability with a 95% confidence?

9.5 Ready-mix concrete is supplied to a building site for construction. In order to ensure that the concrete meets the specified strength requirement, specimens of the ready-mix concrete are subjected to compression tests. Suppose that a batch containing 5% or less under-strength concrete is regarded as good-quality, whereas that containing 20% or more understrength concrete is deemed unsatisfactory. Devise a sampling plan for assuring quality concrete, if the producer's and consumer's risks are 10% and 6%, respectively. *Ans. n = 30; r = 3.*

9.6 Structural-grade lumber is used for falsework in a construction project. Lumber of a given dimension is shipped in truck loads. Suppose that five pieces per truck load are selected at random and subjected to examination and test for quality assurance. Acceptance of a load requires all five pieces to be nondefective.

   (a) If the allowable fraction defective can be as high as 30%, what is the consumer's risk?

   (b) The supplier here is not expected to deliver structural grade wood with less than 10% defective. If the fraction defective in a truck load of structural grade lumber is actually 10%, what is the probability that a truck load will be rejected?

9.7 In an earthdam construction project, the density of the compacted fill is measured at random locations to insure the specified degree of compaction. Suppose that the acceptance criterion requires that the average dry density from 10 locations should be at least 118 lb/ft³. Assume that 120 and 114 lb/ft³ correspond, respectively, to the mean densities of good and poor-quality compaction. Assume also that the density at various locations varies and may be assumed to be Gaussian, with a standard deviation σ of 4 lb/ft³.

   (a) What are the consumer's and producer's risks associated with this sampling plan?

   (b) If the consumer's and producer's risks are specified at 5% and 10%, respectively, devise the appropriate sampling plan.

   (c) Repeat part (b), assuming that σ is unknown, but expected to be around 4 lb/ft³.

9.8   The thickness of a finished pavement relative to the specified thickness is a measure of the quality of construction. Suppose that a thickness of 8 in. or more is considered adequate; otherwise the pavement will be rated as poorly constructed. The engineer estimates that for satisfactory performance of the entire pavement, no more than 10% of the pavement can be less than 8 in. thick, whereas up to 2% of the pavement with less than 8-in. thickness is considered acceptable. Assume that the producer's and consumer's risks are both 10%. Devise a sampling plan assuming that the thickness of pavement has a Gaussian distribution. *Ans. n = 11; M = 0.047.*

9.9   The acceptance of a newly designed solid waste treatment plant will be based on its performance under various types of waste load input. The quality of performance is measured by its efficiency in converting the solid waste to non-polluting materials. Suppose that a plant with a mean efficiency of at least 90% is considered "acceptable," whereas one with a mean efficiency of less than 80% is considered inferior. Assume that the risks of accepting an inferior-quality plant and of rejecting a good-quality plant are both 5%. Devise an acceptance sampling plan, whereby the efficiency in each trial operation is measured, that will fulfill the above requirements. Assume further that the efficiency of each treatment operation is Gaussian with a standard deviation of 4%.

Appendix A

Probability Tables

Table A.1. Table of Standard Normal Probability $\Phi(x) = \dfrac{1}{\sqrt{2\pi}} \displaystyle\int_{-\infty}^{x} \exp\left(-\tfrac{1}{2}\xi^2\right)$

x	$\Phi(x)$	x	$\Phi(x)$	x	$\Phi(x)$
0.0	0.500000	0.50	0.691463	1.00	0.841345
0.01	0.503989	0.51	0.694975	1.01	0.843752
0.02	0.507978	0.52	0.698468	1.02	0.846136
0.03	0.511966	0.53	0.701944	1.03	0.848495
0.04	0.515954	0.54	0.705401	1.04	0.850830
0.05	0.519939	0.55	0.708840	1.05	0.853141
0.06	0.523922	0.56	0.712260	1.06	0.855428
0.07	0.527904	0.57	0.715661	1.07	0.857690
0.08	0.531882	0.58	0.719043	1.06	0.859929
0.09	0.535857	0.59	0.722405	1.09	0.862143
0.10	0.539828	0.60	0.725747	1.10	0.864334
0.11	0.543796	0.61	0.729069	1.11	0.866500
0.12	0.547759	0.62	0.732371	1.12	0.868643
0.13	0.551717	0.63	0.735653	1.13	0.870762
0.14	0.555671	0.64	0.738914	1.14	0.872857
0.15	0.559618	0.65	0.742154	1.15	0.874928
0.16	0.563560	0.66	0.745374	1.16	0.876976
0.17	0.567494	0.67	0.748572	1.17	0.878999
0.18	0.571423	0.68	0.751748	1.18	0.881000
0.19	0.575345	0.69	0.754903	1.19	0.882977
0.20	0.579260	0.70	0.758036	1.20	0.884930
0.21	0.583166	0.71	0.761148	1.21	0.886860
0.22	0.587064	0.72	0.764238	1.22	0.888767
0.23	0.590954	0.73	0.767305	1.23	0.890651
0.24	0.594835	0.74	0.770350	1.24	0.892512
0.25	0.598706	0.75	0.773373	1.25	0.894350
0.26	0.602568	0.76	0.776373	1.26	0.896165
0.27	0.606420	0.77	0.779350	1.27	0.897958
0.28	0.610262	0.78	0.782305	1.28	0.899727
0.29	0.614092	0.79	0.785236	1.29	0.901475
0.30	0.617912	0.80	0.788145	1.30	0.903199
0.31	0.621720	0.81	0.791030	1.31	0.904902
0.32	0.625517	0.82	0.793892	1.32	0.906583
0.33	0.629301	0.83	0.796731	1.33	0.908241
0.34	0.633072	0.84	0.799546	1.34	0.909877
0.35	0.636831	0.85	0.802337	1.35	0.911492
0.36	0.640576	0.86	0.805105	1.36	0.913085
0.37	0.644309	0.87	0.807850	1.37	0.914656
0.38	0.648027	0.88	0.810570	1.38	0.916207
0.39	0.651732	0.89	0.813267	1.39	0.917735
0.40	0.655422	0.90	0.815940	1.40	0.919243
0.41	0.659097	0.91	0.818589	1.41	0.920730
0.42	0.662757	0.92	0.821214	1.42	0.922196
0.43	0.666402	0.93	0.823815	1.43	0.923641
0.44	0.670032	0.94	0.826391	1.44	0.925066
0.45	0.673645	0.95	0.828944	1.45	0.926471
0.46	0.677242	0.96	0.831473	1.46	0.927855
0.47	0.680823	0.97	0.833977	1.47	0.929219
0.48	0.684387	0.98	0.836457	1.48	0.930563
0.49	0.687933	0.99	0.838913	1.49	0.931888

Table A.1. (Cont'd)

x	$\Phi(x)$	x	$\Phi(x)$	x	$\Phi(x)$
1.50	0.933193	2.00	0.977250	2.50	0.993790
1.51	0.934478	2.01	0.977784	2.51	0.993963
1.52	0.935744	2.02	0.978308	2.52	0.994132
1.53	0.936992	2.03	0.978822	2.53	0.994297
1.54	0.938220	2.04	0.979325	2.54	0.994457
1.55	0.939429	2.05	0.979818	2.55	0.994614
1.56	0.940620	2.06	0.980301	2.56	0.994766
1.57	0.941792	2.07	0.980774	2.57	0.994915
1.58	0.942947	2.08	0.981237	2.58	0.995060
1.59	0.944083	2.09	0.981691	2.59	0.995201
1.60	0.945201	2.10	0.982136	2.60	0.995339
1.61	0.946301	2.11	0.982571	2.61	0.995473
1.62	0.947384	2.12	0.982997	2.62	0.995604
1.63	0.948449	2.13	0.983414	2.63	0.995731
1.64	0.949497	2.14	0.983823	2.64	0.995855
1.65	0.950529	2.15	0.984223	2.65	0.995975
1.66	0.951543	2.16	0.984614	2.66	0.996093
1.67	0.952540	2.17	0.984997	2.67	0.996207
1.68	0.953521	2.18	0.985371	2.68	0.996319
1.69	0.954486	2.19	0.985738	2.69	0.996427
1.70	0.955435	2.20	0.986097	2.70	0.996533
1.71	0.956367	2.21	0.986447	2.71	0.996636
1.72	0.957284	2.22	0.986791	2.72	0.996736
1.73	0.958185	2.23	0.987126	2.73	0.996833
1.74	0.959071	2.24	0.987455	2.74	0.996928
1.75	0.959941	2.25	0.987776	2.75	0.997020
1.76	0.960796	2.26	0.988089	2.76	0.997110
1.77	0.961636	2.27	0.988396	2.77	0.997197
1.78	0.962462	2.28	0.988696	2.78	0.997282
1.79	0.963273	2.29	0.988989	2.79	0.997365
1.80	0.964070	2.30	0.989276	2.80	0.997445
1.81	0.964852	2.31	0.989556	2.81	0.997523
1.82	0.965621	2.32	0.989830	2.82	0.997599
1.83	0.966375	2.33	0.990097	2.83	0.997673
1.84	0.967116	2.34	0.990358	2.84	0.997744
1.85	0.967843	2.35	0.990613	2.85	0.997814
1.86	0.968557	2.36	0.990863	2.86	0.997882
1.87	0.969258	2.37	0.991106	2.87	0.997948
1.88	0.969946	2.38	0.991344	2.88	0.998012
1.89	0.970621	2.39	0.991576	2.89	0.998074
1.90	0.971284	2.40	0.991802	2.90	0.998134
1.91	0.971933	2.41	0.992024	2.91	0.998193
1.92	0.972571	2.42	0.992240	2.92	0.998250
1.93	0.973197	2.43	0.992451	2.93	0.998305
1.94	0.973810	2.44	0.992656	2.94	0.998359
1.95	0.974412	2.45	0.992857	2.95	0.998411
1.96	0.975002	2.46	0.993053	2.96	0.998462
1.97	0.975581	2.47	0.993244	2.97	0.998511
1.98	0.976148	2.48	0.993431	2.98	0.998559
1.99	0.976705	2.49	0.993613	2.99	0.998605

Table A.1.   (Cont'd)

x	$\Phi(x)$	x	$\Phi(x)$	x	$1 - \Phi(x)$
3.00	0.998650	3.50	0.999767	4.00	0.316712E-04
3.01	0.998694	3.51	0.999776	4.05	0.256088E-04
3.02	0.998736	3.52	0.999784	4.10	0.206575E-04
3.03	0.998777	3.53	0.999792	4.15	0.166238E-04
3.04	0.998817	3.54	0.999800	4.20	0.133458E-04
3.05	0.998856	3.55	0.999807	4.25	0.106885E-04
3.06	0.998893	3.56	0.999815	4.30	0.853906E-05
3.07	0.998930	3.57	0.999821	4.35	0.680688E-05
3.08	0.998965	3.58	0.999828	4.40	0.541254E-05
3.09	0.998999	3.59	0.999835	4.45	0.429351E-05
3.10	0.999032	3.60	0.999841	4.50	0.339767E-05
3.11	0.999065	3.61	0.999847	4.55	0.268230E-05
3.12	0.999096	3.62	0.999853	4.60	0.211245E-05
3.13	0.999126	3.63	0.999858	4.65	0.165968E-05
3.14	0.999155	3.64	0.999864	4.70	0.130081E-05
3.15	0.999184	3.65	0.999869	4.75	0.101708E-05
3.16	0.999211	3.66	0.999874	4.80	0.793328E-06
3.17	0.999238	3.67	0.999879	4.85	0.617307E-06
3.18	0.999264	3.68	0.999883	4.90	0.479183E-06
3.19	0.999289	3.69	0.999888	4.95	0.371067E-06
3.20	0.999313	3.70	0.999892	5.00	0.286652E-06
3.21	0.999336	3.71	0.999896	5.10	0.169827E-06
3.22	0.999359	3.72	0.999900	5.20	0.996443E-07
3.23	0.999381	3.73	0.999904	5.30	0.579013E-07
3.24	0.999402	3.74	0.999908	5.40	0.333204E-07
3.25	0.999423	3.75	0.999912	5.50	0.189896E-07
3.26	0.999443	3.76	0.999915	5.60	0.107176E-07
3.27	0.999462	3.77	0.999918	5.70	0.599037E-08
3.28	0.999481	3.78	0.999922	5.80	0.331575E-08
3.29	0.999499	3.79	0.999925	5.90	0.181751E-08
3.30	0.999516	3.80	0.999928	6.00	0.986588E-09
3.31	0.999533	3.81	0.999931	6.10	0.530343E-09
3.32	0.999550	3.82	0.999933	6.20	0.282316E-09
3.33	0.999566	3.83	0.999936	6.30	0.148823E-09
3.34	0.999581	3.84	0.999938	6.40	0.77688 E-10
3.35	0.999596	3.85	0.999941	6.50	0.40160 E-10
3.36	0.999610	3.86	0.999943	6.60	0.20558 E-10
3.37	0.999624	3.87	0.999946	6.70	0.10421 E-10
3.38	0.999637	3.88	0.999948	6.80	0.5231 E-11
3.39	0.999650	3.89	0.999950	6.90	0.260 E-11
3.40	0.999663	3.90	0.999952	7.00	0.128 E-11
3.41	0.999675	3.91	0.999954	7.10	0.624 E-12
3.42	0.999687	3.92	0.999956	7.20	0.301 E-12
3.43	0.999698	3.93	0.999958	7.30	0.144 E-12
3.44	0.999709	3.94	0.999959	7.40	0.68 E-13
3.45	0.999720	3.95	0.999961	7.50	0.32 E-13
3.46	0.999730	3.96	0.999963	7.60	0.15 E-13
3.47	0.999740	3.97	0.999964	7.70	0.70 E-14
3.48	0.999749	3.98	0.999966	7.80	0.30 E-14
3.49	0.999758	3.99	0.999967	7.90	0.15 E-14

Table A.2. *p*-Percentile Values of the *t*-Distribution* (*After Brownlee, 1960*)

f \ *p*	0.750	0.900	0.950	0.975	0.990	0.995	0.999
1	1.000	3.078	6.314	12.706	31.821	63.657	318
2	0.816	1.886	2.920	4.303	6.965	9.925	22.3
3	0.765	1.638	2.353	3.182	4.541	5.841	10.2
4	0.741	1.533	2.132	2.776	3.747	4.604	7.173
5	0.727	1.476	2.015	2.571	3.365	4.032	5.893
6	0.718	1.440	1.943	2.447	3.143	3.707	5.208
7	0.711	1.415	1.895	2.365	2.998	3.499	4.785
8	0.706	1.397	1.860	2.306	2.896	3.355	4.501
9	0.703	1.383	1.833	2.262	2.821	3.250	4.297
10	0.700	1.372	1.812	2.228	2.764	3.169	4.144
11	0.697	1.363	1.796	2.201	2.718	3.106	4.025
12	0.695	1.356	1.782	2.179	2.681	3.055	3.930
13	0.694	1.350	1.771	2.160	2.650	3.012	3.852
14	0.692	1.345	1.761	2.145	2.624	2.977	3.787
15	0.691	1.341	1.753	2.131	2.602	2.947	3.733
16	0.690	1.337	1.746	2.120	2.583	2.921	3.686
17	0.689	1.333	1.740	2.110	2.567	2.898	3.646
18	0.688	1.330	1.734	2.101	2.552	2.878	3.610
19	0.688	1.328	1.729	2.093	2.539	2.861	3.579
20	0.687	1.325	1.725	2.086	2.528	2.845	3.552
21	0.686	1.323	1.721	2.080	2.518	2.831	3.527
22	0.686	1.321	1.717	2.074	2.508	2.819	3.505
23	0.685	1.319	1.714	2.069	2.500	2.807	3.485
24	0.685	1.318	1.711	2.064	2.492	2.797	3.467
25	0.684	1.316	1.708	2.060	2.485	2.787	3.450
26	0.684	1.315	1.706	2.056	2.479	2.779	3.435
27	0.684	1.314	1.703	2.052	2.473	2.771	3.421
28	0.683	1.313	1.701	2.048	2.467	2.763	3.408
29	0.683	1.311	1.699	2.045	2.462	2.756	3.396
30	0.683	1.310	1.697	2.042	2.457	2.750	3.385
40	0.681	1.303	1.684	2.021	2.423	2.704	3.307
60	0.679	1.296	1.671	2.000	2.390	2.660	3.232
120	0.677	1.289	1.658	1.980	2.358	2.617	3.160
∞	0.674	1.282	1.645	1.960	2.326	2.576	3.090

* Abridged from Table 12 of *Biometrika Tables for Statisticians*, vol. I, edited by E. S. Pearson and H. O. Hartley, Cambridge University Press, Cambridge (1954), and Table III of *Statistical Tables for Biological, Agricultural, and Medical Research*, R. A. Fisher and F. Yates, Oliver & Boyd, Edinburgh, 1953.

Table A.3. α-Percentile Values of the χ^2 Distribution* (*After Brownlee, 1960*)

f \ α	0.005	0.025	0.050	0.900	0.950	0.975	0.990	0.995	0.999
1	0.04393	0.03982	0.02393	2.71	3.84	5.02	6.63	7.88	10.8
2	0.0100	0.0506	0.103	4.61	5.99	7.38	9.21	10.6	13.8
3	0.0717	0.216	0.352	6.25	7.81	9.35	11.3	12.8	16.3
4	0.207	0.484	0.711	7.78	9.49	11.1	13.3	14.9	18.5
5	0.412	0.831	1.15	9.24	11.1	12.8	15.1	16.7	20.5
6	0.676	1.24	1.64	10.6	12.6	14.4	16.8	18.5	22.5
7	0.989	1.69	2.17	12.0	14.1	16.0	18.5	20.3	24.3
8	1.34	2.18	2.73	13.4	15.5	17.5	20.1	22.0	26.1
9	1.73	2.70	3.33	14.7	16.9	19.0	21.7	23.6	27.9
10	2.16	3.25	3.94	16.0	18.3	20.5	23.2	25.2	29.6
11	2.60	3.82	4.57	17.3	19.7	21.9	24.7	26.8	31.3
12	3.07	4.40	5.23	18.5	21.0	23.3	26.2	28.3	32.9
13	3.57	5.01	5.89	19.8	22.4	24.7	27.7	29.8	34.5
14	4.07	5.63	6.57	21.1	23.7	26.1	29.1	31.3	36.1
15	4.60	6.26	7.26	22.3	25.0	27.5	30.6	32.8	37.7
16	5.14	6.91	7.96	23.5	26.3	28.8	32.0	34.3	39.3
17	5.70	7.56	8.67	24.8	27.6	30.2	33.4	35.7	40.8
18	6.26	8.23	9.39	26.0	28.9	31.5	34.8	37.2	42.3
19	6.84	8.91	10.1	27.2	30.1	32.9	36.2	38.6	43.8
20	7.43	9.59	10.9	28.4	31.4	34.2	37.6	40.0	45.3
21	8.03	10.3	11.6	29.6	32.7	35.5	38.9	41.4	46.8
22	8.64	11.0	12.3	30.8	33.9	36.8	40.3	42.8	48.3
23	9.26	11.7	13.1	32.0	35.2	38.1	41.6	44.2	49.7
24	9.89	12.4	13.8	33.2	36.4	39.4	43.0	45.6	51.2
25	10.5	13.1	14.6	34.4	37.7	40.6	44.3	46.9	52.6
26	11.2	13.8	15.4	35.6	38.9	41.9	45.6	48.3	54.1
27	11.8	14.6	16.2	36.7	40.1	43.2	47.0	49.6	55.5
28	12.5	15.3	16.9	37.9	41.3	44.5	48.3	51.0	56.9
29	13.1	16.0	17.7	39.1	42.6	45.7	49.6	52.3	58.3
30	13.8	16.8	18.5	40.3	43.8	47.0	50.9	53.7	59.7
35	17.2	20.6	22.5	46.1	49.8	53.2	57.3	60.3	66.6
40	20.7	24.4	26.5	51.8	55.8	59.3	63.7	66.8	73.4
45	24.3	28.4	30.6	57.5	61.7	65.4	70.0	73.2	80.1
50	28.0	32.4	34.8	63.2	67.5	71.4	76.2	79.5	86.7
75	47.2	52.9	56.1	91.1	96.2	100.8	106.4	110.3	118.6
100	67.3	74.2	77.9	118.5	124.3	129.6	135.8	140.2	149.4

* Abridged from Table V of *Statistical Tables and Formulas* by A. Hald, John Wiley & Sons, New York, New York (1952).

Table A.4. Critical Values of $D_n{}^\alpha$ in the Kolmogorov-Smirnov Test (*After Hoel, 1962*)

n \ α	0.20	0.10	0.05	0.01
5	0.45	0.51	0.56	0.67
10	0.32	0.37	0.41	0.49
15	0.27	0.30	0.34	0.40
20	0.23	0.26	0.29	0.36
25	0.21	0.24	0.27	0.32
30	0.19	0.22	0.24	0.29
35	0.18	0.20	0.23	0.27
40	0.17	0.19	0.21	0.25
45	0.16	0.18	0.20	0.24
50	0.15	0.17	0.19	0.23
>50	$1.07/\sqrt{n}$	$1.22/\sqrt{n}$	$1.36/\sqrt{n}$	$1.63/\sqrt{n}$

Appendix B

Combinatorial Formulas

In probability problems involving discrete and finite sample spaces, the definition of events and the underlying sample space entails the enumeration of sets or subsets of sample points. For this purpose, the techniques of *combinatorial analysis* are often useful. We summarize here the basic elements of combinatorial analysis.

B.1 THE BASIC RELATION

If there are k positions in a sequence, and n_1 distinguishable elements can occupy position 1, n_2 can occupy position 2, ... and n_k can occupy position k, the number of distinct sequences of k elements each is given by

$$N(k \mid n_1, n_2, \ldots, n_k) = n_1 n_2 \cdots n_k \qquad (B.1)$$

Examples

(a) If a design involves three parameters ϕ_1, ϕ_2, ϕ_3 and there are, respectively, 2, 3, 4 values of these parameters, the number of feasible designs is

$$N(3 \mid 2, 3, 4) = (2)(3)(4) = 24$$

(b) In a three-dimensional Cartesian coordinate system x, y, z, if 10 discrete values are specified for each of the axes, for example, $x = 0, 1, 2, \ldots, 9$, then the total number of coordinate positions is

$$N(3 \mid 10, 10, 10) = (10)^3 = 1000$$

B.2 ORDERED SEQUENCES

In a set of n distinct elements, the number of k-element ordered sequences, or *arrangements*, is

$$(n)_k = n(n-1)(n-2) \cdots (n-k+1) = \frac{n!}{(n-k)!} \qquad (B.2)$$

For the first position in the sequence of k elements, there are n elements available to occupy it; but there are only $(n - 1)$ elements available for the second position since one of the n has been used for the first position, and only $(n - 2)$ elements available for the third position, and so on. Thence, by virtue of the basic relation, Eq. B.1,

$$(n)_k = N (k \mid n, n - 1, \ldots, n - k + 1)$$

thus obtaining Eq. B.2.

Examples

(a)   If no digits are repeated, the number of four-digit figures is $(10)_4 = (10) (9) (8) (7) = 5040$, whereas, if the digits can be repeated the number of four-digit figures would be $(10)^4 = 10,000$—this latter case would include 0000 as one of the figures.

(b)   In taking samples sequentially from a discrete sample space (population), the sampling may be done either *with replacements* or *without replacements*; that is, when an element is drawn from the population it is either returned or not returned to the population before another element is drawn. In a population of n elements, the number of ordered samples of size r then is n^r for sampling with replacements, whereas in sampling without replacements, the corresponding number of ordered samples is $(n)_r$.

B.3   THE BINOMIAL COEFFICIENT

In a set of n distinguishable elements, the number of possible subsets of k different elements each (regardless of order) is given by the *binomial coefficient*

$$\binom{n}{x} = \frac{(n)_k}{k!} \tag{B.3}$$

It may be emphasized that in Eq. B.2 the ordering of the k elements is of significance (that is, different orderings of the same elements constitute different sequences, or arrangements), whereas in Eq. B.3 the order is irrelevant. In a set of k elements, the position of the elements can be permuted $k!$ times; hence, by virtue of Eq. B.2, we obtain Eq. B.3 as the number of different k-element subsets (disregarding order).

Equation B.3 is defined only for $k \leq n$. Using Eq. B.2 for $(n)_k$, we have

$$\binom{n}{k} = \frac{n!}{k!(n - k)!} \tag{B.3a}$$

from which it is clear that

$$\binom{n}{k} = \binom{n}{n-k}$$

(B.4)

Equation B.3 or B.3a is known as the *binomial coefficient* because this is precisely the coefficient in the binomial expansion of $(x + y)^n$; namely,

$$(x + y)^n = \binom{n}{0} x^n y^0 + \binom{n}{1} x^{n-1} y + \binom{n}{2} x^{n-2} y^2 + \cdots + \binom{n}{n} x^0 y^n$$

Examples

(a) From a population of n elements, the number of different samples of size r is $\binom{n}{r}$. This is the same as sampling without replacement, except that the order is disregarded; therefore the number is

$$\frac{(n)_r}{r!} = \binom{n}{r}$$

(b) Among 25 concrete cylinders marked $1, 2, \ldots, 25$; the number of possible samples of five cylinders each is

$$\binom{25}{5} = \frac{25!}{5!20!} = 53{,}130$$

B.4 THE MULTINOMIAL COEFFICIENT

If n distinguishable elements are divided into r different groups of $k_1, k_2, \ldots, k_r$ elements, respectively, so that $k_1 + k_2 + \cdots + k_r = n$, the number of ways to form such r groups is given by the *multinomial coefficient*

$$\binom{n}{k_1, k_2, \ldots, k_r} = \frac{n!}{k_1! k_2! \ldots k_r!}$$

(B.5)

Among the n elements, the first group of k_1 elements can be chosen in $\binom{n}{k_1}$ ways. The second group of k_2 elements can be chosen from the remaining $(n - k_1)$ elements in $\binom{n-k_1}{k_2}$ ways, and so on. The total number of ways of dividing the n elements into r groups of $k_1, k_2, \ldots, k_r$, therefore, is

$$\binom{n}{k_1} \binom{n-k_1}{k_2} \cdots \binom{n - k_1 - k_2 - \cdots - k_{r-2}}{k_{r-1}} \binom{k_r}{k_r} = \frac{n!}{k_1! k_2! \ldots k_r!}$$

Example

In a given region, six earthquakes of intensities V, VI, VII may occur in the next 10 years. Three earthquakes of intensity V, two of VI, and one of VII can occur in

$$\frac{6!}{3!2!1!} = 60 \text{ different sequences}$$

B.5 STIRLING'S FORMULA

An important formula for computing the factorial of large numbers approximately is the *Stirling's formula*:

$$n! \simeq \sqrt{2\pi}\,(n)^{n+\frac{1}{2}}e^{-n} \tag{B.6}$$

A proof of the formula can be found in Feller (1957). The approximation is good even for n as small as 10 (error less than 1%).

Appendix C

Derivation of the Poisson Distribution

The Poisson distribution describes the probability mass function for the number of occurrences of an event within a specified interval of time or space. It is the result of an underlying counting process $X(t)$, known as a *Poisson process*, which is a model of the random occurrences of an event in time (or space) t.

The Poisson process model is based on the following assumptions.

1. At any instant of time (or point in space), there can be at most one occurrence of an event; in other words, the probability of n occurrences of the event over a small interval Δt is of order $\mathbf{o}(\Delta t)^n$.

2. The occurrences of an event in nonoverlapping time (or space) intervals are statistically independent; this is the assumption of *independent increments*.

3. The probability of an occurrence in $(t, t + \Delta t)$ is proportional to Δt; that is,

$$P[X(\Delta t) = 1] = \nu \Delta t$$

where ν is a positive proportionality constant.

On the basis of assumption 2, we obtain, with the theorem of total probability,

$$P[X(t + \Delta t) = x] = P[X(t) = x] P[X(\Delta t) = 0]$$
$$+ P[X(t) = x - 1] P[X(\Delta t) = 1]$$
$$+ P[X(t) = x - 2] P[X(\Delta t) = 2]$$
$$+ \cdots$$

Then on the basis of assumptions 1 and 3, we obtain, using the notation $p_x(t) \equiv P[X(t) = x]$,

$$p_x(t + \Delta t) = [1 - \nu \Delta t - \mathbf{o}(\Delta t)^2 - \cdots]p_x(t) + (\nu \Delta t) p_{x-1}(t)$$
$$+ \mathbf{o}(\Delta t)^2 p_{x-2}(t) + \cdots$$

Neglecting the higher-order terms, the preceding equation becomes

$$\frac{p_x(t + \Delta t) - p_x(t)}{\Delta t} = -\nu p_x(t) + \nu p_{x-1}(t)$$

Therefore, in the limit as $\Delta t \to 0$, we obtain the following differential equation for $p_x(t)$:

$$\frac{dp_x(t)}{dt} = -\nu p_x(t) + \nu p_{x-1}(t) \tag{C.1}$$

It should be recognized that Eq. C.1 applies for any $x \geq 1$. For $x = 0$, the preceding derivation leads to the following:

$$\frac{dp_0(t)}{dt} = -\nu p_0(t) \tag{C.2}$$

If the counting process starts from zero, the initial conditions associated with Eqs. C.1 and C.2 are

$$p_0(0) = 1.0 \quad \text{and} \quad p_x(0) = 0$$

The solution of Eq. C.2 with the first of the initial conditions yields, for $x = 0$,

$$p_0(t) = e^{-\nu t}$$

For $x \geq 1$, the solutions to Eq. C.1 are

$$p_1(t) = \nu t e^{-\nu t}$$

$$p_2(t) = \frac{(\nu t)^2}{2!} e^{-\nu t}$$

and for a general x,

$$p_x(t) = \frac{(\nu t)^x}{x!} e^{-\nu t} \tag{C.3}$$

Equation C.3 is the Poisson probability mass function, in which the parameter ν is the *mean occurrence rate* of the event.

References

1. Aiken, H. H., *Tables of the Cumulative Binomial Probability Distribution*, Harvard Univ. Press, Cambridge, Mass., 1955.
2. Allen, D. E., "Statistical Study of the Mechanical Properties of Reinforcing Bars," *Building Research Note No. 85*, National Research Council, Ottawa, Canada, April 1972.
3. Ang, A. H-S., "Structural Risk Analysis and Reliability-Based Design," *Jour. of the Structural Division*, ASCE, Vol. 99, No. ST9, September 1973, pp. 1891–1910.
4. Bachmann, W. K., "Estimation Stochastique de la Précision des Mesures," *Mensuration Photogrammétrie Génie Rural*, Vol. 71, December 1973, pp. 107–118.
5. Bagnold, R. A., "Interim Report on Wave Pressure Research," *Jour. Institution of Civil Engineers*, London, England, Vol. 12, June 1939, pp. 202–206.
6. Barry, B. A., *Engineering Measurements*, J. Wiley and Sons, New York, 1964.
7. Barsom, J. M., "Relationships Between Plane-Strain Ductility and K_{I_c} for Various Steel," *ASME 1st National Congress on Pressure Vessels and Piping, Paper No. 71-PVP-13*, May 1971.
8. Benjamin, J. R., "Probabilistic Models for Seismic Force Design," *Jour. of the Structural Division*, ASCE, Vol. 94, No. ST5, May 1968, pp. 1175–1196.
9. Brownlee, K. A., *Statistical Theory and Methodology in Science and Engineering*, J. Wiley and Sons, New York, 1960.
10. Bureau of Public Works, "Traffic Assignment and Distribution for Small Urban Areas," U.S. Dept. of Commerce, September 1965.
11. Butts, T. A., Schnepper, D. H., and Evans, R. L., "Statistical Assessment of DO in Navigation Pool," *Jour. of the Sanitary Engineering Division*, ASCE, Vol. 96, No. SA2, April 1970, p. 48.
12. Cartwright, D. E., and Longuet-Higgins, M. S., "The Statistical Distribution of the Maxima of a Random Function," *Proc. Royal Society*, Series A, Vol. 237, 1956.
13. Clopper, C. J., and Pearson, E. S., "The Use of Confidence or Fiducial

Limits illustrated in the Case of the Binomial," *Biometrica*, Vol. 26, 1934, pp. 404–413.

14. Cornell, C. A., "A Normative Second-Moment Reliability Theory for Structural Design," Solid Mechanics Division, Univ. of Waterloo, Waterloo, Ontario, Canada, 1969.

15. Cox, E. A., "Information Needs for Controlling Equipment Costs," *Highway Research Record*, No. 278, Highway Research Board, National Research Council, 1969, pp. 35–48.

16. Cusens, A. R., and Wettern, J. H., "Quality Control in Factory-Made Precast Concrete," *Civil Engineering and Public Works Review*, Vol. 54, 1959.

17. Donovan, N. C., "A Stochastic Approach to the Seismic Liquefaction Problem," *Proc. 1st Int. Conf. on Application of Statistics and Probability to Soil and Structural Engineering*, Hong Kong Univ. Press, 1972, pp. 513–535.

18. Elderton, W. P., *Frequency Curves and Correlation*, 4th Ed., Cambridge Univ. Press, Cambridge, England, 1953.

19. Eisenhart, C., *Tables of the Binomial Probability Distribution*, Applied Mathematics Series 6, National Bureau of Standards, Washington, D.C., 1950.

20. Feller, W., *An Introduction to Probability Theory and Its Applications*, Vol. I, 2nd Ed., J. Wiley and Sons, New York, 1957.

21. Fisher, J. W., Frank, K. H., Hirt, M. A., and McNamee, M., "Effects of Weldments on the Fatigue Strength of Steel Beams," *NCHRP Rept. No. 102*, Highway Research Board, National Research Council, 1970.

22. Forbes, W. S., "A Survey of Progress in House Building," *Building Technology and Management*, Vol. 7 (4), April 1969, pp. 88–91.

23. Freund, J. E., *Mathematical Statistics*, Prentice-Hall, Englewood Cliffs, N. J., 1962.

24. Galligan, W. L., and Snodgrass, D. V., "Machine Stress Rated Lumber: Challenge to Design," *Jour. of the Structural Division*, ASCE, Vol. 96, No. ST12, December 1970.

25. Gerlough, D. L., "Use of Poisson Distribution in Highway Traffic," *Poisson and Traffic*, The Eno Foundation for Highway Traffic Control, Saugatuck, Conn., 1955, p. 38.

26. Goldman, J. L., and Ushijima, T., "Decrease in Hurricane Winds After Landfall," *Jour. of the Structural Division*, ASCE, Vol. 100, No. ST1, January 1974, pp. 129–141.

27. Grandage, A., "Acceptance Sampling by Variables," *Proceeding, National Conf. on Statistical Quality Control Methodology on Highway and Airfield Construction*, May 1966, Univ. of Virginia, Charlottesville, Virginia.

28. Gumbel, E. J., "Statistical Theory of Extreme Values and Some Practical Applications," *Applied Mathematics Series 33*, National Bureau of Standards, Washington, D.C. February 1954.

29. Hald, A., *Statistical Theory with Engineering Applications*, J. Wiley and Sons, New York, 1952.

30. Hardy, G. H., Littlewood, J. E., and Polya, G., *Inequalities*, Cambridge Univ. Press, Cambridge, England, 1959.

31. Harter, H. L., *New Tables of the Incomplete Gamma Function Ratio and of Percentage Points of the Chi-square and Beta Distributions*, Aerospace Research Laboratories, U.S. Air Force (U.S. Gov't. Printing Office, Washington, D.C.), 1963.

32. Hazen, A., *Flood Flows, a Study in Frequency and Magnitude*, J. Wiley and Sons, New York, 1930.

33. Heathington, K. W., and Tutt, P. R., "Traffic Volume Characteristics on Urban Freeway," *Transportation Engineering Jour.*, ASCE, Vol. 97, TE1, February 1971, p. 108.

34. Hoel, P. G., *Introduction to Mathematical Statistics*, 3rd Ed., J. Wiley and Sons, New York, 1962.

35. Hoffman, D., and Lewis, E. V., "Analysis and Interpretation of Full-Scale Data on Midship Bending Stresses of Dry Cargo Ships," Ship Structure Committee, SSC-196, U.S. Coast Guard Hqtrs., Washington, D.C., June 1969.

36. Hognestad, E., "A Study of Combined Bending and Axial Load in Reinforced Concrete Members," *Engineering Experiment Station Bulletin No. 399*, Univ. of Illinois, Urbana, Ill., 1951.

37. Jordan, W., Eggert, O., and Kneissl, M., *Handbuch der Vermessungskunde*, Vol. 1, J. B. Metzlersche Verlagsbuchhandlung, Stuttgart, 1961.

38. Julian, O. G., "Synopsis of First Progress Report of Committee on Factors of Safety," *Jour. of the Structural Division*, ASCE, Vol. 83, No. ST4, July 1957, p. 1316.

39. Kanafani, A. K., "Location Model for Parking Facilities," *Transportation Engineering Jour.*, ASCE, Vol. 98, No. TE1, February 1972, pp. 117–129.

40. Kies, J. A., Smith, H. L., Romine, H. E., and Bernstein, M., "Fracture Testing of Weldments," *ASTM Special Publ. No. 381*, 1965, pp. 328–356.

41. Kimball, B. F., "Assignment of Frequencies to a Completely Ordered Set of Sample Data," *Trans. American Geophysical Union*, Vol. 27, 1946, pp. 843–846.

42. Kothandaraman, V., "A Probabilistic Analysis of Dissolved Oxygen–Biochemical Oxygen Demand Relationship in Streams," *Ph.D.*

dissertation, Dept. of Civil Engineering, Univ. of Illinois at Urbana-Champaign, 1968.

43. Kothandaraman, V., and Ewing, B. B., "A Probabilistic Analysis of Dissolved Oxygen–Biochemical Oxygen Demand Relationship in Streams," *Jour. Water Resources Control Federation*, Part 2, February 1969, pp. 73–90.

44. Kulak, G. L., "Statistical Aspects of Strength of Connections," *Proc. ASCE Specialty Conf. on Safety and Reliability of Metal Structures*, November 1972, pp. 83–105.

45. Lam Put, R., "Dynamic Response of a Tall Building to Random Wind Loads," *3rd Int. Conf. on Wind Effects on Buildings and Structures*, Tokyo, September 1971.

46. Lambe, T. W., and Whitman, R. V., *Soil Mechanics*, J. Wiley and Sons, New York, 1969, p. 375.

47. Linsley, R. K., and Franzini, J. B., *Water Resources Engineering*, McGraw-Hill Book Company, New York, 1964, p. 68.

48. Lipson, C., and Sheth, N. J., *Statistical Design and Analysis of Engineering Experiments*, McGraw-Hill Book Company, New York, 1973.

49. Littleford, T. W., "A Comparison of Flexural Strength-Stiffness Relationships for Clear Wood and Structural Grades of Lumber," *Information Report VP-X-30*, Forest Products Lab., Vancouver, B.C., Canada, December 1967.

50. Loucks, D. P., and Lynn, W. R., "Probabilistic Models for Predicting Stream Quality," *Water Resources Research*, Vol. 2, No. 3, September, 1966, pp. 593–605.

51. Malhotra, V. M., and Zoldners, N. G., "Some Field Experience in the Use of an Accelerated Method of Estimating 28-day Strength of Concrete," *Journal of American Concrete Institute*, November 1969, p. 895.

52. Martin, B. V., Memmott, F. W., and Bone, A. J., "Principles and Techniques of Predicting Future Demand for Urban Area Transportation," *Research Report No. 38*, Dept. of Civil Engineering, Massachusetts Institute of Technology, Cambridge, Mass., January 1963.

53. Martin, J. R., Ledbetter, W. B., Ahmad, H., and Britton, S. C., "Nonbloated Burned Clay Aggregate Concrete," *Jour. of Materials*, *JMLSA*, Vol. 7, No. 4, December 1972, pp. 555–563.

54. Mathematical Association of America, "Introductory Statistics without Calculus," *Report of the Panel on Statistics, Committee on the Undergraduate Program in Mathematics*, June 1972, p. 15.

55. Meadows, D. H., Meadows, D. L., Randers, J., and Behrens, W. W., *The Limits of Growth*, Universe Books, New York, 1972.

During Peak Periods," *Transportation Science*, ORSA, Vol. 4, 1970, pp. 409–411.

57. MIL-STD-414, *Sampling Procedures and Tables for Inspection by Variables for Percent Defective*, Dept. of Defense, June 1957.

58. MIL-STD-105D, *Sampling Procedures and Tables for Inspection by Attributes*, Dept. of Defense, April 1963.

59. Mitchell, G. R., and Woodgate, R. W., "A Survey of Floor Loadings in Office Buildings," *CIRIA Report 25*, London, England, August 1970.

60. Mood, A. M., and Graybill, F. A., *Introduction to the Theory of Statistics*, McGraw-Hill Book, Co., New York, 1963.

61. Morse, W. L., "Stream Temperature Prediction Under Reduced Flow," *Jour. of the Hydraulics Division*, ASCE, Vol. 98, HY6, June 1972.

62. Moulton, L. K., and Schaub, J. H., "Estimations of Climatic Parameters for Frost Depth Predictions," *Transportation Engineering Jour.*, ASCE, Vol. 95, TE4, November 1969.

63. Murdock, J. W., and Kesler, C. E., "Effects of Range of Stress on Fatigue Strength of Plain Concrete Beams," *Jour. of American Concrete Institute*, Vol. 30, No. 2, August 1958.

64. Nishida, Y., "A Brief Note on Compression Index of Soil," *Jour. of the Soil Mechanics and Foundations Div.*, ASCE, Vol. SM3, July 1956.

65. Packman, P. F., Pearson, H. S., Owens, J. S., and Marchese, G. B., "The Applicability of a Fracture Mechanics—Nondestructive Testing Design Criteria," *Technical Report, AFML-TR-68-32*, Air Force Materials Laboratory, Wright-Patterson Air Force Base, Ohio, May 1968.

66. Parratt, L. G., *Probability and Experimental Errors in Science*, J. Wiley and Sons, New York, 1961.

67. Payne, H. J., "Freeway Traffic Control and Surveillance Model," *Transportation Engineering Jour.*, ASCE, Vol. 99, No. TE4, November 1973, pp. 767–783.

68. Pearson, E. S., and Johnson, N. L., *Tables of the Incomplete Beta-Function*, 2nd Ed., Cambridge Univ. Press, Cambridge, England, 1968.

69. Pearson, K., *Tables of the Incomplete B-Function*, Cambridge Univ. Press, Cambridge, England, 1934.

70. Peck, R. B., "Sampling Methods and Laboratory Tests for Chicago Subway Soils," Proceedings, *Purdue Conference on Soil Mechanics and Its Applications*, Lafayette, Indiana, 1940.

71. Pettitt, J. H. D., "Statistical Analysis of Density Tests," *Jour. of the Highway Div.*, ASCE, Vol. No. HW2, November 1967.

72. Proctor, R. P. M., and Paxton, H. W., "Stress Corrosion of Aluminum Alloy in Organic Liquids," *Jour. of Materials*, Vol. 4, No. 3, September 1969, p. 747.

73. Pugsley, A. G., "Structural Safety," *Jour. Royal Aeronautical Society*, Vol. 58, 1955.

74. Richardus, P., *Project Surveying*, J. Wiley and Sons, New York, 1966.

75. Shull, R. D., and Gloyna, E. F., "Transport of Dissolved Water in Rivers," *Jour. of the Sanitary Engineering Division*, ASCE, Vol. 95, SA6, December 1969, p. 1001.

76. Smeed, R. J., "Traffic Studies and Urban Congestion," *Jour. Transportation Economics and Policy*, Vol. 2, No. 1, 1968, pp. 33–70.

77. Tang, W. H., "A Bayesian Evaluation of Information for Foundation Engineering Design," *Proceedings, First Intl. Conf. on Application of Statistics and Probability to Soil and Structural Engineering*, Hong Kong Univ. Press, September 1971, pp. 173–185.

78. Tang, W. H., "Probabilistic Updating of Flaw Information," *Jour. of Testing and Evaluation, JTEVA*, Vol. 1, No. 6, November 1973, pp. 459–467.

79. Thayer, R. P., and Krutchkoff, R. G., "A Stochastic Model for Pollution and Dissolved Oxygen in Streams," Water Resources Research Center, Virginia Polytechnic Inst., Blacksburg, Virginia, 1966.

80. Todd, D. K., and Meyer, C. F., "Hydrology and Geology of the Honolulu Aquifer," *Jour. of the Hydraulics Div.*, ASCE, Vol. 97, No. HY2, February 1971, p. 251.

81. Viner, J. G., "Recent Developments in Roadside Crush Cushions," *Transportation Engineering Jour.*, ASCE, Vol. 98, No. TE1, February 1972, pp. 71–87.

82. Voorhees, A. M., "What Happens When Metropolitan Area Meet," *Jour. of the Urban Planning Division*, ASCE, Vol. 92, UPI, May 1966, p. 16.

83. Ward, E. J., "Systems Approach to Choice in Transport Technology," *Transportation Engineering Jour.*, ASCE, Vol. 96, No. TE4, November 1970.

84. Wynn, F. H., "Shortcut Modal Split Formula," *Highway Research Record*, No. 283, Highway Research Board, National Research Council, 1969.

index

of highway acquisition, 11
initial cost, 186
of material and labor, 169
Covariance, 140, 193
Crack, 147, 215, 341
detection, 147, 341
growth, 215
propagation, 341
Crosswalk, 118
Crushed rock, 74
Culvert, 13
Cumulative distribution function (CDF), 81,
84, 86, 278, 279
cumulative probability, 234, 251
properties of CDF, 82
Cusens and Wettern, 276
Cyclic load, 230
cyclic stress, 215

Dam, 148, 149
control dam, 158
earth dam, 173, 374, 377
Darcy-Weisbach equation, 204
Decision making, 220
criterion, 186
under uncertainty, 11
Defective item, 127, 360, 361
Degree of belief, 354
Degree of freedom, 173, 177, 236, 237, 244,
250, 251, 274, 275, 277, 372
Delay time, 151
Demand for water, 73, 204
de Morgan's rule, 34, 35, 41
Density function, *see* Probability density
function
Density of fill, 365, 369-371, 374, 377
Dependent variable, 170, 297
Derivation of the Poisson distribution, 390
Derivative, 82
nth derivative, 96
partial derivative, 137
Derived probability distribution, 170, 191
derived density function, 171
Design, 11, *see also* Engineering planning
and design
criterion, 117
of experiments, 220
flood, 107, 157, 159
life, 158
load, 195
under uncertainty, 11
wave height, 111
wind velocity, 111, 157, 163
Detectability, 342, 343
Detection of material defects, 341
Determination of probability distribution,
261, 281
Diameter, 215
Diesel engine, 124, 377
Differential equation, 391
Differential settlement, 51, 101, 295, 296
Diffuse prior distribution, 338, 341, 345
Dike, 159, 203
Dimension, 244
Discrete random variable, 81, 84, 94, 330
Diseased trees, 79

Disjoint sets, 30
Dispersion, 7, 9, 90
measure of, 7, 88-90
Dissolved oxygen (DO), 7, 17, 52, 165, 235,
238, 251, 255, 256, 300, 302, 306,
321, 365, 366, 375
Dissolved solid, 319, 320
Distance, 214, 243-247, 258, 260, 315, 358
distance downstream, 313, 315
Distribution function, *see* Cumulative
distribution function
Distributive rule, 33
Drainage system, 71
drainage area, 11
drainage water, 211
Drill hole, 119
Drought, 30
Duality relation, 35

Earthquake, 30, 69, 73, 77, 114, 148, 161,
164, 168, 177, 340, 341, 377, 389
intensity, 73, 121, 389
occurrence, 69, 121, 161, 164, 340, 341
Efficiency, 135, 378
of estimator, 220, 223
Effluent, 165
Eisenhart, 361, 376
Elasticity, theory of, 184
Elderton, 133
Electrical power, 377
Electronic distance measurement, 233
electronic ranging instrument, 245
Elevation, 349, 350
Elongation, 322
Embankment, 16, 168, 252, 254, 364, 368,
369, 374, 375
Emergency control system, 124
Emergency power, 124, 377
Empirical relation, 307, 310
Energy consumption, 288, 322
Energy line, 217
Engineering planning and design, 1, 12, 17,
330, 354
Engineering system, 1, 360
Equipment, 19, 108
construction equipment, 106, 154, 161,
167, 359
Erlang distribution, *see* Gamma distribution
Error, 11, 15, 221, 324
of estimation, 221, 255, 329, 332, 347-
349
of measurement, 15, 194, 325, 349, 357,
358
prediction or model error, 11, 324
propagation of error, 15, 199, 245, 246
standard error, 233, 244-247, 258, 260
systematic error, 243
Esopus Creek, 255
Estimation of correlation coefficient, 315
Estimation of parameter, 254, 255, 281,
329, 330, 337, 349, 354
interval estimation, 133, 254
maximum likelihood method, 222, 228-
231, 252, 254
method of moments, 222, 223, 254, 255
point estimation, 220, 254

SI METRIC UNITS

CONVERSION FACTORS

		Customary to SI
inches (in.)	meters (m)	0.0254
inches (in.)	centimeters (cm)	2.54
inches (in.)	millimeters (mm)	25.4
feet (ft)	meters (m)	0.305
yards (yd)	meters (m)	0.914
miles (miles)	kilometers (km)	1.609
degrees (°)	radians (rad)	0.0174
acres (acre)	hectares (ha)	0.405
acre-feet (acre-ft)	cubic meters (m³)	1233
gallons (gal)	cubic meters (m³)	3.79×10^{-3}
gallons (gal)	liters (l)	3.79
pounds (lb)	kilograms (kg)	0.4536
tons (ton, 2000 lb)	kilograms (kg)	907.2
pound force (lbf)	newtons (N)	4.448
pounds per sq in. (psi)	newtons per sq m (N/m²)	6895
pounds per sq ft (psf)	newtons per sq m (N/m²)	47.88
foot-pounds (ft-lb)	joules (J)	1.356
horsepowers (hp)	watts (W)	746
British thermal units (BTU)	joules (J)	1055
British thermal units (BTU)	kilowatt-hours (kwh)	2.93×10^{-4}